ENCYCLOPEDIA OF MATHEMATICS

EDITED BY G.-C.

Volume 66

# Eigenspaces of Graphs

# ENCYCLOPEDIA OF MATHEMATICS AND ITS APPLICATIONS

ENCYCLOPEDIA OF MATHEMATICS AND ITS APPLICATIONS

# Eigenspaces of graphs

D. Cvetković
*University of Belgrade*

P. Rowlinson
*University of Stirling*

S. Simić
*University of Belgrade*

CAMBRIDGE UNIVERSITY PRESS
Cambridge, New York, Melbourne, Madrid, Cape Town, Singapore, São Paulo

Cambridge University Press
The Edinburgh Building, Cambridge CB2 8RU, UK

Published in the United States of America by Cambridge University Press, New York

www.cambridge.org
Information on this title: www.cambridge.org/9780521573528

First published 1997
This digitally printed version 2008

*A catalogue record for this publication is available from the British Library*

*Library of Congress Cataloguing in Publication data*
Cvetković, Dragoš M.
Eigenspaces of graphs / Dragoš Cvetković, Peter Rowlinson, Slobodan Simić.
p.   cm. – (Encyclopedia of mathematics and its applications; 66)
Includes bibliographical references (p.   –   ) and index.
ISBN 0 521 57352 1 (hc)
1. Graph theory.   2. Spectral theory.   I. Rowlinson, Peter.   II. Simić, Slobodan.
III. Title.   IV. Series: Encyclopedia of mathematics and its applications; v. 66.
QA166.C83   1997
511'.2–dc20   96-2860 CIP

ISBN 978-0-521-57352-8 hardback
ISBN 978-0-521-05718-9 paperback

For Nevenka, Carolyn and Vesna

# Contents

# Preface

The foundations of spectral graph theory were laid in the fifties and six-ties, as a result of the work of a considerable number of mathematicians. Most of the early results are, like this book, concerned with the relation between spectral and structural properties of a graph. The investigation of such a relationship was proposed explicitly by Sachs [Sac1] and Hoff-man [Hof5], although in effect it had already been initiated in an earlier article by Collatz and Sinogowitz [CoSi]. This seminal paper appeared in 1957, but our bibliography contains two references prior to this date: the unpublished thesis of Wei [Wei] from 1952, and a summary (also unpublished) of a 1956 paper by Lihtenbaum [Lih] communicated at the 3rd Congress of Mathematicians of the U.S.S.R.

Another origin of the theory of graph spectra lies beyond mathemat-ics. In quantum chemistry, an approximative treatment of non-saturated hydrocarbons introduced by E. Hückel [Hüc] yields a graph-theoretical model of the corresponding molecules in which eigenvalues of graphs represent the energy levels of certain electrons. The connection between Hückel's model of 1931 and the mathematical theory of graph spectra was recognized many years later in [GüPr] and [CvGu1], and thereafter exploited extensively by many authors, both chemists and mathemati-cians.

In his thesis [Cve7], Cvetković identified 83 papers dealing with eigen-values of graphs which had appeared before 1970. Ten years later, almost all of the results related to the theory of graph spectra published before 1978 were summarized in the monograph *Spectra of Graphs* by Cvetković, Doob and Sachs [CvDS], a book which is almost entirely self-contained; only a little familiarity with graph theory and matrix theory is assumed. Its bibliography contains 564 items, most of which were published between 1960 and 1978. It was supplemented in 1988

x

by *Recent Results in the Theory of Graph Spectra* by Cvetković, Doob, Gutman and Torgašev [CvDGT]. This reviews the results in spectral graph theory from the period 1978-1984, and provides over 700 further references from the mathematical and chemical literature. There are additional references from areas such as physics, mechanical engineering, geography and the social sciences. Although many papers contain only minor results, and some present rediscoveries of known results, the large number of references indicates the rapid rate of growth of spectral graph theory. The third edition of *Spectra of Graphs*, published in 1995, contains an appendix which describes recent developments in the subject.

This book deals with eigenspaces of graphs, and although one cannot speak about eigenvectors without mentioning eigenvalues, or vice versa, the emphasis is on those parts of spectral theory where the structure of eigenspaces is a dominant feature, thus complementing the 'eigenvalue part of the theory' described in *Spectra of Graphs*. For the most part, the eigenspaces considered are those of a $(0, 1)$-adjacency matrix of a finite undirected graph.

Chapters 1 and 2 review 'old' results on eigenvalues and eigenvectors respectively, while the remaining chapters are devoted to 'new' results and techniques. The eigenspace corresponding to the largest eigenvalue (or *index*) of a connected graph is one-dimensional, and in Chapter 3 a spanning eigenvector is used to identify the graphs with extremal index in various families of graphs. The discussion of graph spectra in the first chapter reveals the limitations of the spectrum as a means of characterizing a graph, and motivates the search for further algebraic invariants such as the graph angles considered in Chapters 4 and 5. Angles also have a role in Chapter 6, where the theory of matrix perturbations is applied to adjacency matrices: one can then describe the behaviour of the index of a graph when it undergoes a local modification such as the addition or deletion of an edge or vertex. Graph angles arise from a geometric approach to eigenspaces that leads in Chapter 7 to the notion of a star partition of vertices, an important concept which enables one to construct 'natural' bases for the eigenspaces of a graph. Implications for the graph isomorphism problem are the subject of current research, and this is described in Chapter 8. Some miscellaneous results are gathered together in Chapter 9, and there are two appendices: one contains some classical results from matrix theory, and the other is a table of graph angles.

The authors are indebted to Mladen Cvetković for assistance with the preparation of a LaTeX version of the first draft of the text. The contents of the second author's article on graph perturbations in *Surveys in Combinatorics 1991* (ed. A.D. Keedwell, Cambridge University Press, 1991) have been included, without significant change, in Chapters 3 and 6. With few other exceptions, the results in Chapters 2 to 9 have not previously appeared in book form.

Finally, the authors gratefully acknowledge individual financial support from the following sources over the past ten years: the British Council, the Carnegie Foundation, the Mathematical Institute of the Serbian Academy of Sciences, the Science and Engineering Research Council, the University of Belgrade and the University of Stirling.

January 1996                                                    D.C., P.R., S.S.

# Acknowledgements

The authors are grateful to the following publishers for permission to reproduce, without significant change, sections of the articles cited: The Charles Babbage Research Center [Row7], Gordon and Breach Publishers [RoYu], Elsevier Science Inc. [CvRS1], Elsevier Science Publishers BV [CvRS2], Springer-Verlag GmbH [Row11].

# 1

# A background in graph spectra

In Section 1.1 we introduce notation and terminology which will be used throughout the book. The limitations of the spectrum as a graph invariant are illustrated by the discussion of non-isomorphic cospectral graphs in Section 1.2. In Section 1.3 we describe the extent to which certain classes of graphs are characterized by spectral properties, and in Section 1.4 we discuss ways of extending the spectrum to a set of invariants which together are sufficient to characterize a graph.

## 1.1 Basic notions and results

A comprehensive treatment of the theory of graph spectra is given in the monograph [CvDS], while some of the underlying results from matrix theory are given in Appendix A. Here we present only those basic notions and further results which are needed frequently in other chapters. We recommend as general references the texts by Biggs [Big] and Harary [Har2].

The *adjacency matrix* of a (multi)(di)graph $G$, with vertex set $\{1, 2, \ldots, n\}$, is the $n \times n$ matrix $A = (a_{ij})$ whose $(i, j)$-entry $a_{ij}$ is equal to the number of edges, or arcs, originating at the vertex $i$ and terminating at the vertex $j$. Two vertices of $G$ are said to be *adjacent* if they are connected by an edge or arc. Unless we indicate otherwise we shall assume that $G$ is an undirected graph without loops or multiple edges.

As an example, the adjacency matrix of a 4-cycle is illustrated in Fig. 1.1.

The characteristic polynomial $\det(xI - A)$ of the adjacency matrix $A$ of $G$ is called the *characteristic polynomial of* $G$ and denoted by $P_G(x)$. The eigenvalues of $A$ (i.e. the zeros of $\det(xI - A)$) and the spectrum of $A$ (which consists of the $n$ eigenvalues) are also called the *eigenvalues* and

1

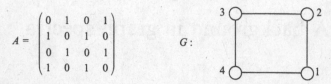

$$A = \begin{pmatrix} 0 & 1 & 0 & 1 \\ 1 & 0 & 1 & 0 \\ 0 & 1 & 0 & 1 \\ 1 & 0 & 1 & 0 \end{pmatrix} \qquad G:$$

Fig. 1.1. A labelled graph $G$ and its adjacency matrix $A$.

the *spectrum* of $G$, respectively. These notions are independent of vertex labelling because a reordering of vertices results in a similar adjacency matrix. The eigenvalues of $G$ are usually denoted by $\lambda_1, \ldots, \lambda_n$; they are real because $A$ is symmetric. Unless we indicate otherwise, we shall assume that $\lambda_1 \geq \lambda_2 \geq \cdots \geq \lambda_n$ and use the notation $\lambda_i = \lambda_i(G)$ for $i = 1, 2, \ldots, n$. Clearly, isomorphic graphs have the same spectrum.

The eigenvalues of $A$ are the numbers $\lambda$ satisfying $A\mathbf{x} = \lambda\mathbf{x}$ for some non-zero vector $\mathbf{x} \in \mathbb{R}^n$. Each such vector $\mathbf{x}$ is called an *eigenvector* of the matrix $A$ (or of the labelled graph $G$) belonging to the eigenvalue $\lambda$. The relation $A\mathbf{x} = \lambda\mathbf{x}$ can be interpreted in the following way: if $\mathbf{x} = (x_1, x_2, \ldots, x_n)^T$ then $\lambda x_u = \sum_{v \sim u} x_v$ where the summation is over all neighbours $v$ of the vertex $u$. If $\lambda$ is an eigenvalue of $A$ then the set $\{\mathbf{x} \in \mathbb{R}^n : A\mathbf{x} = \lambda\mathbf{x}\}$ is a subspace of $\mathbb{R}^n$, called the *eigenspace* of $\lambda$ and denoted by $\mathscr{E}(\lambda)$ or $\mathscr{E}_A(\lambda)$. Such eigenspaces are called eigenspaces of $G$. Of course, relabelling of the vertices in $G$ will result in a permutation of coordinates in eigenvectors (and eigenspaces).

For the eigenvalues $\lambda$ of the graph in Fig. 1.1 we have

$$P_G(\lambda) = \begin{vmatrix} \lambda & -1 & 0 & -1 \\ -1 & \lambda & -1 & 0 \\ 0 & -1 & \lambda & -1 \\ -1 & 0 & -1 & \lambda \end{vmatrix} = \lambda^4 - 4\lambda^2 = 0.$$

The eigenvalues in non-increasing order are $\lambda_1 = 2$, $\lambda_2 = 0$, $\lambda_3 = 0$, $\lambda_4 = -2$ with eigenvectors $\mathbf{x}_1$, $\mathbf{x}_2$, $\mathbf{x}_3$, $\mathbf{x}_4$ where $\mathbf{x}_1 = (1, 1, 1, 1)^T$, $\mathbf{x}_2 = (1, 1, -1, -1)^T$, $\mathbf{x}_3 = (-1, 1, 1, -1)^T$, $\mathbf{x}_4 = (1, -1, 1, -1)^T$. We have $\mathscr{E}(2) = \langle \mathbf{x}_1 \rangle$, $\mathscr{E}(0) = \langle \mathbf{x}_2, \mathbf{x}_3 \rangle$ and $\mathscr{E}(-2) = \langle \mathbf{x}_4 \rangle$, where $\langle \mathbf{y}_1, \mathbf{y}_2, \ldots, \mathbf{y}_k \rangle$ denotes the subspace spanned by the vectors $\mathbf{y}_1, \mathbf{y}_2, \ldots, \mathbf{y}_k$.

The following remarks on matrices will serve to establish more notation.

Since $A$ is a symmetric matrix with real entries there exists an orthogonal matrix $U$ such that $U^T A U$ is a diagonal matrix, $D$ say. Here $D = \text{diag}(\lambda_1, \lambda_2, \ldots, \lambda_n)$ (where $\lambda_1, \lambda_2, \ldots, \lambda_n$ are the eigenvalues of $A$ in some order), and the columns of $U$ are corresponding eigenvectors which form an orthonormal basis of $\mathbb{R}^n$. If this basis is constructed by stringing together orthonormal bases of the eigenspaces of $A$ then $D = \mu_1 E_1 + \cdots + \mu_m E_m$ where $\mu_1, \ldots, \mu_m$ are the distinct eigenvalues of $A$ and each $E_i$ has block diagonal form $\text{diag}(O, \ldots, O, I, O, \ldots O)$ $(i = 1, \ldots, m)$. Then $A$ has the *spectral decomposition*

$$A = \mu_1 P_1 + \cdots + \mu_m P_m \tag{1.1.1}$$

where $P_i = U E_i U^T$ $(i = 1, \ldots, m)$. For fixed $i$, if $\mathscr{E}(\mu_i)$ has $\{\mathbf{x}_1, \ldots, \mathbf{x}_d\}$ as an orthonormal basis then

$$P_i = \mathbf{x}_1 \mathbf{x}_1^T + \cdots + \mathbf{x}_d \mathbf{x}_d^T \tag{1.1.2}$$

and $P_i$ represents the orthogonal projection of $\mathbb{R}^n$ onto $\mathscr{E}(\mu_i)$ with respect to the standard orthonormal basis of $\mathbb{R}^n$. Moreover $P_i^2 = P_i = P_i^T$ $(i = 1, \ldots, m)$ and $P_i P_j = O$ $(i \neq j)$. We shall assume throughout that $\mu_1 > \cdots > \mu_m$. We shall also need the observation that for any polynomial $f$, we have

$$f(A) = f(\mu_1)P_1 + \cdots + f(\mu_m)P_m.$$

In particular, $P_i$ is a polynomial in $A$ for each $i$; explicitly, $P_i = f_i(A)$ where

$$f_i(x) = \frac{\prod_{s \neq i}(x - \mu_s)}{\prod_{s \neq i}(\mu_i - \mu_s)}.$$

The largest eigenvalue $(\mu_1 = \lambda_1)$ of a graph $G$ is called the *index* of $G$; since adjacency matrices are non-negative there is a corresponding eigenvector whose entries are all non-negative (see Theorem A.1 of Appendix A). The index is a simple eigenvalue if and only if $G$ is connected, equivalently if and only if $A$ is irreducible, and in this situation the corresponding eigenspace is spanned by a vector whose entries are all positive (see Theorem A.2 of Appendix A). The unique positive unit eigenvector corresponding to the index of a connected (labelled) graph $G$ is called the *principal eigenvector* of $G$. We may extend this notion as follows to the case in which $G$ is a graph with just one non-trivial component. Without loss of generality the adjacency matrix of $G$ then has the form $\begin{pmatrix} A & O \\ O & O \end{pmatrix}$ where $A$ is irreducible. Since $\mu_1 \neq 0$ we have

$$\begin{pmatrix} A & O \\ O & O \end{pmatrix} \begin{pmatrix} \mathbf{x} \\ \mathbf{y} \end{pmatrix} = \mu_1 \begin{pmatrix} \mathbf{x} \\ \mathbf{y} \end{pmatrix} \text{ if and only if } A\mathbf{x} = \mu_1 \mathbf{x} \text{ and } \mathbf{y} = \mathbf{0}.$$

Accordingly there is a unique non-negative unit eigenvector corresponding to $\mu_1$, and its $i$-th entry is zero if and only if the $i$-th vertex is isolated. Now we call this eigenvector the principal eigenvector of the labelled graph $G$.

**1.1.1 Definition** *For any matrix $M = (\sigma_{i,j})_{m,n}$ we define a weighted bipartite (di)graph $K(M)$ with vertices $r_1, \ldots, r_m$ and $c_1, \ldots, c_n$ in the respective parts as follows: if $\sigma_{ij} \neq 0$, then the vertices $r_i$ and $c_j$ are joined by an edge (arc) whose weight is $\sigma_{ij}$. The (di)graph $K(M)$ is called the König (di)graph of the matrix $M$.*

Here, vertices $r_1, \ldots, r_m$ correspond to rows of $M$ while vertices $c_1, \ldots, c_n$ correspond to the columns of $M$. The term 'König digraph' was introduced in [Cve11] in view of König's use of digraphs in investigating certain problems in matrix theory [Kön].

Next we present certain notation, definitions and results from graph theory.

As usual, $K_n, C_n$ and $P_n$ denote respectively the *complete graph*, the *cycle* and the *path* on $n$ vertices. The *wheel* $W_{n+1}$ is obtained from $C_n$ by adding a vertex $v$ and edges (*spokes*) joining $v$ to each vertex of the $n$-cycle. Further, $K_{m,n}$ denotes the *complete bipartite* graph on $m + n$ vertices. More generally, $K_{n_1,n_2,\ldots,n_k}$ denotes the complete $k$-partite graph with parts of size $n_1, n_2, \ldots, n_k$. The *cocktail-party graph* $CP(n)$ is the unique regular graph with $2n$ vertices of degree $2n - 2$; it is obtained from $K_{2n}$ by deleting $n$ mutually non-adjacent edges.

A connected graph with $n$ vertices is said to be *unicyclic* if it has $n$ edges, *bicyclic* if it has $n + 1$ edges, and *tricyclic* if it has $n + 2$ edges.

Any set of mutually non-adjacent edges in a graph $G$ is called a *matching* of $G$. A matching of $G$ is *perfect* if each vertex of $G$ is the endvertex of an edge from the matching. The weight of a matching in a weighted graph is the sum of weights of edges contained in the matching.

The *complement* of a graph $G$ is denoted by $\overline{G}$, while $mG$ denotes the union of $m$ disjoint copies of $G$. We write $V(G)$ for the vertex set of $G$, and $E(G)$ for the edge set of $G$.

If $uv$ is an edge of $G$ we write $G - uv$ for the graph obtained from $G$ by deleting $uv$. For $v \in V(G)$, $G - v$ denotes the graph obtained from $G$ by deleting the vertex $v$ and all edges incident with $v$. More generally, for $U \subseteq V(G)$, $G - U$ is the subgraph of $G$ induced by $V(G) \setminus U$.

The *join* $G \nabla H$ of (disjoint) graphs $G$ and $H$ is the graph obtained from $G$ and $H$ by joining each vertex of $G$ with each vertex of $H$.

The *coalescence* $G \cdot H$ of (disjoint) rooted graphs $G$ and $H$ is the graph obtained from $G$ and $H$ by identifying the root of $G$ with the root of $H$.

The *line graph* $L(H)$ of any graph $H$ is defined as follows. The vertices of $L(H)$ are the edges of $H$ and two vertices of $L(H)$ are adjacent whenever the corresponding edges of $H$ have a vertex of $H$ in common. Let $N$ denote the vertex-edge $(0,1)$-incidence matrix of $H$. Then the $(0,1)$-adjacency matrices $B$ of $H$ and $A$ of $L(H)$ satisfy

$$NN^T = D + B, \quad N^T N = 2I + A, \tag{1.1.3}$$

where now $D$ is the diagonal matrix whose diagonal entries are the vertex degrees of $H$.

A *generalized line graph* $L(H; a_1, \ldots, a_n)$ is defined for graphs $H$ with vertex set $\{1, \ldots, n\}$ and non-negative integers $a_1, \ldots, a_n$ by taking the graphs $L(H)$ and $CP(a_i)$ $(i = 1, \ldots, n)$ and adding extra edges: a vertex $e$ in $L(H)$ is joined to all vertices in $CP(a_i)$ if $i$ is an endvertex of $e$ as an edge of $H$. We include as special cases an ordinary line graph $(a_1 = a_2 = \cdots = a_n = 0)$ and the cocktail-party graph $CP(n)$ $(n = 1$ and $a_1 = n)$.

Given a subset $U$ of vertices of the graph $G$, the graph $G'$ obtained from $G$ by *switching* with respect to $U$ differs from $G$ as follows: for $u \in U, v \notin U$ the vertices $u, v$ are adjacent in $G'$ if and only if they are non-adjacent in $G$. Note that switching with respect to $U$ is the same as switching with respect to its complement. Switching is described easily in terms of the *Seidel matrix* $S$ of $G$ defined as follows: the $(i, j)$-entry of $S$ is $0$ if $i = j$, $-1$ if $i$ is adjacent to $j$, and $1$ otherwise. The Seidel matrix of $G'$ is $D^{-1}SD$ where $D$ is the (involutory) diagonal matrix whose $i$-th diagonal entry is $1$ if $i \in U$, $-1$ if $i \notin U$. Now it is easy to see that switching with respect to $U$ and then with respect to $V$ is the same as switching with respect to $(U \setminus V) \cup (V \setminus U)$. It follows that switching determines an equivalence relation on graphs; moreover, switching-equivalent graphs have similar Seidel matrices and hence the same Seidel spectrum.

**1.1.2 Example** Let $S_1, S_2, S_3$ be sets of vertices of $L(K_8)$ which induce subgraphs isomorphic to $4K_1$, $C_5 \cup C_3$ and $C_8$, respectively. The graphs $Ch_1, Ch_2, Ch_3$ obtained from $L(K_8)$ by switching with respect to $S_1, S_2, S_3$ respectively are called the *Chang graphs*. The graphs $L(K_8), Ch_1, Ch_2, Ch_3$ are regular of degree 12, cospectral and mutually non-isomorphic (see, for example, [BrCN], p. 105, and also [Sei2]).

If we switch $L(K_8)$ with respect to the set of neighbours of a vertex $v$, we obtain a graph $H$ in which $v$ is an isolated vertex. If we delete $v$ from $H$ we obtain a graph which is called the *Schläfli* graph.     □

The *Laplacian* (or *admittance*) matrix $L$ of $G$ is the $n \times n$ matrix $D - A$ where $A$ is the adjacency matrix of $G$ and $D$ is the diagonal matrix whose $(i, i)$-entry is the degree $d_i$ of the $i$-th vertex $(i = 1, 2, \ldots, n)$. In this book we are concerned primarily with the adjacency matrix $A$, but we note here one property of $L$. For any fixed orientation of the edges of $G$ we may define the corresponding edge-vertex incidence matrix $C$ as the matrix whose $(e, i)$-entry is 1 if $i$ is the endvertex of the edge $e$, $-1$ if $i$ is the initial vertex of $e$, and 0 otherwise. Note that always $L = C^T C$, and so $\mathbf{x}^T L \mathbf{x} = \|C\mathbf{x}\|^2 \geq 0$ for all $\mathbf{x} = (x_1, x_2, \ldots, x_n)^T \in \mathbb{R}^n$. Moreover $C\mathbf{j} = \mathbf{0}$ and if $G$ is connected then conversely $x_1 = x_2 = \cdots = x_n$ whenever $C\mathbf{x} = \mathbf{0}$. It follows that the multiplicity of 0 as an eigenvalue of $L$ is equal to the number of components of $G$. In particular, the second smallest eigenvalue $\lambda$ is zero if and only if $G$ is not connected.

## 1.2 The graph isomorphism problem and cospectral graphs

Since the spectrum of a graph is a graph invariant it is natural to ask whether the spectrum determines a graph to within isomorphism. This attractive but, in a sense, naive conjecture has appealed to many who have encountered graph spectra. If the conjecture were valid, it would provide a polynomial algorithm to decide whether two graphs are isomorphic and thereby solve the *graph isomorphism problem*. As is well known the graph isomorphism problem is not solved in so far as its algorithmic complexity is not known. It belongs to the class NP but it is not known whether it is NP-complete or belongs to the class P.

Graphs with the same spectrum are called *isospectral* or *cospectral* graphs. In this section we review what is known about cospectral graphs.

We first present early results (up to 1971) on cospectral graphs following the review given in [Cve7].

In [CoSi] Collatz and Sinogowitz had already noted that the spectrum of a graph does not determine the graph up to isomorphism. They gave an example of two isospectral trees with eight vertices and different sets of vertex degrees.

The term 'pair of isospectral non-isomorphic graphs' will be denoted by PING. The literature contains various examples of PINGs and a few of the constructions will be described in this section. The importance of PINGs lies in the following observations:

(1) For every pair of non-isomorphic graphs one can find a set of characteristic properties that are different for the two graphs. Therefore, every PING points to properties of graphs that are not uniquely determined by the spectrum.

(2) The existence of a PING rules out various possibilities in the search for families of graphs with the property that different graphs from the same family have different spectra.

We shall restrict our attention to undirected graphs without loops or multiple edges. (It is relatively easy to construct PINGs for other kinds of graphs. All digraphs without cycles have a spectrum containing only numbers equal to zero [Sed]. A further example consisting of directed graphs with seven vertices is cited in [Pon]; see also [Djo].)

In [Har1], Harary states that his conjecture, that isospectrality implies the isomorphism of graphs, was disproved by Bose, who described a PING with 16 vertices. According to [Har1], Bruck and Hoffman also found PINGs with 16 vertices.

There are no PINGs among the connected graphs with at most five vertices – see for example the table of spectra of graphs in [CvDS]. In [Bak2] Baker gives a PING consisting of connected graphs with six vertices, and so the number five above is the best bound possible.

If we consider graphs without the assumption of connectedness, then there exists a PING with five vertices, namely $K_{1,4}$ and $C_4 \cup K_1$. This example has been generalized in [Cve7] as follows. The graph having as components $s$ isolated vertices and one complete bipartite graph $K_{n_1,n_2}$ has eigenvalues $\sqrt{n_1 n_2}, -\sqrt{n_1 n_2}$ and $n_1 + n_2 - 2 + s$ numbers equal to 0. Now consider a graph with spectrum $\sqrt{m}, -\sqrt{m}$ and $n-2$ numbers equal to 0 ($m$ a natural number). This spectrum belongs to each graph of the above type whose parameters $n_1, n_2, s$ satisfy the equations $n_1 + n_2 + s = n$, $n_1 n_2 = m$.

From these examples we see that in general we cannot determine from the spectrum whether or not a graph is connected. If however we consider the narrower class of regular graphs then this information can be extracted from the spectrum (see Theorem 1.3.13). In this case, knowledge of the spectrum is equivalent to knowledge of the Laplacian spectrum, and we recall from Section 1.1 that the Laplacian spectrum of a graph does tell us whether or not the graph is connected.

Turner [Turn2] gives a PING consisting of 12-vertex trees which have the same vertex degree sequence. The author expresses his pessimism concerning the possibility of distinguishing even graphs of restricted type by means of their spectra.

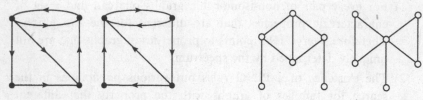

Fig. 1.2. A pair of cospectral digraphs.    Fig. 1.3. A pair of cospectral graphs.

Fisher, who encountered the graph isospectrality problem when investigating the vibration of membranes [Fis], has considered graphs with the following restrictions (among others): (1) the graph does not contain a vertex of degree 1, (2) the graph is planar. He constructed an infinite sequence of PINGs with $5n$ vertices ($n = 3, 4, \ldots$) satisfying conditions (1) and (2). An infinite sequence of sets of mutually non-isomorphic isospectral graphs was also given by Bruck in [Bruc]. It seems that PINGs with a large number of vertices are a common occurrence, and some statistical data are given by Baker in [Bak2].

PINGs can also be found in the family of regular graphs. They can arise from switching-equivalent connected regular graphs of the same degree: examples are provided by the graphs having 16, 28 and 64 vertices which occur as exceptions in Theorems 1.3.5, 1.3.6 and 1.3.27. Further examples arise in the context of Theorems 1.3.9, 1.3.10 and 1.3.11.

If a PING with $n$ vertices is known, then a PING with $m$ vertices ($m > n$) can easily be constructed by adding an arbitrary graph with $m - n$ vertices as a new component in each of the two graphs. Also, from a PING consisting of regular graphs of degree greater than 2, we can construct another PING with more vertices by taking the line graphs of the graphs in question (cf. Theorem 1.3.17).

Another review of cospectral graphs appeared in 1971, written by Harary, King, Mowshowitz and Read [HaKMR]. Among other things they construct the smallest cospectral strongly connected digraphs which are not self-converse (Fig. 1.2), the smallest pair of connected cospectral graphs (Fig. 1.3) and the smallest triplet of connected cospectral graphs (Fig. 1.4).

A third review of cospectral graphs in 1971 appeared in the paper [BaHa], which repeats some of the examples mentioned earlier, and which gives a PING consisting of trees on 12 vertices with the same degrees, the maximal degree being 4. Since these trees are relevant to chemistry the authors justify in this way the main message of the paper, expressed

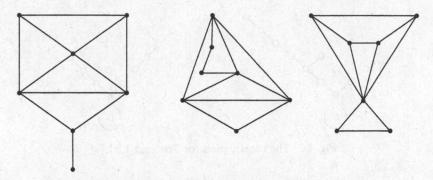

Fig. 1.4. Three cospectral graphs.

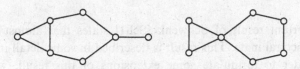

Fig. 1.5. Cospectral graphs with cospectral complements.

by its title: *the characteristic polynomial does not uniquely determine the topology of a molecule.*

The expository article [GoHMK] contains a list of smallest PINGs in various classes of graphs. In addition to the above results the paper gives the smallest cospectral graphs with cospectral complements (Fig. 1.5), the smallest cospectral forests ($K_{1,3} \cup K_2$ and $P_5 \cup K_1$), smallest cospectral regular graphs (two pairs of degree 4 on ten vertices and their complementary pairs; see Fig. 4.1) and some others.

The paper [GoMK1] presents the results of a computational study of spectra of graphs. Characteristic polynomials of all graphs up to nine vertices are computed and the cospectral graphs identified. Statistics are given for cospectral graphs in various classes of graphs.

We quote a theorem which provides a construction for cospectral trees with cospectral complements.

**1.2.1 Theorem** [GoMK1] *Let G be an arbitrary rooted graph. Let S and T be rooted trees as shown in Fig. 1.6. Then G · S and G · T are not isomorphic (unless the root of G is isolated) but are cospectral and have cospectral complements.*

Recall that $G \cdot H$ denotes the coalescence of rooted graphs $G$ and $H$. The proof of Theorem 1.2.1 is based partly on a formula for the

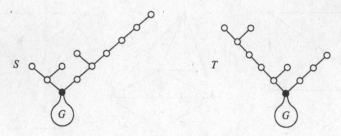

Fig. 1.6. The construction for Theorem 1.2.1.

characteristic polynomial of $G \cdot H$ given in Chapter 4: see equation
(4.3.4).

An important result of Schwenk [Sch1] states that almost all trees
have a cospectral mate. This result is described in some detail in Section
5.1. In order to formulate some extensions of this result, we define
matrix functions called immanants. If $\chi$ is an irreducible character
of the symmetric group $S_n$, if $A = (a_{ij})$ is a square matrix of order
$n$, and if $d_\chi(A) = \sum_{\pi \in S_n} \chi(\pi) a_{1\pi(1)} a_{2\pi(2)} \cdots a_{n\pi(n)}$ then $d_\chi(A)$ is called an
*immanant* of $A$. If $\chi(\pi) = 1$ for all $\pi \in S_n$ then $d_\chi(A)$ is the *permanent*
of $A$, while the alternating character yields the *determinant* of $A$. If
$A$ is the adjacency matrix of a graph $G$ and $\chi$ a fixed character then
$d_\chi(xI - A)$ is the corresponding *immanantal polynomial* of $G$. As a
generalization of a result in [Mer1], it is shown in [BotMe] that almost
every tree has a co-immanantal mate, that is, a tree which shares the
same immanantal polynomials $d_\chi(xI - A)$ for all $\chi$. Indeed the authors
prove a stronger theorem which includes the corresponding result for the
Laplacian matrix $L = D - A$: in the above statement, the set of one-
variable functions $d_\chi(xI - A)$ may be replaced by the set of three-variable
functions $d_\chi(xI - yD - zA)$.

Schwenk [Sch1] found a construction of cospectral graphs which uses
the concept of *cospectral vertices*. This construction will be described in
Section 5.1, along with the notion of *unrestricted vertices*. Both concepts
feature in the general procedures for constructing PINGs described in
[HeEl2]. This paper describes methods for constructing graphs with such
vertices, and discusses cospectral graphs with cospectral complements.

Graphs with cospectral vertices are called *endospectral* graphs [Ran2].
Hence the study of endospectral graphs is closely related to the study
of procedures for constructing cospectral graphs. Some constructions of

endospectral trees are given in [RaKl]. Endospectral trees with up to 16 vertices have been found by a computer search [KnMSTKR].

The paper [DAGT] includes a discussion of some cospectral graphs relevant to chemistry, methods for recognizing cospectrality and certain properties of eigenvectors in cospectral graphs. If the eigenvalues of a graph appear among the eigenvalues of another graph then these graphs are said to be *subspectral*. Several cases of subspectral graphs are reviewed, with an observation that in many cases the smaller graph appears as a fragment of the larger one.

Other references concerning cospectral graphs are [Ach], [Bab1], [Bab3], [Bak1], [Ben], [Chan1], [Chan2], [Chao], [Con], [Cou], [CvGu1], [DAGT], [Din], [Doo4], [Doo6], [FaGr], [GoMK4], [Hei1], [Herm], [Hern1], [Hern2], [HeEl1], [Hof1], [Jia], [KoSu], [KrPa1], [KrPa2], [LiWZ], [Mey], [RaTŽ], [Sch3], [Sei1], [SiMe], [StMa], [ZiTR].

## 1.3 Spectral characterizations of certain classes of graphs

In this section various instances of the following problem are considered:

*Given the spectrum, or some spectral characteristics of a graph, determine all graphs from a given class of graphs having the given spectrum, or the given spectral characteristics.*

In some cases, the solution of such a problem can provide a characterization of a graph up to isomorphism (see below). In other cases we can deduce structural details (see [CvDS], Chapter 3).

The section consists of three subsections. In 1.3.1 we describe some cases in which graphs are characterized by their spectra. Examples of spectral properties equivalent to structural properties are described in 1.3.2. Subsection 1.3.3 deals with cases in which graphs are characterized by a mixture of spectral and structural properties.

It should be noted that the characterization theorems presented here had been proved by the end of the seventies. Almost no further results of this type can be found in papers published subsequently. The theory of graph spectra developed in new directions in the eighties.

### *1.3.1 Characterizations by spectra*

We say that a graph $G$ is characterized by its spectrum if the only graphs cospectral with $G$ are those isomorphic to $G$. Note first that this condition is satisfied by graphs which are characterized by invariants (such as the number of vertices and edges) which can be determined from the spectrum. Examples include the complete graphs and graphs with one edge, together with their complements. Given the spectrum of a graph $G$ we can always establish whether or not $G$ is regular (see Theorem 1.3.13). It follows that if $G$ or $\bar{G}$ is regular of degree 1 then $G$ is characterized by its spectrum. By inspecting eigenvalues of cycles (see [CvDS], p.72) we can show also

**1.3.1 Theorem** [Cve7] *Any regular graph of degree 2 is characterized by its spectrum.*

**1.3.2 Remark** From the spectrum of a regular graph $G$ we can find the spectrum of $\bar{G}$ (see Proposition 4.5.2), and so it follows from Theorems 1.3.1 and 1.3.13 that any $n$-vertex graph which is regular of degree $n - 3$ is characterized by its spectrum. This result was proved for connected multigraphs by Finck [Fin].                                           □

It is straightforward to show that a graph of the form $mK_n$ is characterized by its spectrum, a fact established in complementary form in [Fin]:

**1.3.3 Theorem** *For each positive integer $n$, the complete multipartite graph $K_{n,n,...,n}$ is characterized by its spectrum.*

The next result, however, does not admit a transition to the complement.

**1.3.4 Theorem** [Cve7] *The spectrum of the graph $G$ consists of the natural numbers $n_1 - 1,...,n_k - 1$ together with $s$ numbers equal to $0$ and $n_1 + ... + n_k - k$ numbers equal to $-1$, if and only if $G$ has as its components $s$ isolated vertices together with $k$ complete graphs on $n_1,...,n_k$ vertices respectively.*

In the sixties Hoffman [Hof3] investigated the extent to which regular connected graphs are determined by their distinct eigenvalues. He pointed out that earlier work by Shrikhande [Shr] leads to two characterization theorems which (in view of Theorem 1.3.13) can be stated as follows:

**1.3.5 Theorem** *If $n \neq 8$ then $L(K_n)$ is characterized by its spectrum.*

**1.3.6 Theorem** *If $n \neq 4$ then $L(K_{n,n})$ is characterized by its spectrum.*

We shall now prove a generalization of Theorem 1.3.6 using a number of results from the next subsection. We say that a graph $G$ on $p_1 + p_2$

vertices is *semi-regular bipartite* if there exist distinct positive integers $r_1, r_2$ such that $G$ has an independent set of $p_1$ vertices of degree $r_1$ and an independent set of $p_2$ vertices of degree $r_2$.

**1.3.7 Theorem** ([Cve7], [Doo2]) *If $m + n \geq 19$ and if $\{m, n\} \neq \{2s^2 + s, 2s^2 - s\}$, where $s$ is a positive integer, then $L(K_{m,n})$ is characterized by its spectrum.*

*Proof* We assume that $m > 1$ and $n > 1$ because otherwise $L(K_{m,n})$ is complete and hence determined by its spectrum. Then the eigenvalues of $L(K_{m,n})$ are $m + n - 2$, $m - 2$, $n - 2$, $-2$ with multiplicities 1, $n - 1$, $m - 1$, $mn - m - n + 1$ respectively. (This follows, for example, from Theorem 2.3.4 because, in the notation of Section 2.3, $L(K_{m,n}) = K_m + K_n$.)

Now let $G$ be a graph with the same spectrum as $L(K_{m,n})$. We know from Theorem 1.3.13 that $G$ is a regular connected graph. Since it has degree $\geq 17$ and least eigenvalue $-2$, Theorem 1.3.19 tells us that $G$ is a line graph, say $G = L(H)$ where $H$ has no isolated vertices. Since $G$ is regular, $H$ is either a regular graph or a semi-regular bipartite graph.

Suppose first that $H$ is regular of degree $r$. Then $2(r - 1) = m + n - 2$, whence $m + n$ is even and $r = \frac{1}{2}(m + n)$. The number $q$ of edges in $H$ is the number of vertices in $G$, namely $mn$. Now the number of vertices of $H$ is $\frac{2q}{r}$, or $4\frac{mn}{m+n}$. By considering the multiplicity of $-2$ as a root of $P_{L(H)}(x)$ as given by Theorem 1.3.17 we find that $-\frac{1}{2}(m + n)$ is an eigenvalue of $H$ with multiplicity

$$1 - (mn - 4\frac{mn}{m + n}) = 1 - \frac{(m - n)^2}{m + n}.$$

We deduce that $m = n$, for otherwise $\{m, n\} = \{2s^2 + s, 2s^2 - s\}$, contrary to assumption. Accordingly, the result in this case follows from Theorem 1.3.5.

Secondly suppose that $H$ is a semi-regular bipartite graph with $p_1$ independent vertices of degree $r_1$ and $p_2$ independent vertices of degree $r_2$. Then $p_1 r_1 = p_2 r_2 = mn$ and $r_1 + r_2 = m + n$. By Theorem 1.3.18, $r_1 - 2$ is an eigenvalue of $L(H)$, and a comparison with the eigenvalues of $G$ yields three possibilities: (1) $r_1 = m$, (2) $r_1 = n$, (3) $r_1 = m + n$. The third cannot arise because $r_2 \neq 0$, while in cases (1) and (2) we have $H = K_{m,n}$ as required.     □

In view of further results described in Section 1.3.2, Theorem 1.3.7 can be improved as follows.

**1.3.8 Theorem** ([BuCS1],[BuCS2]) *The graph $L(K_{m,n})$ $(m \geq n)$ is characterized by its spectrum unless*

  (a) $m = n = 4$, *when there is just one exceptional graph, or*

  (b) $m = 6$ *and* $n = 3$, *when there is just one exceptional graph, or*

  (c) $m = 2s^2 + s$, $n = 2s^2 - s$ *and there exists a symmetric Hadamard matrix of order* $4s^2$, *with constant diagonal*.

There are certain families of graphs, defined in terms of graph structure, which have the property that different graphs from the same family have different spectra. In view of Theorem 1.3.1, the regular graphs of degree 2 constitute such a family. A further example consists of the transitive graphs with a prime number of vertices [Turn1] and another is the family $\mathcal{H}$ of all bicyclic Hamiltonian graphs (cycles with one chord). Indeed, different graphs in $\mathcal{H}$ which have the same number of vertices are distinguished by their indices (see [Row3] and [SiKo]).

We now turn to the situation in which a family $\mathcal{G}$ of graphs is spectrally determined in the following (weaker) sense: if $G \in \mathcal{G}$ and $H$ is cospectral with $G$ then $H \in \mathcal{G}$. We describe four such families in terms of structural properties. Recall that the graph of a 2-design $\mathcal{D}$ is a bipartite graph on the points and blocks of $\mathcal{D}$, with adjacency in the graph corresponding to incidence in $\mathcal{D}$. We refer to the corresponding line graph as the line graph of $\mathcal{D}$.

**1.3.9 Theorem** (cf. [Hof3]) *Let $G$ be the line graph of a projective plane of order $n$. If the graph $H$ is cospectral with $G$ then it is the line graph of a projective plane of order $n$.*

**1.3.10 Theorem** (cf. [HoRa1]) *Let $G$ be the line graph of an affine plane of order $n$. If the graph $H$ is cospectral with $G$ then $H$ is the line graph of an affine plane of order $n$.*

**1.3.11 Theorem** (cf. [HoRa2]) *Let $G$ be the line graph of a symmetric balanced incomplete block design with parameters $(v, k, \lambda) \neq (4, 3, 2)$. If the graph $H$ is cospectral with $G$ then $H$ is the line graph of a block design with the same parameters.*

The results described in Subsection 1.3.2 enable us to extend this characterization to 2-designs in general:

**1.3.12 Theorem** ([BuCS1], [BuCS2]) *Let $G_1$ be the line graph of a 2-design with parameters $v, k, b, r, \lambda$. If $G_2$ is cospectral with $G_1$ then one of the following holds:*

*(a)* $G_2$ *is the line graph of a 2-design having the same parameters;*

*(b)* $(v, k, b, r, \lambda) \in \{(3, 2, 6, 4, 2), (4, 3, 4, 3, 2), (4, 4, 4, 4, 4), (3, 3, 6, 6, 6)\};$

*(c)* $v = \frac{1}{2}s(s-1)$, $k = t(s-1)$, $b = \frac{1}{2}s(s+1)$, $r = t(s+1)$, $\lambda = \frac{2t(st-t-1)}{s-2}$,
*where $s$ and $t$ are integers with $st$ even, $t \le \frac{1}{2}s$, $(s-2)|2t(t-1)$,
and $G_2 = L(H)$ where $H$ is a regular graph on $s^2 - 1$ vertices with
the eigenvalues $st$, $\pm\sqrt{ts(s-1-t)(s-2)^{-1}}$, $-t$ of multiplicities $1$,
$\frac{1}{2}(s-2)(s+1)$, $\frac{1}{2}(s-2)(s+1)$, $s$, respectively.*

Just four exceptional graphs arise in case (b) of Theorem 1.3.12: they
are (in order) the graphs numbered 6, 9, 69 and 70 in Table 9.1 of
[BuCS1]. Further examples of spectral characterizations may be found
in [Cve21] and [Doo3].

## *1.3.2 Characterizations by spectral properties*

Here we discuss problems of the following sort. *Some information on
the spectrum of a graph is known; determine all graphs having the given
spectral property.*

Sometimes only limited information on the spectrum determines the
graph completely. For example if $G$ is a graph with $n$ vertices whose
index is equal to $n - 1$, then $G$ is a complete graph. In [Hof2] examples
of characterizations of regular graphs by the so-called *polynomial* of the
graph are given. (This polynomial is the polynomial $f$ of least degree
such that $f(A) = J$, where each entry of $J$ is 1.) In a majority of cases
however the given spectral property determines a class of graphs rather
than a single graph.

The next two theorems provide typical examples of results of this sort.

**1.3.13 Theorem** *Let $\lambda_1 = r, \lambda_2, \ldots, \lambda_n$ $(\lambda_1 \ge \lambda_2 \ge \cdots \ge \lambda_n)$ be the spectrum
of a graph $G$. The graph $G$ is regular (of degree $r$) if and only if $\sum_{i=1}^{n} \lambda_i^2 = n\lambda_1$. In the case of regularity the number of components is equal to the
multiplicity of $\lambda_1$.*

**1.3.14 Theorem** *Let $G$ be a connected graph. Then the following statements
are equivalent:*

*(i) $G$ is bipartite;*

*(ii) if $\lambda$ is an eigenvalue of $G$ then $-\lambda$ is also an eigenvalue of $G$ with the
same multiplicity;*

*(iii) if $r$ is the largest eigenvalue of $G$ then $-r$ is an eigenvalue of $G$.*

See [CvDS], p. 94 and p. 87, for additional information on these theorems.

Next we turn to a special property of line graphs.

**1.3.15 Theorem** [Hof4] *For the smallest eigenvalue q from the spectrum of the line graph L(G) of the arbitrary graph G, we have q ≥ −2. If G has more edges than vertices, then q = −2.*

The first part of this theorem follows from the expression for $N^T N$ in (1.1.3). The following, slightly stronger, version appears as Theorem 6.11 of [CvDS].

**1.3.16 Theorem** ([Doo1], [Doo5]) *Let G be a non-trivial connected graph. The least eigenvalue of L(G) is greater than or equal to −2, equality holding if and only if G has an even cycle or two odd cycles.*

If $G$ is a regular graph, then the characteristic polynomial of $L(G)$ can be expressed in terms of the characteristic polynomial of $G$. The following theorem is proved in [Sac2].

**1.3.17 Theorem** *If G is a regular graph of degree r, with n vertices and $m (= \frac{1}{2}nr)$ edges, then the following relation holds:*

$$P_{L(G)}(x) = (x+2)^{m-n} P_G(x-r+2). \tag{1.3.1}$$

In [Vah] and [Kel] analogous relations are given for the characteristic polynomials of matrices $D+A$ and $D−A$ (where $D$ is the diagonal matrix of vertex degrees, and $A$ is the adjacency matrix).

The next theorem shows that a relation between $P_G(x)$ and $P_{L(G)}(x)$ can be established for certain non-regular graphs.

**1.3.18 Theorem** [Cve7] *Let G be a semi-regular bipartite graph with $n_1$ mutually non-adjacent vertices of degree $r_1$ and $n_2$ mutually non-adjacent vertices of degree $r_2$, where $n_1 \geq n_2$. Then*

$$P_{L(G)}(x) = (x+2)^{\beta} \sqrt{\left(-\frac{\alpha_1}{\alpha_2}\right)^{n_1-n_2} P_G(\sqrt{\alpha_1\alpha_2})P_G(-\sqrt{\alpha_1\alpha_2})},$$

*where $\alpha_i = x - r_i + 2$ $(i = 1, 2)$ and $\beta = n_1 r_1 - n_1 - n_2$.*

The following problem has been discussed by several authors, and is the subject of the expository article [Hof4]: *determine the graphs whose smallest eigenvalue is at least −2.* In [Hof4] and [Hof5] the following theorem from the unpublished paper [HoRa3] is mentioned:

**1.3.19 Theorem** *Let G be a regular connected graph of degree not less* *than 17 and with least eigenvalue* $q \geq -2$. *Then G is either a cocktail-party graph or a line graph. The bound 17 is the best possible.*

An analogous theorem without the assumption of regularity is given below. Here $d(u)$ denotes the degree of vertex $u$, $\delta(G)$ denotes the smallest vertex degree, and $\Delta(u, v)$ denotes the number of vertices adjacent to both of the vertices $u$ and $v$.

**1.3.20 Theorem** [Ray] *Suppose that for the graph G the following hold:* (i) $\delta(G) > 43$, *(ii)* $q = -2$, *(iii) for non-adjacent vertices* $u_1$ *and* $u_2$, $\Delta(u_1, u_2) < d(u_i) - 2$ $(i = 1, 2)$. *Then there exists a graph H such that* $G = L(H)$. *Conversely, if* $G = L(H)$ *with* $\delta(H) > 3$, *then G satisfies conditions (i) and (ii).*

The following observation is crucial to the further investigation of graphs with least eigenvalue $-2$. If $A$ is the adjacency matrix of such a graph $G$ then the symmetric matrix $I + \frac{1}{2}A$ is non-negative and hence expressible as $B^T B$ where $B$ is an $r \times n$ matrix and $r$ is the rank of $I + \frac{1}{2}A$. (To see this, note that $I + \frac{1}{2}A$ is orthogonally diagonalizable.) Thus $I + \frac{1}{2}A$ may be regarded as the Gram matrix of the $n$ vectors in $\mathbb{R}^r$ which form the columns of $B$. From the equation $I + \frac{1}{2}A = B^T B$ we see that these vectors are unit vectors at angles of 60 deg or 90 deg according as the corresponding vertices of $G$ are adjacent or non-adjacent. Now the sets of lines at 60 deg or 90 deg in Euclidean space (called *root systems*) have been determined by Cameron, Goethals, Seidel and Shult [CaGSS]. In order to describe some of the consequences, let $\mathscr{G}$ denote the set of all connected regular graphs which have least eigenvalue $-2$, and which are neither a line graph nor a cocktail-party graph. First we have an improvement of Theorem 1.3.19:

**1.3.21 Theorem** ([HoRa3], [CaGSS]) *Any graph in* $\mathscr{G}$ *has at most 28 vertices and has degree at most 16. These bounds are best possible.*

By investigating the underlying root systems, Bussemaker, Cvetković and Seidel were able to determine the graphs in $\mathscr{G}$, partly by means of a computer search reported in [BuCS2]. There are 187 of them, and they are listed in Table 9.1 of [BuCS1]. As a consequence of this classification, certain characterizations appearing in the literature can be made more precise. The results are described in the following theorems.

**1.3.22 Theorem** *For each* $G \in \mathscr{G}$ *there exists a graph H, with at most eight vertices, such that G is switching equivalent to the line graph of H.*

**1.3.23 Theorem** *Each graph from* $\mathscr{G}$ *is an induced subgraph of one of the* *three Chang graphs or of the Schläfli graph, except for five switching-equivalent graphs on 22 vertices with degree 9.*

The five exceptions are the graphs numbered 148-152 in Table 9.1 of [BuCS1].

**1.3.24 Theorem** *There exist exactly 68 regular graphs which are not line graphs but which are cospectral with a line graph.*

The 68 graphs of Theorem 1.3.24 are those numbered 6, 9-13, 35-54, 59, 60, 69, 70, 108-134, 153-163 in Table 9.1 of [BuCS1]. They have been constructed in [CvRa1] in such a way that the results can be checked without the use of a computer. Further, the following theorem has been proved without recourse to a computer search.

**1.3.25 Theorem** [CvDo1] *The spectrum of a graph G determines whether or not G is a regular, connected line graph except for seventeen cases. In these cases, either G has the spectrum of L(H) where H is one of the 3-connected regular graphs on eight vertices, or H is a connected, semi-regular, bipartite graph on* $6 + 3$ *vertices.*

Finally we can describe how non-isomorphic, cospectral, regular line graphs arise.

**1.3.26 Theorem** *Let* $L(G_1)$, $L(G_2)$ *denote cospectral, connected, regular line graphs of the connected graphs* $G_1$, $G_2$. *Then one of the following holds:*

(a) $G_1$ *and* $G_2$ *are cospectral regular graphs with the same degree;*

(b) $G_1$ *and* $G_2$ *are cospectral semi-regular bipartite graphs with the same parameters;*

(c) $\{G_1, G_2\} = \{H_1, H_2\}$, *where* $H_1$ *is regular and* $H_2$ *is semi-regular bipartite; in addition there exist integers* $s, t$ *with* $0 < t < \frac{1}{2}s$, *and non-negative real numbers* $\lambda_i < t\sqrt{s^2 - 1}$, $i = 2, 3, \ldots, \frac{1}{2}s(s-1)$, *such that* $H_1$ *has* $s^2 - 1$ *vertices, degree* $st$, *and the eigenvalues*

$$st, \; \pm\sqrt{\lambda_i^2 + t^2}, -t \quad (\text{of multiplicity } s);$$

$H_2$ *has* $s^2$ *vertices, parameters* $n_1 = \frac{1}{2}s(s + 1)$, $n_2 = \frac{1}{2}s(s - 1)$, $r_1 = t(s - 1)$, $r_2 = t(s + 1)$, *and the eigenvalues* $\pm t\sqrt{s^2 - 1}$, $\pm\lambda_i$, $0$ *(of multiplicity* $s$).

Theorems 1.3.24 and 1.3.26 represent a spectral characterization of regular connected line graphs. When restricted to special classes of graphs, the theorems have as consequences Theorems 1.3.8 and 1.3.12.

Alternative proofs of the main results of [CaGSS] are also given in [CvDo1]. Valuable commentaries on the problem of characterizing graphs with least eigenvalue $-2$ are contained in [BrCN] and [BuNe]. See also the relevant sections of [CvDGT] and Appendix B of the third edition of [CvDS].

### 1.3.3 Characterizations by a combination of spectral and non-spectral properties

In many cases spectral properties are not sufficient for a characterization. Sometimes one can improve the situation by considering some structural properties of graphs in addition to spectral properties, and then a characterization may be possible. As an example we give some characterizations of the cubic lattice graphs [Cve3]. We omit the proofs and give only a survey of results.

A cubic lattice graph with characteristic $n$ ($n > 1$) is a graph whose vertices are all the $n^3$ ordered triples of $n$ symbols, with two triples adjacent if and only if they differ in exactly one coordinate. Let $d(x, y)$ denote the distance between two vertices $x$ and $y$, $\Delta(x, y)$ the number of vertices adjacent to both $x$ and $y$, and $n_2(x)$ the number of vertices at the distance 2 from $x$. We list some of the properties of the cubic lattice graph $G$:

($P_1$) the number of vertices is $n^3$;
($P_2$) $G$ is connected and regular;
($P_3$) $n_2(x) = 3(n-1)^2$ for all $x$ in $G$;
($P_3'$) $\Delta(x, y) > 1$ for all $x, y$ such that $d(x, y) = 2$;
($P_4$) the distinct eigenvalues of the adjacency matrix of $G$ are $3n - 3$, $2n - 3, n - 3, -3$;
($P$) the adjacency matrix of $G$ has eigenvalues $\lambda_f = 3n - 3 - fn$ ($f = 0, 1, 2, 3$) with multiplicities $p_f = \binom{3}{f}(n-1)^f$.

**1.3.27 Theorem** *For $n \neq 4$ the graph $G$ is the cubic lattice graph with characteristic $n$ if and only if it has properties $(P_1)$, $(P_2)$, $(P_3)$ and $(P_4)$.*

**1.3.28 Theorem** *For $n \neq 4$ the graph $G$ is the cubic lattice graph with characteristic $n$ if and only if it has the properties $(P_1)$, $(P_2)$, $(P_3')$ and $(P_4)$.*

**1.3.29 Theorem** *For $n \neq 4$ the graph $G$ is the cubic lattice graph with characteristic $n$ if and only if it has properties $(P)$ and $(P_3')$.*

It can be proved that properties $(P_1)$, $(P_2)$ and $(P_4)$ are equivalent with $(P)$.

Theorem 1.3.27 was proved in [Las] for $n > 7$. A similar result for the tetrahedral graph can be found in [BoLa]. A method for determining the eigenvalues of similar graphs is described in [Nor].

## 1.4 The search for complete sets of invariants

A graph invariant $\phi$ is said to be *complete* if, for any graphs $G, H$, the equality $\phi(G) = \phi(H)$ implies that $G$ is isomorphic to $H$. Similarly a set of graph invariants is called *complete* if it determines any graph to within isomorphism. For some investigations concerning complete sets of invariants, see [BaPa], [BoMe], [Kri], [Mas], [RiMW], [Turn2].

We have seen that the spectrum of a graph does not, in general, constitute a complete set of invariants; but complete sets of invariants do exist. For instance, it is well known that a graph $G$ can be characterized (determined up to isomorphism) by the largest (or least) binary number obtained by concatenation of the rows (or the rows of the upper triangle) of an adjacency matrix of $G$. An ordering of vertices which yields the characterizing binary number can be considered as a canonical vertex ordering. One can consider several variations of this idea but it turns out that known algorithms for finding such a complete invariant are exponential (see [ReCo], [Bab2]). (This is not to say that the extremal binary number is of no practical use; indeed it has been used successfully in recognizing graphs.)

Nevertheless it would be useful if a complete set of invariants were computable in polynomial time. No such set has been identified to date, and pessimism has been expressed in the literature concerning this question [ReCo]. Optimists may point to the fact (Chapter 2) that a graph is determined by its eigenvalues and eigenspaces, both of which can be found in polynomial time, but this is to ignore the non-invariant nature of eigenspaces: the components of eigenvectors are ordered according to a labelling of vertices. Nevertheless we can reasonably expect that a study of eigenspaces will at least enable us to extend spectral techniques in graph theory, and in this book we present results which have been obtained through efforts in this direction. We can recognize three levels in these efforts: (i) eigenvectors (Chapters 2 and 3), (ii) angles between eigenspaces and coordinate axes (Chapters 4, 5, 6), and (iii) star partitions and star bases of eigenspaces (Chapters 7 and 8).

In Chapters 3 and 6 the emphasis is on one-dimensional eigenspaces, in particular those spanned by a principal eigenvector **x**. The context is the lexicographical ordering of graphs by eigenvalues. Here and in various areas of application (cf. Section 9.3) a natural question is how the largest eigenvalue varies when a graph is modified in some way, for example by the addition or deletion of a vertex or edge. The components of **x** play a key role in answering this question.

In seeking further algebraic invariants which will refine the spectral ordering of graphs we look to geometric attributes of eigenspaces which are coordinate-free. Angles between co-ordinate axes and eigenspaces meet this requirement because they themselves can be ordered naturally. We order these angles by cosines: in the case of an eigenspace spanned by a principal eigenvector **x**, these cosines are just the components of **x**. In Chapter 4 we discuss angles in the general context, and in Chapter 5 we investigate the extent to which a graph is characterized by its eigenvalues and angles.

The geometrical approach leads to the notion of a star partition of vertices. Such a partition is defined in Chapter 7 in terms of orthogonal projections of the coordinate axes onto eigenspaces. Star partitions merit attention because they provide (i) a natural one-to-one correspondence (in general, non-unique) between eigenvalues and vertices, (ii) an explanation of the role of an individual eigenvalue in the structure of a graph, (iii) a means of constructing natural bases for each of the eigenspaces. The import of (iii) is that for a given graph $G$ we obtain only finitely many natural bases for the underlying vector space, since there are are only finitely many star partitions and only finitely many orderings of the vertices. (This is in contrast to the situation for orthonormal bases of eigenvectors: there are infinitely many such bases unless each eigenvalue of $G$ is simple.) A basis constructed from a star partition is called a star basis, and we obtain a canonical star basis by choosing one which is extremal in some sense. Note that a canonical ordering of vertices is associated implicitly with a canonical star basis, and that the eigenvalues together with such a basis constitute a complete set of invariants. The complexity question is now whether a canonical star basis can be constructed in polynomial time, and the current state of play is outlined in Chapter 8. Although the search for a canonical star basis motivated the study of star partitions, recent results relating them directly to graph structure (see Chapter 7) show them to be of further interest in their own right.

# 2

# Eigenvectors of graphs

In situations where the spectrum of a graph does not provide sufficient structural information, a natural way of extending spectral techniques is to bring into consideration the eigenvectors of an adjacency matrix (see, for example, [Cve19] and [Mal]). The basic results presented in this chapter involve eigenvectors in a number of ways; for example they are used to count walks in a graph, to determine the behaviour of spectra under certain graph operations, and to relate the symmetry of a graph to its spectrum.

## 2.1 Some fundamental results

A graph is completely determined by eigenvalues and eigenvectors in the following sense. Let $A$ be the adjacency matrix of a graph $G$ with vertices $1, 2, \ldots, n$ and eigenvalues $\lambda_1, \lambda_2, \ldots, \lambda_n$. If $v_1, v_2, \ldots, v_n$ are linearly independent eigenvectors of $A$ corresponding to $\lambda_1, \lambda_2, \ldots, \lambda_n$ respectively, if $V = (v_1 | v_2 | \cdots | v_n)$ and if $D = \text{diag}(\lambda_1, \lambda_2, \ldots, \lambda_n)$, then

$$A = VDV^{-1}.$$

Since $G$ is determined by $A$, we have proved

**2.1.1 Theorem** *Any graph is determined by its eigenvalues and a basis of corresponding eigenvectors.*

We may construct an orthonormal basis of eigenvectors $u_1, u_2, \ldots, u_n$ by stringing together orthonormal bases of eigenspaces as in Section 1.1. If $U = (u_1 | u_2 | \cdots | u_n)$ then $U^{-1} = U^T$ and we have

$$A = UDU^T, \tag{2.1.1}$$

a relation which we exploit in the next section.

22

Sometimes valuable information about a graph can be obtained from its eigenvectors alone, as in the following (straightforward) result:

**2.1.2 Theorem** *A graph G is regular if and only if its adjacency matrix has an eigenvector all of whose components are equal to 1.*

As is well known, irreducibility of the adjacency matrix of a graph is related to the property of connectedness: a strongly connected digraph has an irreducible adjacency matrix and a digraph with an irreducible adjacency matrix is strongly connected ([DuMe], [Sed]). In undirected graphs, strong connectedness reduces to the property of connectedness.

According to Theorem A.2 of Appendix A, the index of a strongly connected digraph is a simple eigenvalue of the adjacency matrix with a corresponding positive eigenvector. If the adjacency matrix is symmetric, the converse of the last statement also holds, as shown by Theorem A.3. On combining Theorems A.2 and A.3, we obtain

**2.1.3 Theorem** *A graph is connected if and only if its index is a simple eigenvalue with a positive eigenvector.*

Theorem A.4 can also be translated into the language of graph theory:

**2.1.4 Theorem** *If the index r of a graph G has multiplicity p, and if there is a positive eigenvector in the eigenspace corresponding to r, then G has exactly p components, each with index r.*

Theorem 2.1.4 will be used in Section 2.3 in studying the connectedness of graphs obtained by a graph operation called NEPS. The number $p$ of Theorem 2.1.4 is equal to the maximal number of linearly independent non-negative eigenvectors corresponding to $r$; indeed for any eigenvalue $\lambda$, the number of components with index $\lambda$ is equal to the maximal number of linearly independent non-negative eigenvectors corresponding to $\lambda$. This follows from the proof of the next result.

**2.1.5 Theorem** *The number of components of a (labelled) graph G is equal to the maximal number of linearly independent non-negative eigenvectors of G.*

*Proof* Without loss of generality, a non-negative eigenvector has the form $\begin{pmatrix} \mathbf{v} \\ \mathbf{0} \end{pmatrix}$, where each entry of $\mathbf{v}$ is positive. If we partition the adjacency matrix of $G$ accordingly then we obtain an equation

$$\begin{pmatrix} A & B \\ B^T & C \end{pmatrix} \begin{pmatrix} \mathbf{v} \\ \mathbf{0} \end{pmatrix} = \lambda \begin{pmatrix} \mathbf{v} \\ \mathbf{0} \end{pmatrix},$$

from which we see that $\mathbf{v}$ is an eigenvector of $A$ and $B^T\mathbf{v} = \mathbf{0}$. It follows that $B = O$ and $A$ represents a union of components of $G$, say $G_1, G_2, \ldots, G_r$. If $G_i$ has adjacency matrix $A_i$ $(i = 1, 2, \ldots, r)$ then we may take $A = \text{diag}(A_1, A_2, \ldots, A_r)$. It follows that $\mathbf{v}^T = (\mathbf{v}_1^T \,|\, \mathbf{v}_2^T \,|\, \cdots \,|\, \mathbf{v}_r^T)$, where $\mathbf{v}_i$ is a positive eigenvector of $A_i$ $(i = 1, 2, \ldots, r)$. Therefore, $\mathbf{v}_i$ is a scalar multiple of the principal eigenvector of $G_i$ $(i = 1, 2, \ldots, r)$.

Thus if $G_1, G_2, \ldots, G_s$ are all the components of $G$, with principal eigenvectors $\mathbf{x}_1, \mathbf{x}_2, \ldots, \mathbf{x}_s$ respectively, then every non-negative eigenvector of $G$ is a linear combination of the $s$ eigenvectors

$$\begin{pmatrix} \mathbf{x}_1 \\ 0 \\ \vdots \\ 0 \end{pmatrix}, \begin{pmatrix} 0 \\ \mathbf{x}_2 \\ \vdots \\ 0 \end{pmatrix}, \ldots, \begin{pmatrix} 0 \\ 0 \\ \vdots \\ \mathbf{x}_s \end{pmatrix}.$$

In particular, $s$ is the maximal number of linearly independent non-negative eigenvectors of $G$.                                                    □

In the case of an eigenvector $(v_1, v_2, \ldots, v_n)^T$ which has both positive and negative entries, we may investigate the number of components of the subgraph induced by the vertices $i$ for which $v_i \geq 0$; this more general question is pursued in Section 9.1.

## 2.2 The number of walks in a graph

We can use eigenvalues and eigenvectors to count walks in a graph. This is useful since many interesting mathematical problems can be reduced to problems of walk enumeration.

By a *walk of length $k$* in a graph (or digraph) we mean any sequence of (not necessarily different) vertices $x_1, x_2, \ldots, x_k, x_{k+1}$ such that for each $i = 1, 2, \ldots, k$ there is an edge (or arc) from $x_i$ to $x_{i+1}$. The walk is *closed* if $x_{k+1} = x_1$.

Counting walks with specified properties in a graph (or digraph) is related to graph spectra by the following well-known result.

**2.2.1 Theorem** *If $A$ is the adjacency matrix of a graph, then the $(i, j)$-entry $a_{ij}^{(k)}$ of the matrix $A^k$ is equal to the number of walks of length $k$ that originate at vertex $i$ and terminate at vertex $j$.*

Thus, for example, the number of closed walks of length $k$ is equal to the $k$-th spectral moment, since $\sum_{i=1}^{n} a_{ii}^{(k)} = \text{tr}(A^k) = \sum_{i=1}^{n} \lambda_i^k$.

Many combinatorial enumeration problems can be reduced to the enumeration of walks in a suitably chosen graph or digraph. Also, formulas giving the number of walks in terms of eigenvalues and eigenvectors represent a link between spectral and structural properties of a graph, and this is a very useful auxiliary tool in treating many problems on graphs. An important notion related to the number of walks is the main part of the spectrum, described below.

Let $G$ denote a graph with adjacency matrix $A$ and let $U = (u_{ij})$, an orthogonal matrix of eigenvectors of $A$ as described in Section 2.1. Then, according to (2.1.1),

$$a_{ij}^{(k)} = \sum_{s=1}^{n} u_{is} u_{js} \lambda_s^k. \tag{2.2.1}$$

The number $N_k$ of all walks of length $k$ in $G$ is given by

$$N_k = \sum_{i,j} a_{ij}^{(k)} = \sum_{s=1}^{n} \left( \sum_{i=1}^{n} u_{is} \right)^2 \lambda_s^k.$$

Thus we have proved

**2.2.2 Theorem** *The total number $N_k$ of walks of length $k$ in a graph $G$ is given by*

$$N_k = \sum_{s=1}^{n} C_s \lambda_s^k \quad (k = 0, 1, 2, \ldots), \tag{2.2.2}$$

*where $C_s = \left( \sum_{i=1}^{n} u_{is} \right)^2$.*

Let $\mu_1, \mu_2, \ldots, \mu_m$ be the distinct eigenvalues of the graph $G$. Then equation (2.2.2) can be written in the form

$$N_k = D_1 \mu_1^k + D_2 \mu_2^k + \cdots + D_m \mu_m^k \quad (k = 0, 1, 2, \ldots), \tag{2.2.3}$$

where $D_1, D_2, \ldots, D_m$ are numbers with the same sum as $C_1, C_2, \ldots, C_n$. Note that $D_1, D_2, \ldots, D_m$ are determined uniquely by $G$. From the spectral decomposition of $A$ (see Section 1.1) have $A^k = \mu_1^k P_1 + \cdots + \mu_m^k P_m$, and if $\mathbf{j}$ denotes the all-1 vector then $N_k = \mathbf{j}^T A^k \mathbf{j} = \sum_{i=1}^{m} \mu_i^k \mathbf{j}^T P_i \mathbf{j} = \sum_{i=1}^{m} \mu_i^k \|P_i \mathbf{j}\|^2$. This holds for all integers $k \geq 0$ and so $D_i = \|P_i \mathbf{j}\|^2$ $(i = 1, \ldots, m)$. A main eigenvalue of $G$ was originally defined in [Cve7] as an eigenvalue $\mu_i$ for which $D_i \neq 0$. This condition is equivalent in turn to $P_i \mathbf{j} \neq \mathbf{0}$, $\mathbf{j} \notin \mathscr{E}(\mu_i)^\perp$, $\mathscr{E}(\mu_i) \nsubseteq \langle \mathbf{j} \rangle^\perp$ (cf. Section 4.5). The main eigenvalues among $\mu_1, \ldots, \mu_m$ are said to constitute the *main part $\mathscr{M}$* of the spectrum of $G$, which can be characterized as follows. (Note that since $\sum_{i=1}^{m} P_i = I$ we have $\mathbf{j} = \sum_i P_i \mathbf{j}$ where the sum is over those $i$ for which $\mu_i \in \mathscr{M}$.)

**2.2.3 Theorem** [HaSc] *For a graph G, the following statements are equivalent:*

(1) $\mathcal{M}$ *is the main part of the spectrum;*

(2) $\mathcal{M}$ *is the minimum set of eigenvalues the span of whose eigenvectors includes the vector* $(1, 1, \ldots, 1)^T$;

(3) $\mathcal{M}$ *is the set of those eigenvalues which have an eigenvector not orthogonal to* $(1, 1, \ldots, 1)^T$.

The following theorem of Wei [Wei] is noted in [Berg2], p. 131:

**2.2.4 Theorem** *Let* $N_k(i)$ *be the number of walks of length $k$ starting at vertex $i$ of a non-bipartite connected graph $G$ with vertices $1, 2, \ldots, n$. Let* $s_k(i) = N_k(i) \cdot \left( \sum_{j=1}^{n} N_k(j) \right)^{-1}$. *Then, for $k \to \infty$, the vector* $(s_k(1), s_k(2), \ldots, s_k(n))^T$ *tends towards the eigenvector corresponding to the index of $G$.*

*Proof* Let $\{\mathbf{x}_1, \mathbf{x}_2, \ldots, \mathbf{x}_n\}$ be an orthonormal basis of $\mathbb{R}^n$ such that $A\mathbf{x}_i = \lambda_i \mathbf{x}_i$ $(i = 1, \cdots, n)$ and $\lambda_1 \geq \lambda_2 \geq \cdots \geq \lambda_n$; and let $\mathbf{j} = \theta_1 \mathbf{x}_1 + \theta_2 \mathbf{x}_2 + \cdots + \theta_n \mathbf{x}_n$, so that $\theta_i = \mathbf{j}^T \mathbf{x}_i$ $(i = 1, 2, \ldots, n)$. We have $\sum_{j=1}^{n} N_k(j) = \mathbf{j}^T A^k \mathbf{j}$, and so the vector under consideration is $(\mathbf{j}^T A^k \mathbf{j})^{-1} A^k \mathbf{j}$, or

$$\frac{\theta_1 \lambda_1^k \mathbf{x}_1 + \theta_2 \lambda_2^k \mathbf{x}_2 + \cdots + \theta_n \lambda_n^k \mathbf{x}_n}{\theta_1^2 \lambda_1^k + \theta_2^2 \lambda_2^k + \cdots + \theta_n^2 \lambda_n^k}.$$

By Theorem A.2 we have $\theta_1 > 0$ and $\lambda_1 > |\lambda_i|$ for all $i > 1$. Consequently the vector $(\theta_1^2 \lambda_1^k + \cdots + \theta_n^2 \lambda_n^k)^{-1} \theta_i \lambda_i^k \mathbf{x}_i$ approaches $\theta_1^{-1} \mathbf{x}_i$ if $i = 1$, $\mathbf{0}$ if $i > 1$. The results follows.                        □

Note that Theorem 2.2.4 holds also for connected *regular* bipartite graphs because then $\theta_n = 0$ (by Theorem 2.1.2) while $\lambda_1 > |\lambda_i|$ for all $i \in \{2, \ldots, n-1\}$.

The following result has a similar proof.

**2.2.5 Theorem** [LiFe] *Let $G$ be a connected non-bipartite graph with index $\mu_1$ and principal eigenvector $(x_1, x_2, \ldots, x_n)^T$. For fixed vertices $i$ and $j$, the number of $i$-$j$ walks of length $k$ is asymptotic to $\mu_1^k x_i x_j$ as $k \to \infty$.*

Now we turn to some applications. Let us first determine the number of walks of length $k$ in the path $P_n$ with $n$ vertices. The adjacency matrix

Fig. 2.1. A one-dimensional chess-board.

of $P_n$ is of the form

$$\begin{pmatrix} 0 & 1 & & & O \\ 1 & 0 & 1 & & \\ & \ddots & \ddots & \ddots & \\ & & 1 & 0 & 1 \\ O & & & 1 & 0 \end{pmatrix}.$$

It is known that the eigenvalues of this matrix are $\lambda_i = 2\cos\frac{i\pi}{n+1}$ ($i = 1,\ldots,n$). It is easy to verify that the numbers $\sqrt{\frac{2}{n+1}}\sin\frac{ij\pi}{n+1}$ ($j = 1,\ldots,n$) are the coordinates $u_{ij}$ of the normalized eigenvector $\mathbf{u}_i$ belonging to $\lambda_i$ (see [CoSi]). By the use of (2.2.2) we obtain for the number $N_{kn}$ of walks of length $k$ in $P_n$ the expression

$$N_{kn} = \frac{2^{k+1}}{n+1}\sum_{l=1}^{[\frac{n+1}{2}]}\cot^2\frac{2l-1}{n+1}\frac{\pi}{2}\cos^k\frac{2l-1}{n+1}\pi. \qquad (2.2.4)$$

This result is related to the following three problems treated in the literature.

**(1)** In [Doč] the following problem is solved.

Determine the number $N_{kn}$ of all zig-zag lines in the plane which (i) consist of segments of length $\sqrt{2}$ with direction $(\pm 1, 1)^T$, (ii) start from one of the points $(0,0),(0,1),\ldots,(0,k-1)$ and, without leaving the rectangle $0 \le x \le n$, $0 \le y \le k-1$, terminate in one of the points $(n,0),(n,1),\ldots,(n,k-1)$.

This question arises in certain problems of the theory of the function spaces, and the answer is given by (2.2.4).

**(2)** A particular result from [Cve6] reads:

*The number $N_{kn}$ of ways in which a king can make a series of $k$ moves on a one-dimensional chess board (see Fig. 2.1) is given by (2.2.4).*

This is obvious if we use the concept of a graph corresponding to a chess piece on a given chess-board. The vertices of this graph correspond to the squares of the chess-board and two vertices are adjacent if and only if the piece can proceed from one square to the other in one move. In the case considered the corresponding graph is just the path $P_n$.

**(3)** Here we shall obtain a combinatorial identity by counting certain walks in two ways. Let $n$ be a fixed positive integer and let $f_j(n)$ denote the number of sequences of non-negative integers $(a_1, \ldots, a_n)$, such that $a_1 = j$ and

$$|a_i - a_{i+1}| = 1 \quad (i = 1, \ldots, n-1). \tag{2.2.5}$$

Also, let $f_{j,k}(n)$ denote the number of sequences $(a_1, \ldots, a_n)$ which satisfy (2.2.5) with $a_1 = j$, $a_n = k$. Next, let $g_j(n)$ denote the number of sequences $(a_1, \ldots, a_n)$ satisfying $a_1 = j$ and

$$|a_i - a_{i+1}| \le 1 \quad (i = 1, \ldots, n-1). \tag{2.2.6}$$

Finally, let $g_{j,k}(n)$ denote the number of sequences $(a_1, \ldots, a_n)$ which satisfy (2.2.6) with $a_1 = j$, $a_n = k$.

Following [CvSi1] we consider a path $P_m$ where $m = n+j+p$ (arbitrary fixed $p \ge 0$) and the vertices are labelled in a natural manner from 0 to $m-1$.

It is easy to see that the number of walks of length $n-1$ in $P_m$ starting at the vertex $j$ is equal to $f_j(n)$. Moreover $f_{j,k}(n)$ is the number of such walks between vertices $j$ and $k$.

For $g_j(n)$ and $g_{j,k}(n)$ the same remarks hold for the graph $\dot{P}_m$ obtained from $P_m$ by adding a loop at each vertex.

If for each $i$ we associate the vertex labelled $i-1$ with the $i$-th row (or column) of the adjacency matrix, we obtain

$$f_{j,k}(n+1) = a_{j+1,k+1}^{(n)}, \quad f_j(n+1) = \sum_k a_{j+1,k+1}^{(n)},$$

$$g_{j,k}(n+1) = \dot{a}_{j+1,k+1}^{(n)}, \quad g_j(n+1) = \sum_k \dot{a}_{j+1,k+1}^{(n)}.$$

Here $(a_{p,q}^{(n)}) = A^n$ and $(\dot{a}_{p,q}^{(n)}) = \dot{A}^n$, where $A$ and $\dot{A}$ are the adjacency matrices of $P_m$ and $\dot{P}_m$ respectively. Using Theorem 2.2.2 we find that

$$f_{k,j}(n+1) = \frac{2}{n+j+p+2} \sum_{l=1}^{n+j+p+1} \sin \frac{(j+1)l\pi}{n+j+p+2} \sin \frac{(k+1)l\pi}{n+j+p+2}$$

$$\times \left( 2 \cos \frac{l\pi}{n+j+p+2} \right)^n.$$

Now $p$ is an arbitrary non-negative integer, and on letting $p \to +\infty$ we obtain

$$f_{j,k}(n+1) = \frac{1}{\pi} \int_{-\pi}^{\pi} \sin(j+1)x \, \sin(k+1)x \, (2\cos x)^n dx.$$

Similarly for the function $g_{j,k}(n)$ we obtain

$$g_{j,k}(n+1) = \frac{2}{n+j+p+2} \sum_{l=1}^{n+j+p+1} \sin \frac{(j+1)l\pi}{n+j+p+2} \sin \frac{(k+1)l\pi}{n+j+p+2}$$

$$\times \left( 2\cos \frac{l\pi}{n+j+p+2} + 1 \right)^n$$

$$= \frac{1}{\pi} \int_{-\pi}^{\pi} \sin(j+1)x \, \sin(k+1)x \, (2\cos x + 1)^n dx,$$

and so

$$g_{j,k}(n+1) = \sum_{l=0}^{n} \binom{n}{l} f_{j,k}(l+1). \tag{2.2.7}$$

Now, it is easy to show by the calculus of residues that

$$f_{j,k}(n+1) =$$

$$\begin{cases} \binom{n}{(n-j+k)/2} - \binom{n}{(n-j-k-2)/2} & \text{if } n \equiv j+k \;(\mathrm{mod}\,2), \\ 0 & \text{if } n \equiv j+k+1 \;(\mathrm{mod}\,2). \end{cases}$$

There is an analogous expression for $g_{j,k}(n+1)$ (see [Car]) and on substituting in equation (2.2.7) we obtain the following identity:

$$\sum_{s=0}^{\left[\frac{j}{2}\right]} (-1)^s \binom{j-s}{s} \left\{ \binom{2(l-s)+k}{l-s} - \binom{2(l-s)+k}{l-s-1} \right\}$$

$$= \binom{2l+k-j}{l} - \binom{2l+k-j}{l+k+1},$$

Some other identities can be obtained in a similar way (see [CvSi1]) but we shall not deal with them here.

Further references related to the number of walks in a graph are [Ahr], [Bot], [Cve5], [CvGu2], [Ein], [KnMSRT], [Leh], [Ran1], [RaWG].

### 2.3 Eigenspaces of NEPS

In this section we consider a very general graph operation called NEPS (_non-complete extended p-sum_) of graphs.

**2.3.1 Definition** _Let_ $\mathcal{B}$ _be a set of non-zero binary n-tuples, i.e._ $\mathcal{B} \subseteq \{0,1\}^n \backslash \{(0,\ldots,0)\}$. _The NEPS of graphs_ $G_1,\ldots,G_n$ _with basis_ $\mathcal{B}$ _is the graph with vertex set_ $V(G_1) \times \cdots \times V(G_n)$, _in which two vertices, say_ $(x_1,\ldots,x_n)$ _and_ $(y_1,\ldots,y_n)$, _are adjacent if and only if there exists an n-tuple_ $(\beta_1,\ldots,\beta_n) \in \mathcal{B}$ _such that_ $x_i = y_i$ _whenever_ $\beta_i = 0$, _and_ $x_i$ _is adjacent to_ $y_i$ _(in_ $G_i$) _whenever_ $\beta_i = 1$.

The notion of NEPS was introduced in [CvLu] and rediscovered in [She]. It generates a lot of binary graph operations in which the vertex set of the resulting graph is the Cartesian product of vertex sets of graphs on which the operation is performed (see [CvDS], pp. 65-66, and the references cited in [CvDS]).

We now recall some special cases in which a graph is the NEPS of graphs $G_1,\ldots,G_n$ with basis $\mathcal{B}$. In particular, for $n = 2$ we have the following instances of NEPS:

(i) the sum $G_1 + G_2$, when $\mathcal{B} = \{(0,1),(1,0)\}$;

(ii) the product $G_1 \times G_2$, when $\mathcal{B} = \{(1,1)\}$;

(iii) the strong product $G_1 * G_2$, when $\mathcal{B} = \{(0,1),(1,0),(1,1)\}$.

The _p-sum_ of graphs is a NEPS in which the basis consists of all _n_-tuples with exactly $p$ 1s. The _J-sum_ of graphs, where $J$ is a subset of $\{1,\ldots,n\}$, is a NEPS in which the basis consists of all _n_-tuples in which the number of 1s belongs to $J$.

The notion of NEPS arises in a natural way when studying spectral properties of graphs obtained by binary operations of the type mentioned above. The main ideas are described in [Cve1], while the early references [Cve2], [Cve4], [Cve6] are summarized and generalized in [Cve7].

In [CvPe1], the definition of a NEPS of graphs has been extended to digraphs (which may have multiple arcs and/or loops), and in [MoOm1] and [Petr] to infinite graphs in two different ways.

There are some other graph operations in which the resulting graph has as its vertex set the Cartesian product of vertex sets of the starting graphs. In [Cve9], the so-called _Boolean operations_ on graphs are defined, while [Šok] introduces a more general operation called the _generalized direct product of graphs_, containing NEPS and Boolean operations as special cases. The generalized direct product of graphs has been extended to digraphs in [Pet3], [Pet4], [Pet5]. The generalized direct product of graphs does not have such nice properties as a NEPS, and so here we consider only NEPS.

Theorems 2.3.2 and 2.3.4 are taken from [CvLu] (see also [CvDS], pp. 68-69).

**2.3.2 Theorem** *Let $A_1, \ldots, A_n$ be the adjacency matrices of graphs $G_1, \ldots, G_n$ respectively. The NEPS of graphs $G_1, \ldots, G_n$ with basis $\mathscr{B}$ has an adjacency matrix given by*

$$A = \sum_{\beta \in \mathscr{B}} A_1^{\beta_1} \otimes \cdots \otimes A_n^{\beta_n}. \tag{2.3.1}$$

Here $A_i^0 = I$ (the identity matrix of the same size as $A_i$), $A_i^1 = A_i$, and $\otimes$ denotes the *Kronecker product* of matrices defined below. The proof is left to the reader.

**2.3.3 Definition** *The Kronecker product $A \otimes B$ of matrices $A = (a_{ij})_{m,n}$ and $B = (b_{ij})_{p,q}$ is the $mp \times nq$ matrix obtained from $A$ by replacing each entry $a_{ij}$ with the block $a_{ij}B$.*

Thus the entries of $A \otimes B$ consist of all the $mnpq$ possible products of an entry of $A$ with an entry of $B$.

The following relations are well known (see, for example, [MaMi], pp. 18 and 8):

$$\text{tr}\,(A \otimes B) = \text{tr}\,A \cdot \text{tr}\,B, \tag{2.3.2}$$

$$(A \otimes B) \cdot (C \otimes D) = (AC) \otimes (BD). \tag{2.3.3}$$

The relation (2.3.2) holds whenever $A$ and $B$ are square matrices, and (2.3.3) holds whenever the products $AC$ and $BD$ exist. Moreover the Kronecker product is an assosiative operation.

Starting from (2.3.3) and using induction, we obtain

$$(A_1 \otimes \cdots \otimes A_n) \cdot (B_1 \otimes \cdots \otimes B_n) \cdots (M_1 \otimes \cdots \otimes M_n) \tag{2.3.4}$$

$$= (A_1 B_1 \cdots M_1) \otimes \cdots \otimes (A_n B_n \cdots M_n).$$

**2.3.4 Theorem** *If $\lambda_{i1}, \ldots, \lambda_{in_i}$ is the spectrum of $G_i$ $(i = 1, \ldots, n)$, then the spectrum of the NEPS of $G_1, \ldots, G_n$ with basis $\mathscr{B}$ consists of all possible values $\Lambda_{i_1, \ldots, i_n}$ where*

$$\Lambda_{i_1, \ldots, i_n} = \sum_{\beta \in \mathscr{B}} \lambda_{1i_1}^{\beta_1} \cdots \lambda_{ni_n}^{\beta_n} \quad (i_k = 1, \ldots, n_k; \ k = 1, \ldots, n). \tag{2.3.5}$$

*Proof* Since the adjacency matrix $A_s$ of $G_s$ is symmetric, there exist linearly independent vectors $\mathbf{x}_{s1}, \ldots, \mathbf{x}_{sn_s}$ such that $A_s \mathbf{x}_{si_s} = \lambda_{si_s} \mathbf{x}_{si_s}$ ($i_s = 1, 2, \ldots, n_s$; $s = 1, 2, \ldots, n$). Consider the vector

$$\mathbf{x} = \mathbf{x}_{1i_1} \otimes \cdots \otimes \mathbf{x}_{ni_n}.$$

Using Theorem 2.3.2 we see that $A\mathbf{x} = \Lambda_{i_1, \ldots, i_n} \mathbf{x}$. This yields $n_1 n_2 \cdots n_k$ eigenvalues and independent eigenvectors, and so all eigenvalues have been determined.                                                                    $\square$

In particular, if $\lambda_1, \ldots, \lambda_n$ and $\mu_1, \ldots, \mu_m$ are the eigenvalues of $G$ and $H$, respectively, then

$\lambda_i + \mu_j$ ($i = 1, \ldots, n$; $j = 1, \ldots, m$) are the eigenvalues of $G + H$;
$\lambda_i \mu_j$ ($i = 1, \ldots, n$; $j = 1, \ldots, m$) are the eigenvalues of $G \times H$;
$\lambda_i + \mu_j + \lambda_i \mu_j$ ($i = 1, \ldots, n$; $j = 1, \ldots, m$) are the eigenvalues of $G * H$.

Theorem 2.3.4 has been extended in [MoOm1], [MoOm2], [Petr] to infinite digraphs.

We have seen in Section 2.2 that the number $N_k$ of walks of length $k$ in a graph with distinct eigenvalues $\mu_1, \ldots, \mu_m$ is given by (2.2.2). In conjunction with Theorem 2.3.4 this yields

**2.3.5 Theorem** (cf. [Cve4],[Cve7]) *Let* $\sum_{i_j} C_{ji_j} \lambda_{ji_j}^k$ ($j = 1, \ldots, n$) *denote the number of walks of length $k$ for the graph $G_j$. Then the NEPS of graphs $G_1, \ldots, G_n$ with basis $\mathscr{B}$ contains*

$$\sum_{i_1, \ldots, i_n} C_{1i_1} \ldots C_{ni_n} \left( \sum_{\beta \in \mathscr{B}} \lambda_{1i_1}^{\beta_1} \cdots \lambda_{ni_n}^{\beta_n} \right)^k$$

*walks of length $k$.*

A formula for the number of walks of length $k$ between two specified vertices in a NEPS was derived in [Cve8]. It was applied to the NEPS of complete graphs thereby solving the problem of enumerating the ways in which a rook (a chess piece) can make a series of $k$ moves between two specified cells of a chess-board. A similar problem for a knight appeared in the *American Mathematical Monthly* (Problem E2392 of January 1973); here we reproduce the problem and the solution as given in [CvSi2], where additional comments and references can be found.

*Let the distance between the two cells of the (infinite) chess-board be defined as the minimum number of steps for a knight to move from one cell to the other. Determine $D(O, P)$, the distance between cells $O = (0, 0)$ and $P = (a, b)$.*

*Solution* To determine $D(O, P)$, we first determine $N_{(0,0),(a,b)}^k$, the number of ways for a knight to move from the cell $(0,0)$ to the cell $(a, b)$ (or, equivalently, from the cell $(1, 1)$ to the cell $(a+1, b+1)$) in exactly $k$ steps. For this purpose we shall consider, for sufficiently large $n$, a chess-board of dimension $n \times n$ on a torus, and a graph which represents the knight's moves on this chess-board. The adjacency matrix $\mathscr{A}$ of this graph can be represented in the form

$$\mathscr{A} = A \otimes (A^2 - 2I) + (A^2 - 2I) \otimes A, \qquad (2.3.6)$$

where $A$ is the adjacency matrix of a cycle of length $n$. It is well known that the eigenvalues $\lambda_j$ of the matrix $A$ are given by $\lambda_j = 2\cos(2\pi j/n)$ $(j = 1, \ldots, n)$, while $x_{jl} = \frac{1}{n} \exp\left(\frac{2\pi i}{n} l j\right)$ $(l = 1, \ldots, n; \ i^2 = -1)$ are the coordinates of an eigenvector corresponding to $\lambda_j$. Hence the eigenvalues and corresponding eigenvectors of the matrix (2.3.6) are given by

$$\Lambda_{j,l} = (\lambda_j + \lambda_l)(\lambda_j \lambda_l - 2), \quad \mathbf{x}_{jl} = \mathbf{x}_j \otimes \mathbf{x}_l, \quad j, l = 1, \ldots, n.$$

It can easily be checked that, for $0 \le a, b \le n$ and $k < n$, we have

$$N_{(0,0),(a,b)}^k = (\mathscr{A}^k)_{(1,1),(a+1,b+1)} = \sum_{p,q=1}^n x_{p1} x_{q1} \overline{x}_{p,a+1} \overline{x}_{q,b+1} \Lambda_{p,q}^k$$

$$= \sum_{p,q=1}^n \frac{4^k}{n^2} \exp\left(\frac{2\pi}{n}(-pa - qb)j\right)$$

$$\left(\cos\frac{2\pi}{n}p + \cos\frac{2\pi}{n}q\right)^k \left(2\cos\frac{2\pi}{n}p \cos\frac{2\pi}{n}q - 1\right)^k.$$

By letting $n \to +\infty$, we obtain

$$N_{(0,0),(a,b)}^k$$

$$= \frac{4^{k-1}}{\pi^2} \int_0^{2\pi} \int_0^{2\pi} \exp(-j(ax + by))(\cos x + \cos y)^k (2\cos x \cos y - 1)^k dxdy$$

$$= \sum_{t=0}^k (-2)^{k-t} \binom{k}{t} \sum_{s_1 + s_2 = k} \frac{k!}{s_1! s_2!} \binom{s_1 + t}{\frac{1}{2}(s_1 + t - a)} \binom{s_2 + t}{\frac{1}{2}(s_2 + t - b)},$$

$$(2.3.7)$$

where $\binom{m}{\alpha} = 0$ if $a \notin \{0, 1, \ldots, m\}$ and $s_1, s_2$ range between 0 and $k$ (inclusive). It follows that

$$D(O, P) = \min\{k : N_{(0,0),(a,b)}^k \ne 0\}. \qquad (2.3.8)$$

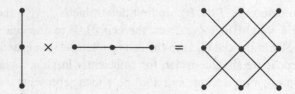

Fig. 2.2. A simple example of a NEPS.

Without loss of generality, we may assume that $a \geq b$ and $a > 2$. By a simple, but tedious, analysis of formula (2.3.7), making use of (2.3.8) we obtain

$$D(O,P) = \begin{cases} a+b-2\lfloor(a+b)/3\rfloor & \text{if } a \leq 2b, \\ a+b-2\lfloor\frac{a}{2}\rfloor-2\lfloor\frac{1}{2}(b-\lfloor\frac{a}{2}\rfloor)\rfloor & \text{if } a \geq 2b. \end{cases}$$

□

Further specific examples of eigenvectors of NEPS of paths and cycles, and of eigenvectors of related graphs, may be found in [LeYe].

As we have seen in Section 2.2, an eigenvalue (of a graph) is called a *main* eigenvalue if its eigenspace contains a vector not orthogonal to $(1,\ldots,1)^T$. The next theorem was proved in [Pet1], [Pet2].

**2.3.6 Theorem** *An eigenvalue of a NEPS of graphs is main if and only if, when expressed in terms of eigenvalues of graphs on which the operation is performed as in Theorem 2.3.4, it depends only on main eigenvalues of these graphs.*

The product of two connected graphs can be a disconnected graph, and such an example is given in Fig. 2.2. Similar situations appear in more general examples of NEPS, and we shall use Theorems 2.3.5 and 2.3.6 to investigate conditions under which certain operations on graphs result in connected graphs. In a NEPS this question is related in a natural way to the question of whether or not the resulting graph is bipartite, and so we address this problem also.

Consider a NEPS of graphs $G_1,\ldots,G_n$ each containing at least one edge. The indices $r_1,\ldots,r_n$ of the graphs are then positive, and we obtain the index of the NEPS from (2.3.5) if we put $i_1 = i_2 = \cdots = i_n = 1$, in accordance with the convention that $\lambda_{ji_j} = \lambda_{j1} = r_j$ $(j = 1,\ldots,n)$. Thus, for the index $r$ of the NEPS we have

$$r = \Lambda_{1,\ldots,1} = \sum_{\beta \in \mathscr{B}} r_1^{\beta_1} \cdots r_n^{\beta_n} \quad (> 0). \tag{2.3.9}$$

We shall consider only those NEPS with a basis $\mathscr{B}$ which for each $j \in \{1, 2, \ldots, n\}$ contains at least one $n$-tuple $(\beta_1, \ldots, \beta_n)$ with $\beta_j = 1$. We denote this condition by (D). This condition implies that the index of a NEPS effectively depends on each $r_j$.

If $G_1, \ldots, G_n$ are connected graphs, there are positive eigenvectors $\mathbf{x}_1, \ldots, \mathbf{x}_n$ corresponding to the indices $r_1, \ldots, r_n$. It can be easily verified (see also Theorem 2.3.4) that the eigenvector $\mathbf{x} = \mathbf{x}_1 \otimes \cdots \otimes \mathbf{x}_n$ belongs to the index $r$ of the NEPS. Moreover, $\mathbf{x}$ itself is a positive vector.

Now, by Theorem 2.1.4, the number of components of the NEPS is equal to the multiplicity of the index $r$. Hence it is necessary to investigate whether or not $\Lambda_{i_1, \ldots, i_n}$ is equal to $r$ for some $n$-tuple $(i_1, \ldots, i_n)$ different from $(1, \ldots, 1)$. For this it is necessary that for at least one $j \in \{1, \ldots, n\}$, the relation $|\lambda_{ji_j}| = |\lambda_{j_1}| = r_j$ holds for some $i_j > 1$. Since $G_1, \ldots, G_n$ are connected graphs, their indices are simple eigenvalues and the above equality can be satisfied only if $\lambda_{ji_j} = -r_j$. According to Theorem 1.3.14 we then have that $G_j$ is a bipartite graph.

It follows that if a NEPS of connected graphs (each containing at least one edge) is disconnected then it is because one or more of these graphs is bipartite. However, the presence of a bipartite graph does not guarantee that a NEPS is disconnected; the structure of the particular NEPS has a certain influence too.

By further analysis we see that the desired $n$-tuple of indices $i_1, \ldots, i_n$ must be such that for every $i_j \neq 1$ the graph $G_j$ is bipartite, i.e. $\lambda_{ji_j} = -r_j$, and such that every summand in (2.3.5) contains an even number of quantities $\lambda_{ji_j}$ $(i_j \neq 1)$.

In order to formulate a theorem with precise conditions for a NEPS to be connected (or bipartite) we introduce the following definition.

**2.3.7 Definition** *A function in several variables is called even (odd) with respect to a given non-empty subset of variables if the function does not change its value (changes only in sign) when all the variables from the subset are simultaneously changed in sign. The function is even (odd) if at least one non-empty subset of variables exists with respect to which the function is even (odd).*

In view of the previous remarks we have the following theorem.

**2.3.8 Theorem** [Cve7] *Let $G_1, \ldots, G_n$ be connected graphs each containing at least one edge. Suppose also that $G_{i_1}, \ldots, G_{i_s}$ ($\{i_1, \ldots, i_s\} \subseteq \{1, \ldots, n\}$) are the bipartite graphs among the $G_i$. Then the NEPS of graphs $G_1, \ldots, G_n$, with basis $\mathscr{B}$ satisfying condition (D), is a connected graph if and only if* *the function*

$$\sum_{\beta \in \mathscr{B}} x_1^{\beta_1} \cdots x_n^{\beta_n} \qquad\qquad (2.3.10)$$

*is never even with respect to any non-empty subset of the set* $\{x_{i_1}, \ldots, x_{i_s}\}$. *In the case of disconnectedness the number of components is equal to the multiplicity of the index of the NEPS.*

We shall now deduce the conditions under which the NEPS of connected graphs is a bipartite graph.

All components of a NEPS of connected graphs have the same index $r$ by Theorem 2.1.4. Thus the number of components of such a NEPS is equal to the multiplicity of its index. A NEPS is bipartite if, naturally, all its components are bipartite. According to Theorem 1.3.14, each component must then contain the number $-r$ in the spectrum. Since no component contains in the spectrum the number $-r$ with multiplicity greater than 1, it follows that a necessary and sufficient condition for a NEPS to be bipartite is that the numbers $r$ and $-r$ have the same multiplicity in the spectrum of the NEPS.

We see from (2.3.5) that the number $-r$ appears in the spectrum of a NEPS only if some of the graphs are bipartite and if there exist subsets of the variables $x_1, \ldots, x_n$ with respect to which the function (2.3.10) is odd. Accordingly we obtain the following theorem.

**2.3.9 Theorem** [Cve7] *Let* $G_1, \ldots, G_n$ *be connected graphs, each containing at least one edge. Suppose also that* $G_{i_1}, \ldots, G_{i_s}$ *(*$\{i_1, \ldots, i_s\} \subseteq \{1, \ldots, n\}$*) are the bipartite graphs among the* $G_i$*. Then a NEPS of graphs* $G_1, \ldots, G_n$*, with basis* $\mathscr{B}$ *satisfying condition (D), is bipartite if and only if the number of non-empty subsets of the set* $\mathscr{L} = \{x_{i_1}, \ldots, x_{i_s}\}$*, with respect to which the function (2.3.10) is even, is smaller by 1 than the number of such subsets with respect to which it is odd.*

This theorem represents the basis for proving the following result which gives the precise conditions under which the $p$-sum is a bipartite graph.

**2.3.10 Theorem** ([Cve2], [Cve7]) *Let* $G_1, \ldots, G_n$ *be connected graphs each containing at least two vertices. Then the $p$-sum of these graphs is a bipartite graph if and only if one of the following conditions holds:*

    *(a) $p$ is equal to $n$ and at least one of the graphs* $G_1, \ldots, G_n$ *is bipartite;*
    *(b) $p$ is odd and less than $n$, and all the graphs* $G_1, \ldots, G_n$ *are bipartite.*

*Proof* Consider first the case in which $p = n$. The function (2.3.10) is then of the form $x_1 \cdots x_n$. If $\mathscr{L} = \emptyset$ (see Theorem 2.3.9), the $p$-sum is not bipartite. Accordingly, let $\mathscr{L}$ contain $l$ elements, where $l \geq 1$. Then the function $x_1 \cdots x_n$ is even with respect to exactly $\binom{l}{2} + \binom{l}{4} + \cdots = 2^{l-1} - 1$ non-empty subsets of $\mathscr{L}$, and is odd with respect to exactly $\binom{l}{1} + \binom{l}{3} + \cdots = 2^{l-1}$ such subsets. According to Theorem 2.3.10, the $p$-sum is then bipartite.

Next let $p$ be odd and less than $n$. The function (2.3.10) is then not even. It is odd only with respect to the set of all variables; for if it were odd with respect to a proper subset of variables then there would be one summand among the summands of (2.3.10) containing an even number of variables from the same subset, contradicting the assumption of the oddness in (2.3.10). Thus, for the $p$-sum to be bipartite in this case it is necessary (and sufficient) that all graphs $G_1, \ldots, G_n$ are bipartite.

Finally, if $p$ is even and less than $n$, using similar reasoning, we see that the function (2.3.10) cannot be odd.

This completes the proof. □

## 2.4 Divisors of a graph

Before we introduce the concept of a *graph divisor* we present some relevant theorems from matrix theory. The first applies not just to real symmetric matrices but to any Hermitian matrix. (Recall that the matrix $A$ with complex entries $a_{ij}$ is called *Hermitian* if $A^T = \overline{A}$, i.e. $a_{ji} = \overline{a}_{ij}$ for all $i, j$.)

**2.4.1 Theorem** (see, for example, [MaMi], p. 119) *Let $A$ be a Hermitian matrix with eigenvalues $\lambda_1 \geq \lambda_2 \geq \cdots \geq \lambda_n$ and let $B$ be one of its principal submatrices. If the eigenvalues of $B$ are $v_1 \geq v_2 \cdots \geq v_m$ then $\lambda_{n-m+1} \leq v_i \leq \lambda_i$ $(i = 1, \ldots, m)$*

The inequalities of Theorem 2.4.1 are known as *Cauchy's inequalities* and the whole theorem is known as the *Interlacing Theorem*. It is used frequently as a spectral technique in graph theory.

**2.4.2 Theorem** (C.C.Sims, see [HeHi]) *Let $A$ be a real symmetric matrix with eigenvalues $\lambda_1 \geq \lambda_2 \geq \cdots \geq \lambda_n$. Given a partition $\{1, \ldots, n\} = \Delta_1 \dot\cup \Delta_2 \dot\cup \cdots \dot\cup \Delta_m$ with $|\Delta_i| = n_i > 0$, consider the corresponding blocking $A = (A_{ij})$, where $A_{ij}$ is an $n_i \times n_j$ block. Let $e_{ij}$ be the sum of the entries in $A_{ij}$ and set $B = (e_{ij}/n_i)$ (Note that $e_{ij}/n_i$ is the average row sum in $A_{ij}$.) Then the spectrum of $B$ is contained in the interval $[\lambda_n, \lambda_1]$.*

Haemers [Hae] has shown that the interlacing properties also hold for the matrices $A$ and $B$ of this theorem. If we assume that in each block $A_{ij}$ from Theorem 2.4.2 all row sums are equal then we can say more:

**2.4.3 Theorem** ([Hay], [PeSa1]) *Let $A$ be any matrix partitioned into blocks as in Theorem 2.4.2. Let the block $A_{ij}$ have constant row sums $b_{ij}$ and let $B = (b_{ij})$. Then the spectrum of $B$ is contained in the spectrum of $A$ (having in view also the multiplicities of the eigenvalues).*

The content of Theorem 2.4.3 justifies the introduction of the following definition.

**2.4.4 Definition** *Given an $s \times s$ matrix $B = (b_{ij})$, let the vertex set of a graph $G$ be partitioned into (non-empty) subsets $X_1, X_2, \ldots, X_s$ so that for any $i, j = 1, 2, \ldots, s$ each vertex from $X_i$ is adjacent to exactly $b_{ij}$ vertices of $X_j$. The multidigraph $H$ with adjacency matrix $B$ is called a front divisor of $G$, or briefly, a divisor of $G$.*

The concept of a divisor of a graph was introduced by Sachs ([Sac1], [Sac2]). The existence of a divisor means that the graph has a certain structure; indeed, a divisor can be interpreted as a homomorphic image of the graph. On the other hand, by Theorem 2.4.3, the characteristic polynomial of a divisor divides the characteristic polynomial of the graph (i.e. the spectrum of a divisor is contained in the spectrum of the graph). In this way the notion of a divisor can be seen as a link between spectral and structural properties of a graph [FiSa].

Divisors have been considered in [Sch2], [HaSc] under the name *equitable partitions*. The reference [Mow] is given in [Sch3] for an implicit proof of Theorem 2.4.3. The notion of a (front) divisor of a multigraph was rediscovered by Yap [Yap] who proved (using different terminology) that, for a digraph, the characteristic polynomial of a divisor divides the characteristic polynomial. The notion of a divisor was extended to infinite graphs in [MoOm2].

We mention in passing two notions related to the concept of a divisor. Let $P = P_1, P_2, \ldots$ be a partition of the vertex set of a graph, and let $Q = Q_1, Q_2, \ldots$ be a partition of the edge set $E$. A pair $(P, Q)$ is called a *colouring* of a graph $(V, E)$. A colouring is a *regular colouring* if any two vertices in $P_i$ are endvertices of the same number of edges in $Q_j$ for all $i, j$, and if any two edges in $Q_j$ have the same number of endvertices in $P_i$ for all $i$ and $j$. The importance of regular colourings in the study of graph spectra is pointed out in [Sie].

*Walk partitions* of the vertex set of a graph are introduced in [PoSu] and studied in relation to spectral properties of graphs. Two vertices belong to the same cell of a walk partition if the numbers of walks of any given length starting at these vertices are the same. The paper includes a discussion of the relationship between walk partitions and divisors of a graph.

In order to prove the next theorem let us consider the generating function $H_G(t) = \sum_{k=0}^{+\infty} N_k t^k$ for the numbers $N_k$ of walks of length $k$ in a graph $G$.

The function $\frac{1}{u} H_G\left(\frac{1}{u}\right)$ has only simple poles and they represent the main part of the spectrum of $G$, since

$$\frac{1}{u} H_G\left(\frac{1}{u}\right) = \frac{1}{u} \sum_{k=0}^{+\infty} N_k \left(\frac{1}{u}\right)^k = \frac{1}{u} \sum_{k=0}^{+\infty} \frac{1}{u^k} \sum_{i=1}^{m} D_i \mu_i^k$$

$$= \frac{1}{u} \sum_{i=1}^{m} D_i \sum_{k=0}^{+\infty} \left(\frac{\mu_i}{u}\right)^k = \sum_{i=1}^{m} \frac{D_i}{u - \mu_i}. \qquad (2.4.1)$$

Next let $n_k^{(i)}$ be the number of walks of length $k$ in a multidigraph $H$ which start from the vertex $i$. Then $n_k^{(i)} = \text{sum}^{(i)} B^k$, the sum of all entries in the $i$-th row of $B^k$, where $B$ is the adjacency matrix of $H$. Consider the function $\frac{1}{u} F_H^{(i)}\left(\frac{1}{u}\right)$, where $F_H^{(i)}(t) = \sum_{k=0}^{+\infty} n_k^{(i)} t^k$. We have

$$\frac{1}{u} F_H^{(i)}\left(\frac{1}{u}\right) = \frac{1}{u} \sum_{k=0}^{+\infty} \frac{n_k^{(i)}}{u^k} = \frac{1}{u} \sum_{k=0}^{+\infty} \frac{1}{u^k} \text{sum}^{(i)} B^k = \frac{1}{u} \text{sum}^{(i)} \sum_{k=0}^{+\infty} \frac{B^k}{u^k}$$

$$= \text{sum}^{(i)} \left(I - \frac{1}{u} B\right)^{-1} = \text{sum}^{(i)}(uI - B)^{-1}. \qquad (2.4.2)$$

**2.4.5 Theorem** [Cve10] *The spectrum of any divisor $H$ of a graph $G$ includes the main part of the spectrum of $G$.*

*Proof* Let $H$ be formed on the basis of the partition $X_1 \dot\cup X_2 \dot\cup \cdots \dot\cup X_s$ of the vertex set of $G$. Let $|X_i| = n_i$ $(i = 1, 2, \ldots, s)$ and let $i$ be the vertex of $H$ corresponding to $X_i$. Since $H$ is a homomorphic image of $G$, the image of any walk in $G$ is a walk in $H$, where of course different walks in $G$ can have the same image in $H$. If a walk $\mathcal{W}$ in $H$ starts at vertex $i$ then the corresponding walks of $G$ start in $X_i$. If we fix a vertex in $X_i$ as the starting point of such a walk, then this walk in $G$ is determined uniquely by $\mathcal{W}$, since there exists a one-to-one correspondence between the edges starting at $i$ and the edges starting at any fixed vertex in $X_i$. Thus, in the notation above, we have $N_k = n_1 n_k^{(1)} + \cdots + n_s n_k^{(s)}$, and from

(2.4.2) we obtain

$$\frac{1}{u}F_G\left(\frac{1}{u}\right) = \sum_{i=1}^{s} n_i \, \text{sum}^{(i)}(uI - B)^{-1}.$$

Since $(uI - B)^{-1} = \text{adj}(uI - B)/P_H(u)$ the function on the right-hand side is expressible as a rational function with the characteristic polynomial $P_H(u)$ of $H$ in the denominator. Some factors of $P_H(u)$ may cancel, but since by (2.4.1) the (simple) poles of $\frac{1}{u}H_G\left(\frac{1}{u}\right)$ form the main part of the spectrum of $G$, all these main eigenvalues must be zeros of $P_H(u)$. □

Since the largest eigenvalue $r$ of a graph always belongs to the main part of the spectrum we have the following result.

**2.4.6 Corollary** *Any divisor of a graph $G$ has the index of $G$ as an eigenvalue.*

It was conjectured in [HaSc] that the spectrum of a divisor $H$ with the smallest number of vertices is just the main part of the spectrum of $G$. Theorem 2.4.5 confirms one of the two inclusions implicit in this conjecture. The reverse inclusion does not hold in general, as shown by the following counterexample from [Cve10]. A counterexample with only seven vertices subsequently appeared in [PoSu].

According to Theorem 1.2.1, the graphs $G_1$ and $G_2$ of Fig. 2.3 are cospectral and have cospectral complements. According to (4.5.5) and (4.5.6), $G_1$ and $G_2$ have the same main part of the spectrum. (Notice that cospectral graphs need not have the same main part of the spectrum.) Because of its symmetry $G_1$ has a divisor with ten vertices. By Theorem 2.4.5 the main part of the spectrum of $G_1$ and $G_2$ has at most ten eigenvalues. But a divisor of $G_2$ with the minimal number of vertices has 19 vertices. Thus $G_2$ is a counterexample to the above conjecture.

**2.4.7 Remark** It can easily be seen that the graphs with only one eigenvalue in the main part of the spectrum are just the regular graphs. It would be interesting to know, for given $k$, which graphs have exactly $k$ eigenvalues in the main part of the spectrum. In particular, semiregular bipartite graphs have exactly two eigenvalues in the main part of the spectrum. Are there other graphs with this property? Note that the eigenvalue $-2$ in the line graphs is never a main eigenvalue (Corollary 2.6.5). □

The concept of a divisor has also featured in coding theory. As

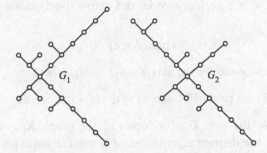

Fig. 2.3. Cospectral graphs with cospectral complements.

an application of the divisor concept in this field we shall outline an elementary proof of *Lloyd's Theorem* due to Cvetković and van Lint [CvLi]. For the general concepts of coding theory, see for example [Lin] (with Lloyd's Theorem on p. 111).

We need some preparations. Consider a set $\mathscr{F}$ of $b$ distinct symbols which we call the *alphabet*. The elements of $\mathscr{F}^n$ will be called *words of length n*. In $\mathscr{F}^n$ the *Hamming distance d* is defined by

$$d(\mathbf{x}, \mathbf{y}) = |\{i : x_i \neq y_i, 1 \leq i \leq n\}|.$$

A subset $\mathscr{S}$ of $\mathscr{F}^n$ is called a *perfect e-code* if $\mathscr{F}^n$ is partitioned by the spheres $\mathscr{S}_e(\mathbf{c})$ ($\mathbf{c} \in \mathscr{S}$), where

$$\mathscr{S}_e(\mathbf{c}) := \{\mathbf{x} \in \mathscr{F}^n : d(\mathbf{x}, \mathbf{c}) \leq e\}.$$

In 1957 Lloyd [Llo] proved a strong necessary condition for the existence of a perfect $e$-code when $b = 2$ (the binary case). In the years since 1972 several authors (see [CvDS], p. 131) have proved that the theorem (always referred to as Lloyd's Theorem) holds for all $b$:

**2.4.8 Theorem** (Lloyd's Theorem) *If a perfect e-code of length n over an alphabet of b symbols exists, then the e zeros $x_i$ of the polynomial*

$$\phi_{enb}(x) := \sum_{i=0}^{e} (-1)^i (b-1)^{e-i} \binom{n-x}{e-i} \binom{x-1}{i}$$

*are distinct positive integers $\leq n$.*

**Sketch of the proof.** From other parts of the theory [Lin] it is well known that the zeros $x_i$ of $\phi_{enb}(x)$ are all distinct; all that matters is to show that $x_j \in \{1, 2, \ldots, n\}$.

Assume that $\mathscr{C}$ is a perfect $e$-code, and define the distance $d(\mathbf{x}, \mathscr{C})$ of $\mathbf{x}$ from $\mathscr{C}$ by

$$d(\mathbf{x}, \mathscr{C}) := \min\{d(\mathbf{x}, \mathbf{c}) : \mathbf{c} \in \mathscr{C}\}.$$

Then $\mathscr{F}^n$ is partitioned by the sets $\mathscr{C}_0, \mathscr{C}_1, \ldots, \mathscr{C}_e$, where

$$\mathscr{C}_i := \{x \in \mathscr{F}^n : d(\mathbf{x}, \mathscr{C}) = i\} \quad (i = 0, 1, \ldots, e).$$

Consider now the sum $G$ of $n$ copies of the graph $K_b$: according to Theorem 2.3.4, the distinct eigenvalues of $G$ are the numbers

$$\lambda_j = (b-1)(h-j) - j = bn - n - bj \quad (j = 0, 1, \ldots, n).$$

The vertices of $G$ can be understood as the elements of $\mathscr{F}^n$: then the partition $\mathscr{C}_0, \mathscr{C}_1, \ldots, \mathscr{C}_e$ defines a front divisor $H$ of $G$ with adjacency matrix

$$B = \begin{pmatrix} 0 & n(b-1) & & & & 0 \\ 1 & b-2 & (n-1)(b-1) & & & \\ & \ddots & \ddots & \ddots & & \\ & & e-1 & (e-1)(b-2) & (n-e+1)(b-1) \\ 0 & & & e & n(k-1)-c \end{pmatrix}.$$

Since the spectrum of $H$ is contained in the spectrum of $G$, each of the $e+1$ eigenvalues $\mu_i$ of $H$, i.e. each of the zeros of the polynomial $P_H(x) = \det(xI - B)$, is equal to one of the numbers $\lambda_j$:

$$\mu_i = bn - n - bj_i, \quad j_i \in \{0, 1, \ldots, n\}.$$

This means that each root $x_i$ of the equation

$$P_H(bn - n - bx) = 0$$

is equal to one of the numbers $0, 1, \ldots, n$. By expanding $\det\{(bn - n - bx)I - B\}$ (see [CvLi]) it can be shown that

$$P_H(bn - n - bx) = e! \cdot x \cdot \phi_{enb}(x),$$

and so we conclude that $\phi_{enb}(x)$ has distinct zeros $x_j$, with $x_j \in \{1, 2, \ldots, n\}$. $\qquad\square$

We now prove a theorem which relates the divisor concept to switching in graphs.

**2.4.9 Theorem** [Cve10] *If a regular graph $G$ of degree $r$ with $n$ vertices can be switched into a regular graph of degree $r^*$, then $r^* - \frac{n}{2}$ is an eigenvalue of $G$.*

*Proof* If $G$ has the stated property then the switching sets form a divisor with the adjacency matrix

$$\begin{pmatrix} r - \frac{1}{2}(n - x - r^* + r) & \frac{1}{2}(n - x - r^* + r) \\ \frac{1}{2}(x - r^* + r) & r - \frac{1}{2}(x - r^* + r) \end{pmatrix},$$

where $x$ is the size of a switching set ($1 \le x < n$). This matrix has the eigenvalues $r$ and $r^* - \frac{n}{2}$, which proves the theorem. $\qquad\qquad\square$

**2.4.10 Corollary** *If $n$ is odd then $G$ cannot be switched into another regular graph since an eigenvalue of a graph cannot be a non-integer rational number.*

**2.4.11 Corollary** *If $G$ can be switched into a regular graph of the same degree and if $q$ is the least eigenvalue of $G$, then $r - \frac{n}{2} \ge q$, i.e. $n \le 2r - 2q$. Since $q \ge -r$, we have $r - \frac{n}{2} \ge -r$, i.e. $r \ge n/4$.*

**2.4.12 Example** There is no cospectral pair of non-isomorphic cubic graphs with less than 14 vertices [BuČCS]. Accordingly it follows from Corollary 2.4.11 that the existence of cospectral cubic graphs cannot be explained by switching. $\qquad\qquad\square$

**2.4.13 Example** If $L(K_s)$ ($s > 1$) can be switched to another regular graph of the same degree then by Theorem 2.4.9, $2s - 4 - \frac{s(s-1)}{4} \ge -2$, whence $s \le 8$. When $s = 8$ there are three different graphs which are switching equivalent to $L(K_8)$, cospectral with $L(K_8)$, but not isomorphic to $L(K_8)$. These three graphs are the Chang graphs of Example 1.1.2.

## 2.5 The automorphism group and eigenvectors

Let $G$ be a graph with vertex set $V = V(G)$. A bijection $\pi : V \to V$ is called an *automorphism* of $G$ if it preserves the adjacency relation of $G$. The set $\Gamma = \Gamma(G)$ of all automorphisms $\pi$ of $G$ is a group with respect to the composition of mappings.

There are many results, described in Chapter 5 of [CvDS] and Chapter 3 of [CvDGT], which relate the automorphism group of a graph to its spectrum and eigenvectors. These results are, generally speaking, of the following two types: first, bounds for the number of simple eigenvalues (or eigenvalues of a small multiplicity), and second, procedures for factoring the characteristic polynomial by means of the automorphism group.

In this section we establish by elementary methods some noteworthy relations between the spectrum of a (multi)graph $G$ and its automorphism group $\Gamma = \Gamma(G)$.

The group $\Gamma$ is realized by the set $\Gamma^*$ of all permutation matrices $P = P(\pi)$ which commute with the adjacency matrix $A$ of $G$. We have

$$P \in \Gamma^* \Leftrightarrow PA = AP.$$

For an arbitrary multigraph $G$ with non-trivial automorphism group $\Gamma$, let $\mathbf{x}$ be an eigenvector of $A$ with corresponding eigenvalue $\lambda$ and let $P \in \Gamma^*$. Then $A\mathbf{x} = \lambda\mathbf{x}$ implies

$$AP\mathbf{x} = PA\mathbf{x} = P\lambda\mathbf{x} = \lambda P\mathbf{x};$$

this means that together with $\mathbf{x}$ all vectors $P\mathbf{x}$ ($P \in \Gamma^*$) are eigenvectors of $G$. If the vectors $\mathbf{x}$ and $P\mathbf{x}$ are linearly independent (and this is, in a sense, the 'general case'), then $\lambda$ must have a multiplicity $m > 1$. This simple observation is crucial to what follows.

Let $G$ be a multigraph and assume that $\lambda$ is a simple eigenvalue of $G$ with corresponding real eigenvector $\mathbf{x} = (x_1, x_2, \ldots, x_n)^T$. Then for any $P = P(\pi) \in \Gamma^*$ the eigenvectors $\mathbf{x}$ and $P\mathbf{x}$ are linearly dependent, and so there is a real number $\mu$ such that $P\mathbf{x} = \mu\mathbf{x}$. Let $\gamma$ be any cycle of $\pi$; assume, without loss of generality, that $\gamma = (12\ldots t)$ and that the partial vector $\mathbf{x}' = (x_1, x_2, \ldots, x_t)^T$ of $\mathbf{x}$ is non-zero. Since $P\mathbf{x} = \mu\mathbf{x}$ we have

$$C\mathbf{x}' = \mu\mathbf{x}', \tag{2.5.1}$$

where $C$ is the permutation matrix corresponding to $\gamma$. Since $C^t\mathbf{x}' = \mathbf{x}' = \mu^t\mathbf{x}'$. We have $\mu^t = 1$, and we conclude that

$$P\mathbf{x} = \mu\mathbf{x}, \text{ where } \begin{cases} \mu = 1 & \text{if } t \text{ is odd,} \\ \mu = \pm 1 & \text{if } t \text{ is even.} \end{cases} \tag{2.5.2}$$

From (2.5.1) and (2.5.2) we deduce

(i) if $t$ is odd, then $x_1 = x_2 = \cdots = x_{t-1} = x_t$,

(ii) if $t$ is even, then $x_1 = x_2 = \cdots = x_{t-1} = x_t$ or $x_1 = -x_2 = \cdots = x_{t-1} = -x_t$,

(iii) if $\mathbf{x}$ is a principal eigenvector then $x_1 = x_2 = \cdots = x_{t-1} = x_t$.

We note two consequences. First, if $\mathbf{x}$ is a principal eigenvector then $x_i = x_j$ whenever $i$ and $j$ are similar vertices, that is, whenever there exists $\pi \in \Gamma$ such that $i = \pi(j)$. Secondly, if $G$ has $n$ distinct eigenvalues then $P^2\mathbf{x} = \mathbf{x}$ for all $P \in \Gamma^*$ and all eigenvectors $\mathbf{x}$ of $A$. Since $\mathbb{R}^n$ is spanned by the eigenvectors of $A$, this means that $P^2 = I$. Thus we have proved the following theorem.

**2.5.1 Theorem** ([Mow], [PeSa2]) *If a multigraph has no repeated eigenvalues then all of its non-trivial automorphisms are involutions; equivalently, its automorphism group is an elementary abelian 2-group.*

It is observed in [Row16] that, more generally, if the $n$-vertex graph $G$ has $n - r$ simple eigenvalues then the exponent of $\Gamma(G)$ divides the least common multiple of $2, 3, \ldots, r + 2$.

Theorem 2.5.1 may be seen in the context of group representations as follows. We know that there exists an orthogonal matrix $U$ such that $U^T A U = D$ where $D$ has block-diagonal form $\text{diag}(\mu_1 I, \mu_2 I, \ldots, \mu_m I)$. Here $\mu_1, \mu_2, \ldots, \mu_m$ are the distinct eigenvalues of $G$ and $\mu_i I$ has size $k_i \times k_i$ where $k_i$ is the multiplicity of $\mu_i$ ($i = 1, 2, \ldots, m$). Now $P(\pi)$ commutes with $A$ if and only if the orthogonal matrix $U^T P(\pi) U$ commutes with $D$; and matrices which commute with $D$ have block-diagonal form $\text{diag}(B_1, \ldots, B_m)$ where $B_i$ has size $k_i \times k_i$ ($i = 1, 2, \ldots, m$). Hence the map $\pi \mapsto U^T P(\pi) U$ is a representation of $\Gamma$ as a subgroup of the direct product of orthogonal groups of degrees $k_1, \ldots, k_m$. In particular if $m = n$ and $k_1 = k_2 = \cdots = k_m = 1$ then $U^T P(\pi) U$ is diagonal with each diagonal entry equal to $\pm 1$. This provides an alternative proof of Theorem 2.5.1. In the general case the representation can be used to establish a sharp upper bound for the number of simple eigenvalues of a connected graph $G$: if the orbits of $\Gamma(G)$ on $V(G)$ have sizes $2^{m_i} r_i$ ($r_i$ odd; $i = 1, 2, \ldots, s$) then $G$ has at most $2^{m_1} + 2^{m_2} + \cdots + 2^{m_s}$ simple eigenvalues. Details may be found in [Row1], but it is worth noting here that to show that the bound is sharp one exploits the fact that always the orbits of $\Gamma(G)$ determine a divisor of $G$ as defined in Section 2.4.

## 2.6 Eigenspaces of line graphs

Of particular interest are the eigenvectors of the line graph $L(G)$ of a connected regular multigraph $G$ of degree $r$.

The relation between the spectrum of $G$ and that of $L(G)$ is given in Theorem 1.3.17. The formula from this theorem establishes a one-to-one correspondence between the eigenvalues $\lambda \neq -r$ of $G$ and the eigenvalues $\mu \neq -2$ of $L(G)$. If $\lambda$ ($\neq -r$) is an eigenvalue of $G$ of multiplicity $p$ then $\mu = \lambda + r - 2$ ($\neq -2$) is an eigenvalue of $L(G)$ of multiplicity $p$; and if $\mu$ ($\neq -2$) is an eigenvalue of $L(G)$ of multiplicity $p$, then $\lambda = \mu - r + 2$ ($\neq -r$) is an eigenvalue of $G$ of multiplicity $p$.

Therefore, we shall call $(\lambda, \mu)$ a pair of corresponding eigenvalues if $\lambda$ is an eigenvalue of $G$, $\mu$ is an eigenvalue of $L(G)$, and $\lambda + r = \mu + 2 \neq 0$. (Recall that $-r$ is an eigenvalue of $G$ if and only if $G$ is bipartite, and in this case its multiplicity is 1; cf. Theorem 1.3.14.)

In the following theorem, $X(\lambda)$ denotes the eigenspace of the eigenvalue $\lambda$ of $G$, and $Y(\mu)$ denotes the eigenspace of the corresponding eigenvalue $\mu$ of $L(G)$.

**2.6.1 Theorem** [Sac2] *Let $G$ be a connected regular multigraph of degree $r$ with $n$ vertices and $m$ edges, let $(\lambda, \mu)$ be a pair of corresponding eigenvalues of $G$ and $L(G)$, and let $R$ denote the $n \times m$ vertex-edge incidence matrix of $G$. Then $R^T$ maps the eigenspace $X(\lambda)$ onto the eigenspace $Y(\mu)$, and $R$ maps $Y(\mu)$ onto $X(\lambda)$.*

*Proof* Let $A$ and $B$ denote the adjacency matrices of $G$ and $L(G)$ respectively.

Suppose that $\mathbf{x} \in X(\lambda)$, $\mathbf{x} \neq \mathbf{0}$, and $\mathbf{y} = R^T\mathbf{x}$. Then,

$$R\mathbf{y} = RR^T\mathbf{x} = (A + D)\mathbf{x} = (A + rI)\mathbf{x} = (\lambda + r)\mathbf{x} \neq \mathbf{0},$$

and so $\mathbf{y} \neq \mathbf{0}$. Further we have

$$B\mathbf{y} = (R^TR - 2I)\mathbf{y} = R^TRR^T\mathbf{x} - 2\mathbf{y} = R^T(\lambda + r)\mathbf{x} - 2\mathbf{y} = (\lambda + r - 2)\mathbf{y} = \mu\mathbf{y},$$

and so $\mathbf{y} \in Y(\mu)$, i.e.

$$\mathbf{x} \in X(\lambda), \mathbf{x} \neq \mathbf{0} \text{ implies } R^T\mathbf{x} \in Y(\mu), R^T\mathbf{x} \neq \mathbf{0}.$$

In a similar way it can be proved that

$$\mathbf{y} \in Y(\mu), \mathbf{y} \neq \mathbf{0} \text{ implies } R\mathbf{y} \in X(\lambda), R\mathbf{y} \neq \mathbf{0}.$$

We deduce that if $\mathbf{y} \in Y(\mu)$ then there is a unique $\mathbf{x} \in X(\lambda)$ such that $R^T\mathbf{x} = \mathbf{y}$, namely

$$\mathbf{x} = \frac{1}{\mu + 2}R\mathbf{y}.$$

Similarly, if $\mathbf{x} \in X(\lambda)$, then there is a unique $\mathbf{y} \in Y(\mu)$ such that $R\mathbf{y} = \mathbf{x}$, namely

$$\mathbf{y} = \frac{1}{\lambda + r}R^T\mathbf{x}.$$

This proves the theorem. $\square$

**2.6.2 Remark** The equations $\mathbf{y} = R^T\mathbf{x}$ and $\mathbf{x} = R\mathbf{y}$ from the proof of Theorem 2.6.1 can be given a more intuitive form. If we write $x(v)$ for the component of **x** that corresponds to the vertex $v$ and $y(e)$ for the component of **y** that corresponds to the edge $e$ then we have

$$y(e) = \sum_{v \in e} x(v) \text{ and } x(v) = \sum_{v \in e} y(e).$$

$\square$

Next we consider the eigenvectors **x** of $G$ and **y** of $L(G)$, corresponding to $-r$, $-2$, respectively.

**2.6.3 Theorem** *Let $G$ be any connected graph. Then $G$ is bipartite if and only if the system of equations $\sum_{v \in e} x(v) = 0$ ($e \in E(G)$), equivalent to $R^T x = 0$, has a solution space of dimension 1.*

*In particular, if $G$ is bipartite and regular of degree $r$, then the solution space is the eigenspace of $G$ corresponding to $-r$.*

The simple proof may be left to the reader.

**2.6.4 Theorem** ([Doo1], [Doo3]); for regular multigraphs see [Sac2]) *Let $G$ be any connected multigraph. Then **y** is an eigenvector of $L(G)$ corresponding to the eigenvalue $-2$ if and only if $\sum_{v \in e} y(e) = 0$ for each vertex $v$ of $G$, equivalently $R y = 0$.*

**2.6.5 Corollary** *If **y** is an eigenvector of $L(G)$ corresponding to the eigenvalue $-2$, then $\sum_{i=1}^{m} y_i = 0$, i.e. the eigenspace $Y(-2)$ is orthogonal to the vector $(1, 1, \ldots, 1)^T$. Hence, in line graphs, the eigenvalue $-2$ never belongs to the main part of the spectrum.*

Note that, for a regular multigraph of degree $r$, each eigenspace $X(\lambda)$ ($\lambda \neq r$) is orthogonal to $(1, 1, \ldots, 1)^T$.

We formulate next some results concerning the multiplicity of the eigenvalue $-2$ in line graphs and generalized line graphs.

**2.6.6 Theorem** ([Doo1], [Doo5]) *Let $G$ be a connnected graph with $n$ vertices and $m$ edges such that the least eigenvalue of $L(G)$ is $-2$. Then the multiplicity of this eigenvalue is $m - n + 1$ if $G$ is bipartite, and $m - n$ otherwise.*

For regular and semi-regular bipartite graphs this theorem follows from Theorems 1.3.17 and 1.3.18.

**2.6.7 Theorem** [CvDSi] *Let $G$ be a graph with $n$ vertices and $m$ edges, and let $H = L(G; a_1, \ldots, a_n)$ where not all $a_i$ are zero. Then the multiplicity of $-2$ as an eigenvalue of $H$ is $m - n + \sum_{i=1}^{n} a_i$.*

Recall that a regular spanning submultigraph (of degree $s$) of a regular multigraph $G$ is called a *factor* (*s-factor*) of $G$. We shall now establish an interesting relation between the factors of $G$ and the eigenvectors of $L(G)$.

A spanning submultigraph $G'$ of a multigraph $G$ can be represented by a vector $\mathbf{c} = (c_1, c_2, \ldots, c_m)^T$, with $c_j = 1$ if the $j$-th edge of $G$ belongs to $G'$, and $c_j = 0$ otherwise; the vector $\mathbf{c}$ is called the *characteristic vector of $G'$* (with respect to $G$).

Let $G$ be a connected regular multigraph of degree $r$. The largest eigenvalue of $L(G)$ is $2r - 2$ and the least eigenvalue is not smaller than $-2$. Let $\{\mathbf{y}_1, \mathbf{y}_2, \ldots, \mathbf{y}_p\}$ be a maximal set of linearly independent eigenvectors corresponding to the eigenvalues of $L(G)$ which are greater than $-2$ and smaller than $2r - 2$. It is easy to see that $p = n - 2$ if $G$ is bipartite and $p = n - 1$ otherwise; thus if $n > 2$ then $p > 0$. Denote by $M$ the $m \times p$ matrix with columns $\mathbf{y}_1, \mathbf{y}_2, \ldots, \mathbf{y}_p$. The next theorem provides a means of investigating the existence of an $s$-factor in $G$, provided that $M$ is known.

**2.6.8 Theorem** ([Sac3], [Sac4]) *Let $G$ be a connected regular multigraph with $n$ vertices and $m$ edges. A vector $\mathbf{z}$ with $m$ components is the characteristic vector of an $s$-factor of $G$ if and only if it satisfies the following conditions:*

(1) *$\frac{1}{2}sn$ components of $\mathbf{z}$ are equal to 1, and the other components are equal to zero;*
(2) *$M^T \mathbf{z} = 0$.*

A proof can be found in the original papers or in [CvDS]; see also the comments immediately following Theorem 4.3.17.

# 3

# Eigenvector techniques

Each of the first four sections of this chapter is devoted to an eigenvector technique and its use in proving results concerning the eigenvalues of a graph. The emphasis is on the problem of finding the graphs with maximal or minimal index in a given class of graphs. In the final section we survey further results of this sort which may be proved using a variety of the techniques that have been identified.

## 3.1 Rayleigh quotients

Suppose first that $A$ is any symmetric $n \times n$ matrix with real entries. A *Rayleigh quotient* for $A$ is a scalar of the form $\mathbf{y}^T A \mathbf{y} / \mathbf{y}^T \mathbf{y}$ where $\mathbf{y}$ is a non-zero vector in $\mathbb{R}^n$.. The supremum of the set of such scalars is the largest eigenvalue $\mu_1$ of $A$, equivalently

$$\mu_1 = \sup\{\mathbf{x}^T A \mathbf{x} : \mathbf{x} \in \mathbb{R}^n, \ \|\mathbf{x}\| = 1\}. \tag{3.1.1}$$

This well-known fact follows immediately from the observation that if $\{\mathbf{x}_1, \ldots, \mathbf{x}_n\}$ is an orthonormal basis of eigenvectors of $A$ and if $\mathbf{x} = \alpha_1 \mathbf{x}_1 + \cdots + \alpha_n \mathbf{x}_n$ then $\alpha_1^2 + \cdots + \alpha_n^2 = 1$, while

$$\mathbf{x}^T A \mathbf{x} = \lambda_1 \alpha_1^2 + \cdots + \lambda_n \alpha_n^2, \tag{3.1.2}$$

where $A \mathbf{x}_i = \lambda_i \mathbf{x}_i$ $(i = 1, \ldots, n)$. We take $\mu_1 = \lambda_1 \geq \lambda_2 \geq \cdots \geq \lambda_n$ except where stated otherwise.

Note that for $\mathbf{x} \neq \mathbf{0}$, $\mathbf{x}^T A \mathbf{x} = \mu_1$ if and only if $A \mathbf{x} = \mu_1 \mathbf{x}$. Moreover each eigenvalue $\lambda_k$ $(k = 1, \ldots, n)$ can be characterized in terms of subspaces of $\mathbb{R}^n$ as follows. Let $\mathcal{U}$ be an $(n - k + 1)$-dimensional subspace of $\mathbb{R}^n$, so that $\langle \mathbf{x}_1, \ldots, \mathbf{x}_k \rangle \cap \mathcal{U} \neq \{\mathbf{0}\}$. If $\mathbf{x}$ is a unit vector in this intersection of subspaces then $\alpha_{k+1} = \cdots = \alpha_n = 0$ and so $\mathbf{x}^T A \mathbf{x} \geq \lambda_k$. It follows that $\sup\{\mathbf{x}^T A\mathbf{x} : \|\mathbf{x}\| = 1\} \geq \lambda_k$. On the other hand, by (3.1.2) again, this lower bound is attained when $\mathcal{U} = \langle \mathbf{x}_k, \ldots, \mathbf{x}_n \rangle$ because in this case $\alpha_1 = \cdots = \alpha_{k-1} = 0$ for every unit vector in $\mathcal{U}$. Hence for each $k \in \{1, \ldots, n\}$ we have

$$\lambda_k = \inf\{\sup\{\mathbf{x}^T A\mathbf{x} \; : \; \mathbf{x} \in \mathcal{U}, \; \|\mathbf{x}\| = 1\} \; : \; \mathcal{U} \in \Lambda_{n-k+1}\} \qquad (3.1.3)$$

where $\Lambda_{n-k+1}$ denotes the set of all $(n-k+1)$-dimensional subspaces of $\mathbb{R}^n$.

Clearly equation (3.1.1) may be used to obtain lower bounds for $\mu_1$ by making appropriate choices for $\mathbf{x}$. For example if $G + uv$ denotes the graph obtained from the connected graph $G$ by adding the edge $uv$ then we may take $A$ to be the adjacency matrix of $G + uv$ and $\mathbf{x}$ the principal eigenvector $(x_1, \ldots, x_n)^T$ of $G$. In this case we find that $\mu_1(G + uv) \geq \mu_1(G) + 2x_u x_v$. Upper bounds involving the components of the principal eigenvector are discussed in Section 6.4 in the context of graph perturbations.

We note here however that if the graphs $G, G'$ differ in only one edge then $|\mu_1(G) - \mu_1(G')| \leq 1$. We may assume that $G$ is connected and that $G'$ is obtained from $G$ by deleting the edge $ij$, say. In this case we have $\mu_1(G) > \mu_1(G') \geq \mu_1(G) - 2x_i x_j \geq \mu_1(G) - x_i^2 - x_j^2 \geq \mu_1(G) - 1$ and the required inequality follows. We describe without proof a consequence for random graphs (see Theorem 5.2 of [CvRo5]): if $\mu_1$ is the index of a random graph $G^{n,p}$ on $n$ vertices then for $t > 0$,

$$\Pr(|\mu_1 - E(\mu_1)| \geq t) \leq 2\exp\{-2t^2 / \binom{n}{2}\}. \qquad (3.1.4)$$

Here $E(\mu_1)$ is the expected value of $\mu_1$, and each edge of $G^{n,p}$ is present with the same probability $p$ $(0 < p < 1)$. The inequality (3.1.4) is proved by the method of bounded differences (see Lemma 3.3 of [McD]) and may be seen as an analogue of the Chernoff bound [Cher] on the tails of a binomial distribution.

For a second application of (3.1.1) we turn to the class $\mathscr{H}_n$ of all maximal outerplanar graphs with $n$ vertices $(n \geq 4)$. Thus if $G \in \mathscr{H}_n$ then $G$ has a plane representation as an $n$-gon triangulated by $n - 3$ chords, and the boundary of this $n$-gon is the unique Hamiltonian cycle $Z$ of $G$. A triangle of $G$ with no edge in common with $Z$ is called an *internal triangle* of $G$. We denote by $\mathscr{K}_n$ the class of all graphs in $\mathscr{H}_n$ without internal triangles. Recall that $K_1 \nabla P_{n-1}$ is the graph obtained

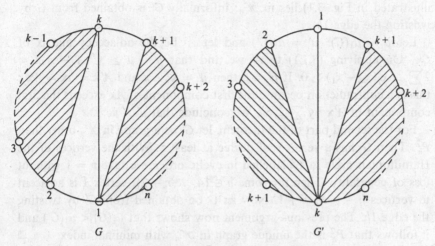

Fig. 3.1. The graphs $G$ and $G'$ of Theorem 3.1.1

from the path $P_{n-1}$ by adding a new vertex of degree $n-1$, and that $P_n^2$ is the graph obtained from $P_n$ by adding new edges joining all pairs of vertices at distance 2 in $P_n$. Note that both $K_1 \triangledown P_{n-1}$ and $P_n^2$ lie in $\mathscr{K}_n$. Computer evidence suggests that among the graphs in $\mathscr{K}_n$ the graph $K_1 \triangledown P_{n-1}$ alone has maximal index and $P_n^2$ alone has minimal index. We use (3.1.1) to prove that these statements are true of the class $\mathscr{K}_n$.

**3.1.1 Theorem** [Row7] *For $n \geq 4$, let $\mathscr{K}_n$ be the class of maximal outerplanar graphs with $n$ vertices and no internal triangles. A graph in $\mathscr{K}_n$ with maximal index is isomorphic to $K_1 \triangledown P_{n-1}$ and a graph in $\mathscr{K}_n$ with minimal index is isomorphic to $P_n^2$.*

*Proof* We assume that $n \geq 6$ because each of $\mathscr{K}_4$ and $\mathscr{K}_5$ contains just one graph. We show first that if $G$ is a graph in $\mathscr{K}_n$ other than $K_1 \triangledown P_{n-1}$ then $\mathscr{K}_n$ contains a graph $G'$ such that $\mu_1(G') > \mu_1(G)$. We take the vertices of the Hamiltonian cycle in $G$ to be labelled $1, 2, \ldots, n$ in cyclic order. Since $G \neq K_1 \triangledown P_{n-1}$ the labelling may be chosen so that some vertex $k$ $(3 < k < n-1)$ is adjacent to vertices 1 and 2, and vertex $k+1$ is adjacent to vertex 1.

The graph $G$ is illustrated in Fig. 3.1 where the symmetry shows that without loss of generality we may assume $x_k \geq x_1$, where $(x_1, \ldots, x_n)^T$ is the principal eigenvector $\mathbf{x}$ of $G$. Let $J$ be the set of neighbours $j$ of vertex 1 such that $k+2 \leq j \leq n$, and let $G'$ be the graph obtained from $G$ by replacing edge $1j$ with edge $kj$ for each $j \in J$. Then $G'$ (also illustrated in Fig. 3.1) lies in $\mathcal{H}_n$: informally $G'$ is obtained from $G$ by twisting the edge $1k$.

Let $\mu = \mu_1(G)$, $\mu' = \mu_1(G')$ and let $A'$ be the adjacency matrix of $G'$. On applying (3.1.1) to $A'$ we find that $\mu' - \mu \geq \mathbf{x}^T(A' - A)\mathbf{x} = 2\sum_{j\in J} x_j(x_k - x_1) \geq 0$. If $\mu' = \mu$ then $\mu' = \mathbf{x}^T A'\mathbf{x}$ and $A'\mathbf{x} = \mu\mathbf{x} = A\mathbf{x}$ : this is a contradiction because the first component of $A\mathbf{x}$ exceeds the first component of $A'\mathbf{x}$ by $\sum_{j\in J} x_j$. We conclude that $\mu' > \mu$.

For the second part of the theorem, let $G'$ be a graph in $\mathcal{H}_n$ other than $P_n^2$. Then $G'$ has a vertex $v$ of degree at least 5. With the vertices of the Hamiltonian cycle in $G'$ labelled in cyclic order, we take $v = 1$ without loss of generality. Then for some $h \in \{4, \ldots, n-2\}$, vertex 1 is adjacent to vertices $h - 1$, $h$, $h + 1$. Now let $G$ be obtained from $G'$ by twisting the edge $1h$. The previous argument now shows that $\mu_1(G) < \mu_1(G')$ and it follows that $P_n^2$ is the unique graph in $\mathcal{H}_n$ with minimal index. $\quad\square$

It is noted in [Row7] that the index of $K_1 \triangledown P_{n-1}$ lies between 4 and $4 - \frac{6}{n}$, while the index of $P_n^2$ lies between $1 + \sqrt{n}$ and $1 + \sqrt{n} - \frac{3}{n}$.

In order to involve eigenvalues other than the largest we need the following generalization of equation (3.1.1).

**3.1.2 Lemma** *Suppose that the symmetric matrix $A$ has eigenvalues $\lambda_1, \lambda_2, \ldots, \lambda_n$ in arbitrary order, and let $\mathbf{x}_1, \ldots, \mathbf{x}_r$ be linearly independent eigenvectors of $A$ corresponding to $\lambda_1, \ldots, \lambda_r$ respectively. If $r < n$ then the largest eigenvalue $\lambda \in \{\lambda_{r+1}, \ldots, \lambda_n\}$ is equal to $\sup\{\mathbf{x}^T A\mathbf{x}/\mathbf{x}^T\mathbf{x} : 0 \neq \mathbf{x} \in \langle\mathbf{x}_1, \ldots, \mathbf{x}_r\rangle^\perp\}$; moreover, for $0 \neq \mathbf{x} \in \langle\mathbf{x}_1, \ldots, \mathbf{x}_r\rangle^\perp, \lambda = \mathbf{x}^T A\mathbf{x}/\mathbf{x}^T\mathbf{x}$ if and only if $A\mathbf{x} = \lambda\mathbf{x}$.*

*Proof* We may assume that $\|\mathbf{x}_i\| = 1$ $(i = 1, \ldots, r)$, so that we may extend $\{\mathbf{x}_1, \ldots, \mathbf{x}_r\}$ to an orthonormal basis $\{\mathbf{x}_1, \ldots, \mathbf{x}_n\}$ of $\mathbb{R}^n$. If $\mathbf{x}$ is a unit vector in $\langle\mathbf{x}_1, \ldots, \mathbf{x}_r\rangle^\perp$ then $\mathbf{x} = \alpha_{r+1}\mathbf{x}_{r+1} + \cdots + \alpha_n\mathbf{x}_n$ where $\alpha_{r+1}^2 + \cdots + \alpha_n^2 = 1$. Then $\mathbf{x}^T A\mathbf{x} = \lambda_{r+1}\alpha_{r+1}^2 + \cdots + \lambda_n\alpha_n^2$ and the result follows. $\quad\square$

We use Lemma 3.1.2 to characterize graph isomorphisms in terms of eigenvalues. To be precise, let $\theta$ be a bijection $V(G) \to V(H)$, where $G, H$ are disjoint finite graphs. We define the *recognition graph* $\Gamma(G, \theta, H)$ as the graph consisting of $G, H$ and the edges $\{v, \theta(v)\}$ $(v \in V(G))$. (The terminology is suggested by a graph-theoretical model for pattern recognition formulated in [BiMa].) With a suitable ordering of vertices, $\Gamma(G, \theta, H)$ has adjacency matrix $\begin{pmatrix} A & I \\ I & B \end{pmatrix}$, where $A, B$ are adjacency matrices for $G, H$ respectively. If $\theta$ is an isomorphism then $A = B$ and $\Gamma(G, \theta, H)$ has characteristic polynomial

$$\det((x + 1)I - A) \det((x - 1)I - A);$$

hence if $\lambda_1, \ldots, \lambda_n$ are the eigenvalues of $G$ then those of $\Gamma(G, \theta, H)$ are $\lambda_1 \pm 1, \lambda_2 \pm 1, \ldots, \lambda_n \pm 1$. We show that the converse holds for cospectral graphs $G, H$. (Of course if $G$ and $H$ are not cospectral then there is no isomorphisim $\theta : V(G) \to V(H)$.) We note in passing that if $\theta$ is an isomorphism then $\Gamma(G, \theta, H)$ is a NEPS (see Section 2.3).

**3.1.3 Theorem** [Row10] *Suppose that $G, H$ are cospectral graphs, with common eigenvalues $\lambda_1, \ldots, \lambda_n$, and let $\theta$ be a bijection $V(G) \to V(H)$. Then $\theta$ is an isomorphisim if and only if the eigenvalues of $\Gamma(G, \theta, H)$ are $\lambda_1 \pm 1, \ldots, \lambda_n \pm 1$.*

*Proof* It remains to prove sufficiency. We may suppose that $\lambda_1 \geq \lambda_2 \geq \cdots \geq \lambda_n$, so that $(\mathbf{x}^T \mathbf{y}^T) \begin{pmatrix} A & I \\ I & B \end{pmatrix} \begin{pmatrix} \mathbf{x} \\ \mathbf{y} \end{pmatrix} = \mathbf{x}^T A \mathbf{x} + \mathbf{y}^T B \mathbf{y} + 2\mathbf{x}^T \mathbf{y} \leq \lambda_1 \mathbf{x}^T \mathbf{x} + \lambda_1 \mathbf{y}^T \mathbf{y} + \mathbf{x}^T \mathbf{x} + \mathbf{y}^T \mathbf{y} = (\lambda_1 + 1)(\mathbf{x}^T \mathbf{x} + \mathbf{y}^T \mathbf{y})$.

If now $\begin{pmatrix} \mathbf{x} \\ \mathbf{y} \end{pmatrix}$ is an eigenvector corresponding to $\lambda_1 + 1$ then this upper bound is attained; and in this case $A\mathbf{x} = \lambda_1 \mathbf{x}$, $B\mathbf{y} = \lambda_1 \mathbf{y}$, $\mathbf{x} = \mathbf{y}$. Accordingly a unit eigenvector corresponding to $\lambda_1 + 1$ has the form $\begin{pmatrix} \mathbf{x}_1 \\ \mathbf{x}_1 \end{pmatrix}$, and we see straightaway that $\begin{pmatrix} \mathbf{x}_1 \\ -\mathbf{x}_1 \end{pmatrix}$ is a unit eigenvector corresponding to $\lambda_1 - 1$. Among the eigenvalues of $\Gamma(G, \theta, H)$ other than $\lambda_1 \pm 1$ the largest is $\lambda_2 + 1$; and by Lemma 3.1.2 this is the supremum of the Rayleigh quotients $\dfrac{(\mathbf{x}^T \mathbf{y}^T) \begin{pmatrix} A & I \\ I & B \end{pmatrix} \begin{pmatrix} \mathbf{x} \\ \mathbf{y} \end{pmatrix}}{\mathbf{x}^T \mathbf{x} + \mathbf{y}^T \mathbf{y}}$ taken over non-zero $\begin{pmatrix} \mathbf{x} \\ \mathbf{y} \end{pmatrix}$ orthogonal to both $\begin{pmatrix} \mathbf{x}_1 \\ \mathbf{x}_1 \end{pmatrix}$ and $\begin{pmatrix} \mathbf{x}_1 \\ -\mathbf{x}_1 \end{pmatrix}$. This last condition is satisfied if and only if $\mathbf{x}^T \mathbf{x}_1 = 0 = \mathbf{y}^T \mathbf{x}_1$ and so by Lemma 3.1.2 (applied to $A$ and $B$) we have $(\mathbf{x}^T \mathbf{y}^T) \begin{pmatrix} A & I \\ I & B \end{pmatrix} \begin{pmatrix} \mathbf{x} \\ \mathbf{y} \end{pmatrix} \leq (\lambda_2 + 1)(\mathbf{x}^T \mathbf{x} + \mathbf{y}^T \mathbf{y})$. Again equality holds if and only if $A\mathbf{x} = \lambda_2 \mathbf{x}$, $B\mathbf{y} = \lambda_2 \mathbf{y}$ and $\mathbf{x} = \mathbf{y}$. Hence as before eigenvectors of $\begin{pmatrix} A & I \\ I & B \end{pmatrix}$ corresponding to $\lambda_2 \pm 1$ have the form $\begin{pmatrix} \mathbf{x}_2 \\ \pm \mathbf{x}_2 \end{pmatrix}$.

We may repeat the process, beginning with vectors $\begin{pmatrix} \mathbf{x} \\ \mathbf{y} \end{pmatrix}$ orthogonal to the four vectors $\begin{pmatrix} \mathbf{x}_1 \\ \pm\mathbf{x}_1 \end{pmatrix}, \begin{pmatrix} \mathbf{x}_2 \\ \pm\mathbf{x}_2 \end{pmatrix}$, to obtain $n$ non-zero pairwise orthogonal vectors $\mathbf{x}_1,\dots,\mathbf{x}_n$ such that $A\mathbf{x}_i = \lambda_i\mathbf{x}_i = B\mathbf{x}_i$ $(i=1,\dots,n)$. It follows that $A = B$ and that $\theta$ is an isomorphism.                         $\square$

Finally we prove two results which relate the structure of a graph $G$ to its Laplacian spectrum, and note their implications in terms of the usual spectrum. Recall from Section 1.1 that the second smallest eigenvalue $\lambda$ of the Laplacian matrix of $G$ is zero if and only if $G$ is not connected. For an arbitrary graph $G$, $\lambda$ is called the *algebraic connectivity* of $G$ [Fie1] and we describe how it is related to the vertex connectivity $\kappa(G)$ and edge connectivity $\kappa'(G)$.

**3.1.4 Theorem** [Fie1] *If $G$ is not a complete graph and if $\lambda$ is the second smallest eigenvalue in the Laplacian spectrum of $G$ then $\lambda \le \kappa(G)$.*

*Proof* It suffices to show that if the graph $H$ is obtained from $G$ by removing a vertex then the second smallest eigenvalue of the Laplacian $L(H)$ satisfies the inequality $\lambda(H) \ge \lambda - 1$; for then if we remove $\kappa(G)$ vertices in succession to obtain a disconnected graph $H_0$ we have $0 = \lambda(H_0) \ge \lambda - \kappa(G)$.

Let $H = G - v$ and let $G'$ be the graph obtained from $G$ by adding all the edges $vu$ $(u \in V(H))$ for which $u$ is not adjacent to $v$ in $G$. With appropriate labelling of vertices we have

$$L(G') = \begin{pmatrix} L(H)+I & -\mathbf{j} \\ -\mathbf{j}^T & n-1 \end{pmatrix}$$

where $n = |V(G)|$. Now $L(H)\mathbf{v} = \lambda(H)\mathbf{v}$ for some unit vector $\mathbf{v}$ orthogonal to $\mathbf{j}$, and if $\mathbf{v}' = \begin{pmatrix} \mathbf{v} \\ 0 \end{pmatrix}$ then $L(G')\mathbf{v}' = (\lambda(H)+1)\mathbf{v}'$.

If we apply Lemma 3.1.2 to $-L$ we find that $\lambda = \inf\{\mathbf{x}^T L\mathbf{x} \; : \; \mathbf{x} \in U\}$ where

$$U = \{(x_1,\dots,x_n)^T \in \mathbb{R}^n \; : \sum_{i=1}^{n} x_i^2 = 1 \text{ and } \sum_{i=1}^{n} x_i = 0\}.$$

Equivalently $\lambda = \inf_{\mathbf{x}\in U}\{\sum_{j\sim k} (x_j - x_k)^2\}$, from which it is clear that $\lambda \le \lambda(G')$. On the other hand, $\mathbf{v}' \in U$ and so $\lambda(G') \le \mathbf{v}'^T L(G')\mathbf{v}' = \lambda(H)+1$. It follows that $\lambda(H) \ge \lambda - 1$, as required.                         $\square$

For any graph $G$, $\kappa(G) \le \kappa'(G)$ and so by Theorem 3.1.4, $\lambda \le \kappa'(G)$ for

any non-complete graph $G$. (If $G = K_n$ then $\lambda = n$, $\kappa(G) = \kappa'(G) = n-1$.)
In [Fie3], Fiedler establishes a sharp lower bound for $\lambda$ in terms of $\kappa'(G)$,
namely

$$\lambda \geq 2(1 - \cos\frac{\pi}{n})\kappa'(G) \tag{3.1.5}$$

for any $n$-vertex graph $G$. Further, equality holds in (3.1.5) if and only if
$G = P_n$.

**3.1.5 Theorem** [AlMi] *Let $\lambda$ be the second smallest eigenvalue in the Laplacian spectrum of the $n$-vertex graph $G$. For any non-trivial bipartition $S \dot\cup T$
of $V(G)$, the number of edges between $S$ and $T$ is at least $\frac{\lambda}{n}|S||T|$.*

*Proof* We know from Section 1.1 that the smallest eigenvalue of $L$ is
zero, with $\mathbf{j}$ as a corresponding eigenvector. From equation (3.1.2) (with
$\mathbf{x}_n = n^{-\frac{1}{2}}\mathbf{j}$, $\lambda_n = 0$, $\lambda_{n-1} = \lambda$) we see that

$$\lambda = \inf\{\mathbf{x}^T L\mathbf{x}/\mathbf{x}^T\mathbf{x} : \mathbf{x} \neq \mathbf{0} \text{ and } \mathbf{x}^T\mathbf{j} = 0\}.$$

Now let $s = |S|$, $t = |T|$, and let $E(S, T)$ denote the set of edges between
$S$ and $T$. We define $\mathbf{x} = (x_1,\ldots,x_n)^T$ where $x_i = -s^{-1}$ if $i \in S$, $+t^{-1}$ if
$i \in T$. Then $\mathbf{x}^T\mathbf{j} = 0$ and so $\lambda \leq \mathbf{x}^T L\mathbf{x}/\mathbf{x}^T\mathbf{x}$. We choose an orientation of
edges in which all edges between $S$ and $T$ are oriented from $S$ to $T$, and
we let $C$ be the corresponding edge-vertex incidence matrix. Then the
$e$-entry of $C\mathbf{x}$ is $s^{-1} + t^{-1}$ if $e \in E(S, T)$ and 0 otherwise. Consequently
$\mathbf{x}^T L\mathbf{x} = \|C\mathbf{x}\|^2 = |E(S, T)|/(s^{-1} + t^{-1})^2$, while $\mathbf{x}^T\mathbf{x} = s^{-1} + t^{-1}$. Hence
$\lambda \leq |E(S, T)|(s^{-1} + t^{-1}) = |E(S, T)|n/st$, and the result follows. $\square$

Theorem 3.1.5 is a special case of the following result, with a similar
proof [AlMi], which applies to any two disjoint non-empty subsets $S, T$.
Here $|S| = s$, $|T| = t$; $e$ is the distance between $S$ and $T$; $E$ is the set of
all edges in $G$, $E_S$ is the set of edges with both ends in $S$ and $E_T$ is the
set of edges with both ends in $T$:

$$\lambda \leq e^{-2}(s^{-1} + t^{-1})(|E| - |E_S| - |E_T|). \tag{3.1.6}$$

The *isoperimetric number* $i(G)$, a measure of expansion in $G$, is defined
as the minimum of $|E(S, T)|/\min\{|S|,|T|\}$ taken over all non-trivial
bipartitions $S \dot\cup T$ of $V(G)$. It follows from Theorem 3.1.5 that

$$i(G) \geq \frac{1}{2}\lambda. \tag{3.1.7}$$

This is noted by Mohar [Moh2], who goes on to derive an upper
bound for $i(G)$ when $G \neq K_1, K_2, K_3$ :

$$i(G) \leq \sqrt{\lambda(2\Delta - \lambda)} \tag{3.1.8}$$

where $\Delta$ denotes the maximal degree of $G$. For (weaker) inequalities in terms of the second largest eigenvalue of $G$, we compare $L$ with the matrices $\Delta I - A$, $\delta I - A$ where $\delta$ denotes the minimal degree of $G$. Each of $(\Delta I - A) - L$, $L - (\delta I - A)$ is a diagonal matrix with non-negative entries and so it follows from (3.1.3) that $\delta - \lambda_2(G) \leq \lambda \leq \Delta - \lambda_2(G)$. Now suppose that $G$ is connected, so that $\lambda_2(G) = \mu_2$. In this case it follows from Theorem 3.1.4 that

$$\kappa(G) \geq \delta - \mu_2, \tag{3.1.9}$$

and it follows from (3.1.7) and (3.1.8) that (for $G \neq K_1, K_2$)

$$\frac{1}{2}(\delta - \mu_2) \leq i(G) \leq \sqrt{\Delta^2 - \mu_2^2}. \tag{3.1.10}$$

Analogous results for the isoperimetric number of a locally finite graph are given in [Moh1]. For further results on the Laplacian spectrum, the reader is referred to [Gro], [GrMe1], [GrMe2], [GrMe3], [GrMS], [GrMW], [GrZi], [Maa1], [Maa3], [Mer2] and [Mer3].

## 3.2 Comparing vectors

For $\mathbf{x} = (x_1, \ldots, x_n)^T$, $\mathbf{y} = (y_1, \ldots, y_n)^T \in \mathbb{R}^n$, we write $\mathbf{x} \leq \mathbf{y}$ if $x_i \leq y_i$ for all $i \in \{1, \ldots, n\}$, and $\mathbf{x} < \mathbf{y}$ if $\mathbf{x} \leq \mathbf{y}$ but $\mathbf{x} \neq \mathbf{y}$. Now let $A$ be any symmetric $n \times n$ matrix with real entries and largest eigenvalue $\lambda_1$. It is clear from (3.1.1) that

$$\text{if } \mathbf{y} > 0 \text{ and } A\mathbf{y} > \rho\mathbf{y} \text{ then } \lambda_1 > \rho. \tag{3.2.1}$$

On the other hand if $A$ is irreducible, in particular if $A$ is the adjacency matrix of a connected graph, we have also

$$\text{if } \mathbf{y} > 0 \text{ and } A\mathbf{y} < \rho\mathbf{y} \text{ then } \lambda_1 < \rho. \tag{3.2.2}$$

For if $\mathbf{x}$ is the principal eigenvector of $A$ then $\mathbf{x}^T\mathbf{y} > 0$ while $\lambda_1\mathbf{x}^T\mathbf{y} = \mathbf{x}^T A\mathbf{y} < \rho\mathbf{x}^T\mathbf{y}$. We give two applications of (3.2.2): in each case a local modification of a connected graph $G$ is shown to have smaller index than $G$.

First suppose that $G'$ is obtained from $G$ by splitting a vertex $v$: if the edges incident with $v$ are $vw$ ($w \in W$) then $G'$ is obtained from $G - v$ by adding two new vertices $v_1, v_2$ and edges $v_1w_1$ ($w_1 \in W_1$), $v_2w_2$ ($w_2 \in W_2$) where $W_1 \dot\cup W_2$ is a non-trivial bipartition of $W$.

**3.2.1 Theorem** [Sim1] *If $G'$ is obtained from the connected graph $G$ by splitting a vertex then $\lambda_1(G') < \lambda_1(G)$.*

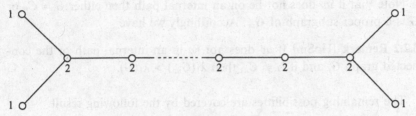

Fig. 3.2. The tree $T_n$ $(n \geq 6)$.

*Proof* We may suppose that $G'$ is connected for otherwise $G'$ has two components each of which is a proper subgraph of $G$, with index less than $\lambda_1(G)$. Accordingly we may apply (3.2.2) to the adjacency matrix $A'$ of $G'$.

Suppose that vertices are numbered so that vertex 1 of $G$ splits into vertices $0, 1$ of $G'$. Let $\lambda_1 = \lambda_1(G)$, let $(x_1, \ldots, x_n)^T$ be the principal eigenvector of $G$ and define **y** as $(y_0, y_1, y_2, \ldots, y_n)^T$ where $y_0 = x_1$ and $y_i = x_i$ $(i = 1, \ldots, n)$. Then $A'\mathbf{y} = (y_0', y_1', y_2', \ldots, y_n')^T$ where $y_0' < \lambda_1 y_0$, $y_1' < \lambda_1 y$ and $y_i' = \lambda_1 y_i$ $(i = 2, \ldots, n)$. Thus $A'\mathbf{y} < \lambda_1 \mathbf{y}$ and by (3.2.2) we have $\lambda_1(G') < \lambda_1$. □

Secondly we consider the graph $G_{uv}$ obtained from the connected graph $G$ by subdividing the edge $uv$, that is by replacing $uv$ with edges $uw, wv$ where $w$ is an additional vertex. Note that no change in index results when $G$ is an $n$-cycle $C_n$ $(n \geq 3)$ or when $uv$ is a non-pendant edge of the tree $T_n$ $(n \geq 6)$ shown in Fig. 3.2: in both cases the index remains 2. (For $T_n$, Fig. 3.2 shows the components of an eigenvector corresponding to 2.)

In all other cases, as we shall see, the index strictly decreases or increases according as $uv$ is or is not on an *internal path* of $G$ [HoSm]. Such a path is of one of two types:

(a) a sequence of vertices $v_0, v_1, \ldots, v_{k+1}$ $(k \geq 2)$ where $v_0, v_1, \ldots, v_k$ are distinct, $v_{k+1} = v_0$ of degree at least 3, $\deg(v_i) = 2$ for $i = 1, \ldots, k$, and $v_{i-1} \sim v_i$ $(i = 1, \ldots, k+1)$;

(b) a sequence of distinct vertices $v_0, v_1, \ldots, v_{k+1}$ $(k \geq 0)$ such that $v_{i-1} \sim v_i$ $(i = 1, \ldots, k+1), \deg(v_0) \geq 3, \deg(v_{k+1}) \geq 3$ and $\deg(v_i) = 2$ whenever $1 \leq i \leq k$.

Thus in case (a) the vertices $v_0, \ldots, v_k$ induce a subgraph isomorphic to $C_{k+1}$, while in case (b) the vertices $v_0, \ldots, v_{k+1}$ lie in an induced subgraph isomorphic to $T_{k+6}$.

Note that if $uv$ does not lie on an internal path then either $G = C_n$ or $G$ is a proper subgraph of $G_{uv}$. Accordingly we have

**3.2.2 Remark** [HoSm] If $uv$ does not lie in an internal path of the connected graph $G$, and if $G \neq C_n$, then $\lambda_1(G_{uv}) > \lambda_1(G)$. □

The remaining possibilities are covered by the following result.

**3.2.3 Theorem** [HoSm] *If $uv$ lies on an internal path of the connected graph $G$, and if $G \neq T_n$, then $\lambda_1(G_{uv}) < \lambda_1(G)$.*

*Proof* Suppose first that $uv$ lies on an internal path $v_0, v_1, \ldots, v_{k+1}$ of type (a) where the vertices $v_0, \ldots, v_k$ are labelled $0, 1, \ldots, k$. Let $x_0, x_1, \ldots, x_k$ be the corresponding components of the principal eigenvector $\mathbf{x}$ of $G$ and let $\lambda_1 = \lambda_1(G)$. (It is helpful to picture the components of $\mathbf{x}$ ascribed to the relevant vertices of $G$, as in Fig. 3.2.) We have $\lambda_1 x_j = x_{j-1} + x_{j+1}$ $(i = 1, \ldots, k)$ where $x_{k+1} = x_0$ and by symmetry $x_j = x_{k+1-j}$ $(j = 1, \ldots, k)$. If $k$ is even then without loss of generality we may take $u = \frac{1}{2}k$, $v = \frac{1}{2}k + 1$. Now let $\mathbf{y}$ be obtained from $\mathbf{x}$ by inserting the additional component $x_w$ equal to $x_u$ and $x_v$. If $A$ is the adjacency matrix of $G_{uv}$ then $A\mathbf{y}$ and $\lambda_1 \mathbf{y}$ differ only in the components corresponding to $w$, for which $x_u + x_v = 2x_w < \lambda_1 x_w$. Hence $A\mathbf{y} < \lambda_1 \mathbf{y}$ and we have $\lambda_1(G_{uv}) < \lambda_1$ by (3.2.2). If $k$ is odd then we may take $u = \frac{1}{2}(k-1)$, $v = \frac{1}{2}(k+1)$ and construct $\mathbf{y}$ as before with $x_w = x_v$. In this situation, $\lambda_1 x_v = 2x_u$ whence $x_u > x_v, \lambda_1 x_v > x_u + x_w$ and $\lambda_1 x_w > x_u + x_v$. It follows that $A\mathbf{y} < \lambda_1 \mathbf{y}$ and so again $\lambda_1(G_{uv}) < \lambda_1$ by (3.2.2).

Secondly suppose that $uv$ lies on an internal path of type (b), whose vertices are labelled $0, 1, \ldots, k + 1$, and let $x_0, x_1, \ldots, x_{k+1}$ be the corresponding components of the principal eigenvector $\mathbf{x}$ of $G$. Reversing the path if necessary we may assume that $x_0 \leq x_{k+1}$. Let $t$ be least such that $x_t$ is the least of $x_0, x_1, \ldots, x_{k+1}$, so that $t < k + 1$. Without loss of generality we let $u = t$, $v = t + 1$.

Consider first the case $t > 0$: here we take $\mathbf{y}$ to be the vector obtained from $\mathbf{x}$ by inserting an additional component $x_w$ equal to $x_t$. We have $x_{t-1} + x_w \leq x_{t-1} + x_{t+1} = \lambda_1 x_t$ and $x_t + x_{t+1} < x_{t-1} + x_{t+1} = \lambda_1 x_w$, whence $A\mathbf{y} < \lambda_1 \mathbf{y}$ and $\lambda_1(G_{uv}) < \lambda_1$ by (3.2.2). Accordingly we suppose that $t = 0$. Let $S$ be the set of neighbours of $0$ other than $1$, and let $s = \sum_{j \in S} x_j$. If $s \geq x_0$ then we construct $\mathbf{y}$ as above with $x_w = x_0$. We have $s + x_w \leq s + x_1 = \lambda_1 x_0$ and $x_0 + x_1 \leq s + x_1 = \lambda_1 x_w$; moreover one of these inequalities is strict for otherwise $\lambda_1 = 2$, contradicting the fact that $T_{k+6}$ is a proper subgraph of $G$. Hence $A\mathbf{y} < \lambda_1 \mathbf{y}$. Finally suppose that $s < x_0$. In this case we construct $\mathbf{y}$ from $\mathbf{x}$ by replacing $x_0$ with $s$ and inserting $x_w$ equal to $x_0$. Now $x_0 + s < 2x_0 \leq |S|x_0$, while $|S|x_0 \leq \lambda_1 s$ because $x_0 \leq \lambda_1 x_j$ for all $j \in S$. Hence $x_0 < \lambda_1 s$ and it follows that $A\mathbf{y} < \lambda_1 \mathbf{y}$. A final application of (3.2.2) completes the proof of the theorem. □

The proof of Theorem 3.2.3 makes use of the relation

$$\lambda_1 x_j = x_{j-1} + x_{j+1} \quad (j = 1, 2, \ldots, k) \qquad (3.2.3)$$

for an induced cyclic subgraph with vertices $0, 1, \ldots, k$. Here the $x_j$ are components of a principal eigenvector, the vertices $0, k + 1$ are one and the same, and $x_j = x_{k+1-j}$ by symmetry. In particular $2x_1 = \lambda_1 x_0$ and it follows by induction on $j$ that when $\lambda_1 > 2$ we have $x_{j-1} < x_j$ for $j = 1, 2, \ldots, \lfloor \frac{1}{2}k \rfloor$. This fact makes it possible to use the 'comparison of vectors' method to compare indices of graphs of the following type. Let $G$ be a connected rooted graph and for $3 \leq k \leq m - 3$ let $G_{k,m-k}$ be the graph obtained from $G$ by identifying its root with a vertex of $C_k$ and a vertex of $C_{m-k}$.

**3.2.4 Theorem** [Sim1] *For fixed $m \geq 6$ the index of $G_{k,m-k}$ is strictly decreasing for $3 \leq k \leq \lfloor \frac{1}{2}m \rfloor$.*

If two similar vertices of a graph are connected by an induced path with vertices $0, 1, \ldots, k + 1$ then again (3.2.3) holds with the same constraints, because similarity of the vertices $0, k + 1$ ensures that $x_0 = x_{k+1}$. Accordingly, near-identical arguments apply to the graphs $H_{k,m-k}$ obtained from a connected graph $H$ by adding edge-disjoint paths of lengths $k, m - k$ between vertices which are similar in $H$. As noted in [Sim1], for fixed $m \geq 3$ the index of $H_{k,m-k}$ is strictly decreasing for $1 \leq k \leq \lfloor \frac{m}{2} \rfloor$. In the case that $H = P_2$, the graphs $H_{k,m-k}$ are just the bicyclic Hamiltonian graphs with $m$ vertices, graphs which are treated in a different context in Section 3.4.

Li and Feng [LiFe] proved the analogue of Theorem 3.2.4 for graphs obtained from the rooted graph $G$ by attaching paths of lengths $k$ and $m - k$ at the root. Their method was to compare characteristic polynomials, and their result is presented in Section 6.2 in the context of graph perturbations. The result is clearly pertinent to the problem of ordering unicyclic graphs by index, and their methods were extended by Simić [Sim2] in this context. Among unicyclic graphs with $n$ vertices, the unique graph with largest index is obtained from the star $K_{1,n-1}$ by adding an edge (see also Section 3.5). That with smallest index is necessarily $C_n$ by a Rayleigh quotient argument: if the unicyclic graph $G$ has index $\lambda_1$ and adjacency matrix $A$ then $\lambda_1 \geq \mathbf{j}^T A \mathbf{j}/\mathbf{j}^T \mathbf{j} = 2$ with equality if and only if $\mathbf{j}$ is an eigenvector corresponding to 2 (cf. Theorem 2.1.2).

## 3.3 Biquadratic forms

Let $G, G'$ be $n$-vertex graphs with adjacency matrices $A, A'$ and indices $\mu$, $\mu'$ respectively. Suppose that each of $G, G'$ has a unique non-trivial component, and let $\mathbf{x}, \mathbf{x}'$ be the principal eigenvectors of $G, G'$ respectively. Then

$$(\mu - \mu')\mathbf{x}^T \mathbf{x}' = \mathbf{x}^T (A - A')\mathbf{x}'. \tag{3.3.1}$$

We suppose that the vertices of $G$ and $G'$ are labelled so that in both $\mathbf{x}$ and $\mathbf{x}'$, the non-zero components precede the zero components. Then $\mathbf{x}^T \mathbf{x}' > 0$ and so the sign of $\mu - \mu'$ is the same as that of $\mathbf{x}^T(A - A')\mathbf{x}'$, which may be regarded as a biquadratic form $\sum_{i=1}^n \sum_{j=1}^n x_i(a_{ij} - a'_{ij})x'_j$. In some circumstances the equation (3.3.1) is more useful than the inequalities (3.2.1), (3.2.2) and we give two applications in this section.

The first relates to the class $\mathcal{G}(n, m)$ of $n$-vertex graphs with $m$ edges, where to exclude trivial cases we suppose that $0 < m < \binom{n}{2}$. In Section 3.5 we describe for arbitrary $m$ the graphs in $\mathcal{G}(n, m)$ with maximal index, while here we show that when $m = \binom{d}{2}$ such a graph consists of $K_d$ together with $n - d$ isolated vertices. We show first that a graph in $\mathcal{G}(n, m)$ with maximal index has a *stepwise* adjacency matrix, that is an adjacency matrix $(a_{ij})$ which satisfies the condition

if $i < j$ and $a_{ij} = 1$ then $a_{hk} = 1$ whenever $h < k \leq j$ and $h \leq i$.

Thus all non-diagonal entries above or to the left of a 1 are also 1s, and the boundary between 1s and non-diagonal 0s has a stepwise form.

**3.3.1 Lemma** [BrHo] *A graph in $\mathcal{G}(n, e)$ with maximal index has a stepwise adjacency matrix.*

*Proof* Let $G$ be a graph in $\mathcal{G}(n, e)$ with maximal index. Note first that $G$ has a unique non-trivial component, for otherwise the index may be increased by relocating an edge $uv$ not lying in a given component $C$ with largest index. (We may replace $uv$ with an edge $u'v$ where $u'$ is any vertex of $C$.) Now let $A$ be the adjacency matrix $(a_{ij})$ of $G$ with vertices ordered so that the components of the principal eigenvector $\mathbf{x}$ satisfy the relation $x_1 \geq x_2 \geq \cdots \geq x_n$.

Suppose by way of contradiction that $a_{pq} = 0$ and $a_{p,q+1} = 1$ for some $p, q$ such that $p < q$. Let $A'$ be the matrix obtained from $A$ by interchanging the $(p, q)$- and $(p, q+1)$-entries and interchanging the $(q, p)$- and $(q+1, p)$-entries. On applying (3.1.1) to $A'$ we have $\mu_1(A') - \mu_1(A) \geq \mathbf{x}^T(A' - A)\mathbf{x} = 2x_p(x_q - x_{q+1}) \geq 0$. Since $A'$ is the adjacency matrix of a graph in $\mathscr{G}(n, m)$ (obtained from $G$ by relocating an edge) we have $\mu_1(A') = \mu_1(A)$, and then $\mu_1(A')\mathbf{x} = A'\mathbf{x}$ because $\mu_1(A') = \mathbf{x}^T A'\mathbf{x}$. Thus $(A' - A)\mathbf{x} = \mathbf{0}$. But the $q$-th entry of $(A' - A)\mathbf{x}$ is $x_p$ and this is non-zero because the vertex $p$ is not isolated in $G$ (see Section 1.1). We obtain a similar contradiction in the case that $a_{pq} = 0$, $a_{p+1,q} = 1$, $p + 1 < q$. Accordingly $A$ is a stepwise matrix and the lemma is proved. $\square$

Using the relation (3.3.1) it is now a simple matter to deduce the following result, which was proved in a different way by Brualdi and Hoffman.

**3.3.2 Theorem** [BrHo] *Let $d$ be an integer greater than 1. Among the graphs with $\binom{d}{2}$ edges, those with maximal index are the graphs with $K_d$ as a unique non-trivial component.*

*Proof* We may suppose that $n > d$. In view of Lemma 3.3.1 it suffices to show that $\mathbf{x}^T(A - A')\mathbf{x}' > 0$ when $A = \begin{pmatrix} J - I & 0 \\ 0 & 0 \end{pmatrix}$ and $A'$ is any other $n \times n$ stepwise adjacency matrix with $d(d - 1)$ non-zero entries. (Here $J$ is the $d \times d$ matrix with each entry 1.) In this situation there exists a positive integer $r$ such that $A - A'$ has $2r$ entries equal to $+1$ and $2r$ entries equal to $-1$. Thus $\mathbf{x}^T(A - A')\mathbf{x}'$ is expressible as $\alpha - \beta$ where $\alpha$ is a sum of $r$ terms of the form $x_i x'_j + x'_i x_j$ $(1 \leq i \leq d, 1 \leq j \leq d)$ and $\beta$ is a sum of $r$ terms of the form $x_i x'_j + x'_i x_j$ $(1 \leq i \leq d, d + 1 \leq j \leq n)$. Now $x_1 = x_2 = \cdots = x_d > 0$ while $x_{d+1} = x_{d+2} = \cdots = x_n = 0$; moreover, from the relation $A'\mathbf{x}' = \mu'\mathbf{x}'$ it is straightforward to check that (because $A'$ is a stepwise matrix) $x'_1 \geq x'_2 \geq \cdots \geq x'_n$ (see Lemma 2 of [Row4]). Note that $x'_d > 0$ because the corresponding vertex is not isolated. It follows that $\alpha \geq 2rx_d x'_d$, $\beta \leq rx_d x'_{d+1}$ and hence that $\mathbf{x}^T(A - A')\mathbf{x}' = \alpha - \beta \geq rx_d (2x'_d - x'_{d+1}) > 0$ as required. $\square$

Next we show how further analysis of the biquadratic form contributes to the solution of another maximal index problem. In this case the graphs in question constitute the class $\mathscr{G}_n$ of tricyclic Hamiltonian graphs with $n$ vertices $(n \geq 5)$. Thus a graph $G$ in $\mathscr{G}_n$ consists of an $n$-cycle together with two chords which may or may not have a vertex in common. The next result (in which $\Delta(G)$ denotes maximal degree) shows that the two chords do have a common vertex when the index of $G$ is maximal.

**3.3.3 Theorem** [BeRo2] *If $G \in \mathcal{G}_n$, $n \geq 5$ and $\Delta(G) = 3$ then there exists $G' \in \mathcal{G}_n$ such that $\Delta(G') = 4$ and $\mu_1(G') > \mu_1(G)$.*

*Proof* We take the vertices of a Hamiltonian cycle $Z$ in $G$ to be labelled $1, 2, \ldots, n$ in cyclic order, and we let $A$ be the corresponding adjacency matrix of $G$. Let $(x_1, x_2, \ldots, x_n)^T$ be the principal eigenvector of $G$ and suppose that the two chords of $Z$ join $h$ to $i$ and $j$ to $k$ ($h, i, j, k$ distinct). Without loss of generality, we take $x_i$ to be smallest among $x_h, x_i, x_j, x_k$. If $h$ is not adjacent to $j$ let $G'$ be the graph obtained from $G$ by replacing edge $hi$ with edge $hj$. Then $\Delta(G') = 4$ and by (3.1.1) we have, in the usual notation, $\mu' - \mu \geq \mathbf{x}^T A' \mathbf{x} - \mathbf{x}^T A \mathbf{x} = 2x_h(x_j - x_i) \geq 0$. If $\mu' = \mu$ then $\mathbf{x}^T A' \mathbf{x} = \mu'$ and $A' \mathbf{x} = \mu \mathbf{x} = A \mathbf{x}$: this is a contradiction because $A' \mathbf{x}$ has $i$-th component $x_{i-1} + x_{i+1}$ (subscripts reduced modulo $n$) while $A \mathbf{x}$ has $i$-th component $x_{i-1} + x_{i+1} + x_h$. Thus $\mu' > \mu$ and we have the required result when $h$ and $j$ are non-adjacent.

Now suppose that $h$ and $j$ are adjacent. If $h$ is not adjacent to $k$ then we may repeat the above argument, this time replacing edge $hi$ with edge $hk$. It remains to deal with the case in which $j, h, k$ are consecutive points of $Z$. Without loss of generality, we take $k = 1$, $h = 2$, $j = 3$; and since $n \geq 5$ we may invoke symmetry to assume that $i \neq n$. Now let $G'$ be obtained from $G$ by replacing edge $2i$ with edge $1i$, so that $\Delta(G') = 4$.

In the usual notation we have $\mu' x_1' = x_2' + x_3' + x_i' + x_n'$ and $\mu' x_2' = x_1' + x_3'$. On subtracting and rearranging we obtain

$$\frac{x_1' - x_2'}{x_i'} = \frac{1}{\mu' + 1}\left(1 + \frac{x_n'}{x_i'}\right). \tag{3.3.2}$$

On the other hand, $\mu x_1 = x_2 + x_3 + x_n$ and $\mu x_2 = x_1 + x_3 + x_i$, whence

$$\frac{x_2 - x_1}{x_i} = \frac{1}{\mu + 1}\left(1 - \frac{x_n}{x_i}\right). \tag{3.3.3}$$

If $\mu' \leq \mu$ then it follows from (3.3.2) and (3.3.3) that $(x_1' - x_2')/x_i' > (x_2 - x_1)/x_i$. But now $\mathbf{x}^T(A - A')\mathbf{x}' = x_i(x_1' - x_2') - x_i'(x_2 - x_1) > 0$ and by (3.3.1) we have $\mu' > \mu$. This contradiction completes the proof. $\square$

The characteristic polynomial of a graph in $\mathcal{G}_n$ with maximal degree 4 can be expressed in terms of Chebyshev polynomials [BeRo2]. It is then possible to show that a graph in $\mathcal{G}_n$ ($n \geq 5$) with maximal index is isomorphic to the graph obtained from the cycle $123 \ldots n1$ by adding the edges 13 and 14. The graphs in $\mathcal{G}_n$ with minimal index are discussed in Section 3.5.

## 3.4 Implicit functions

The idea here is to solve recurrence relations for the components of a principal eigenvector of a connected graph $G$ in order to obtain the index of $G$ as an implicit function of certain parameters of $G$. It is then a matter of calculus to determine the values of the parameters for which the index attains its extremal values. We illustrate the technique as used by Simić and Kocić [SiKo] in the case that $G$ belongs to the class $\mathscr{P}(n,k)$ of $n$-vertex graphs which are homeomorphic to the multigraph consisting of $k$ edges ($k \geq 3$) between two vertices $u,v$. Thus a graph in $\mathscr{P}(n,k)$ consists of $k$ edge-disjoint $u$-$v$ paths, say of lengths $m_1, m_2, \ldots, m_k$, such that $m_1 + m_2 + \cdots + m_k = n + k - 2$. Here $n$ and $k$ are fixed and the index $\mu$ of a graph $G$ in $\mathscr{P}(n,k)$ depends on the parameters $m_1, m_2, \ldots, m_k$: we write $G = G(m_1, m_2, \ldots, m_k)$. Since $k \geq 3$, $G$ contains a cycle as a proper subgraph and so $\mu > 2$.

If the vertices in a $u$-$v$ path are labelled $0, 1, \ldots, m$ then corresponding components of the principal eigenvector satisfy the relation

$$\mu x_j = x_{j-1} + x_{j+1} \quad (j = 1, \ldots, m-1). \tag{3.4.1}$$

It is convenient to write $\mu = 2\cosh(2t)$ ($t > 0$), so that the solution of the recurrence relation (3.4.1) is $x_j = a\mathrm{e}^{2jt} + b\mathrm{e}^{-2jt}$ ($j = 0, 1, \ldots, m$).

By symmetry, $x_j = x_{m-j}$ ($j = 0, \ldots, m$) and so

$$x_j = a(\mathrm{e}^{2jt} + \mathrm{e}^{2(m-j)t}) \quad (j = 0, 1, \ldots, m). \tag{3.4.2}$$

An equation of the form (3.4.2) holds in respect of $u$-$v$ paths of lengths $m = m_1, m_2, \ldots, m_k$, say with corresponding constants $a = a_1, a_2, \ldots, a_k$. In each case $x_0$ is the component of the principal eigenvector corresponding to the vertex $u$, and so we have

$$x_0 = a_1(1 + \mathrm{e}^{2m_1 t}) = a_2(1 + \mathrm{e}^{2m_2 t}) = \cdots = a_k(1 + \mathrm{e}^{2m_k t}). \tag{3.4.3}$$

Considering now the neighbours of $u$, we have

$$\mu x_0 = \sum_{i=1}^{k} a_i(\mathrm{e}^{2t} + \mathrm{e}^{2(m-1)t}). \tag{3.4.4}$$

If we now use (3.4.3) to substitute for $a_1, \ldots, a_k$ in (3.4.4) then on cancelling $x_0$ we obtain $F(m_1, m_2, \ldots m_k, t) = 0$ where

$$F(m_1, m_2, \ldots, m_k, t) = 2\cosh(2t) - \sum_{i=1}^{k} \frac{\cosh((m_i - 2)t)}{\cosh(m_i t)}. \qquad (3.4.5)$$

Now suppose that all but two of the parameters $m_i$ are fixed, say $m_3, \ldots, m_k$. Then $m_1 + m_2$ have constant sum and it is a straightforward matter of calculus to check that $\frac{\partial F}{\partial m_1} > 0$ for $m_1 < m_2$, $\frac{\partial F}{\partial m_1} < 0$ for $m_1 > m_2$.

Since $\frac{\partial F}{\partial t}$ is always positive and $\frac{\partial F}{\partial m_1} + \frac{\partial F}{\partial t}\frac{\partial t}{\partial m_1} = 0$ we conclude that $\frac{\partial t}{\partial m_1} < 0$ for $m_1 < m_2$ and $\frac{\partial t}{\partial m_1} > 0$ for $m_1 > m_2$.

Therefore $t$, and hence $\mu$, is an increasing function of $|m_1 - m_2|$. Consequently we have the following result.

**3.4.1 Theorem** [SiKo] *A graph in $\mathscr{P}(n,k)$ ($k \geq 3$) with maximal index is isomorphic to $G(n - k + 1, 2, 2, \ldots, 2, 1)$. A graph in $\mathscr{P}(n,k)$ ($k \geq 3$) with minimal index is isomorphic to $G(m_1, m_2, \ldots, m_k)$ with $m_1 = \cdots = m_r = q+1$, $m_{r+1} = \cdots = m_k = q$, where $q = \lfloor (n+k-2)/k \rfloor$ and $r = n+k-2-kq$.*

Note that $\mathscr{P}(n, 3)$ is the class of bicyclic Hamiltonian graphs with $n$ vertices which received passing mention in Section 3.2. In this case, Theorem 3.4.1 confirms a conjecture of Cvetković [Cve13]. In particular, a graph in $\mathscr{P}(n, 3)$ is characterized by its index, a result proved for even $n$ by different methods in [Row3]. An analogue of Theorem 3.4.1, for $n$-vertex graphs homoeomorphic to $k$ loops with a common vertex, may be proved with only minor changes to the arguments given above [SiKo]. The method of implicit functions was also used by Simić [Sim3] to determine the graphs with minimal index among the unicyclic graphs with a prescribed number of vertices. A variation of the technique is described in the next section, devoted to a survey of further maximal and minimal index problems.

The equations (3.4.1) are also the starting point for an investigation [BeSi] of the indices of *broken wheels*, which are obtained from wheels by deleting spokes. Let $\mathscr{W}(n,k)$ denote the family of broken wheels with $n + 1$ vertices and $k$ spokes. Solutions of (3.4.1), subject to appropriate boundary conditions, are used to find an expression for the difference between characteristic polynomials of two graphs in $\mathscr{W}(n,k)$. After some analysis it can be shown that the index of a graph in $\mathscr{W}(n,k)$ is maximal

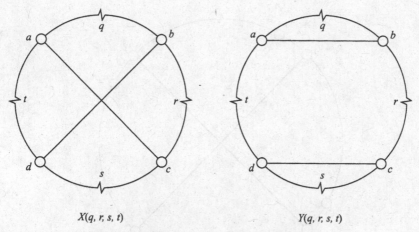

Fig. 3.3. The graphs $X(q,r,s,t)$, $Y(q,r,s,t)$.

when the spokes join the central vertex to $k$ successive vertices of the $n$-cycle. It turns out that the minimal index arises when the spokes are distributed as evenly as possible (in a unique way); but it remains an open problem to find an algorithm to determine in general which of two graphs in $\mathscr{W}(n,k)$ has the larger index.

## 3.5 More extremal index problems

In this section we review further results concerning graphs with maximal or minimal index in a given class of graphs. We do not attempt to give complete proofs but we indicate the role of the techniques which have been illustrated in the preceding sections.

We begin by showing how to find the graphs with minimal index in the class $\mathscr{G}_n$ of tricyclic Hamiltonian $n$-vertex graphs considered in Section 3.3. First the techniques employed in the proof of Theorem 3.3.3 may be used to show that $\Delta(G) = 3$ for such a graph $G$. Thus $G$ has one of the forms $X(q,r,s,t)$, $Y(q,r,s,t)$ shown in Fig. 3.3, where $a,b,c,d$ denote vertices and $q,r,s,t$ denote numbers of edges. Thus $n = q+r+s+t$. We note the following property of $X(q,r,s,t)$.

**3.5.1 Lemma** [RoYu] *Let* $(x_1, x_2, \ldots, x_n)^T$ *be the principal eigenvector of* $X(q,r,s,t)$; *and let* $a,b,c,d$ *be the vertices of degree 3 as illustrated in Fig. 3.3. Suppose that* $t+q$ *is smallest among* $q+r$, $r+s$, $s+t$, $t+q$ *(equivalently,* $s \geq q$ *and* $r \geq t$*). Then*

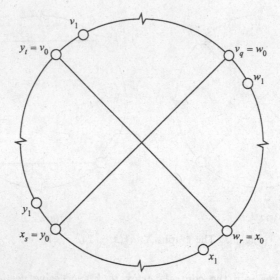

Fig. 3.4. Components of an eigenvector of $X(q,r,s,t)$.

(i) $x_a \geq x_b$, *with equality if and only if* $r = t$,

(ii) $x_b \geq x_c$, *with equality if and only if* $q = s$,

(iii) $x_a \geq x_d$, *with equality if and only if* $q = s$,

(iv) $x_d \geq x_c$, *with equality if and only if* $r = t$.

The proof of Lemma 3.5.1 uses the fact (see (4.2.8)) that $x_i^2 = P_{X-i}(\mu_1)/P'_X(\mu_1)$ [LiFe], where $X = X(q,r,s,t)$ and $\mu_1 = \mu_1(X)$. Accordingly for $i \in \{a,b,c,d\}$ the values of $x_i$ may be compared by comparing the characteristic polynomials of the unicyclic graphs $X - i$, which are expressible in terms of Chebyshev polynomials.

The next step is to find the graphs of the form $X(q,r,s,t)$ with least index, and this can be done using implicit functions (cf. Section 3.4). It is convenient to relabel the components of the principal eigenvector $v_1, v_2, \ldots, v_q, w_1, w_2, \ldots, w_r, x_1, x_2, \ldots, x_s, y_1, y_2, \ldots, y_t$ as shown in Fig. 3.4, and to set $v_0 = y_t$, $w_0 = v_q$, $x_0 = w_r$, $y_0 = x_s$. Thus

$$\left.\begin{array}{l} \mu_1 v_0 = y_{t-1} + v_1 + x_0, \\ \mu_1 w_0 = v_{q-1} + w_1 + y_0, \\ \mu_1 x_0 = w_{r-1} + x_1 + v_0, \\ \mu_1 y_0 = x_{s-1} + y_1 + w_0, \end{array}\right\} \qquad (3.5.1)$$

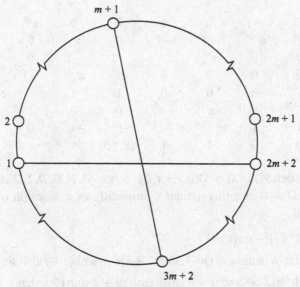

Fig. 3.5. A graph isomorphic to $X(m, m+1, m, m+1)$.

and

$$\left.\begin{array}{ll}
\mu_1 v_h = v_{h-1} + v_{h+1} & (1 \le h \le q-1), \\
\mu_1 w_i = w_{i-1} + w_{i+1} & (1 \le i \le r-1), \\
\mu_1 x_j = x_{j-1} + x_{j+1} & (1 \le j \le s-1), \\
\mu_1 y_k = y_{k-1} + y_{k+1} & (1 \le k \le t-1).
\end{array}\right\} \quad (3.5.2)$$

If we write $\mu_1 = 2 \cosh \theta$ ($\theta > 0$) then for $q \ge 2$, the recurrence relation $\mu_1 v_h = v_{h-1} + v_{h+1}$ $(1 \le h \le q-1)$ has the general solution $v_h = Ae^{h\theta} + Be^{-h\theta}$ $(0 \le h \le q)$, where $A$ and $B$ are constants. Similarly $w_i = Ce^{i\theta} + De^{-i\theta}$ $(0 \le i \le r)$, $x_j = Ee^{j\theta} + Fe^{-j\theta}$ $(0 \le j \le s)$ and $y_k = Ge^{k\theta} + He^{-k\theta}$ $(0 \le k \le t)$. Since $v_0 = y_t, w_0 = v_q, x_0 = w_r$ and $y_0 = x_s$ we have

$$\left.\begin{array}{l}
A + B = Ge^{t\theta} + He^{-t\theta}, \\
C + D = Ae^{q\theta} + Be^{-q\theta}, \\
E + F = Ce^{r\theta} + De^{-r\theta}, \\
G + H = Ee^{s\theta} + Fe^{-s\theta}.
\end{array}\right\} \quad (3.5.3)$$

Note that if (say) $q = 1$ then we may simply require $A, B$ to be solutions of $A + B = Ge^{t\theta} + He^{-t\theta}, Ae^\theta + Be^{-\theta} = C + D$. Now equations (3.5.1) and (3.5.3) may be written in the form

$$\mathbf{M}(A, B, C, D, E, F, G, H)^T = \mathbf{0}, \quad (3.5.4)$$

where

$$
\mathbf{M} = \begin{pmatrix}
-1 & -1 & 0 & 0 & 0 & 0 & e^{t\theta} & e^{-t\theta} \\
e^{q\theta} & e^{-q\theta} & -1 & -1 & 0 & 0 & 0 & 0 \\
0 & 0 & e^{r\theta} & e^{-r\theta} & -1 & -1 & 0 & 0 \\
0 & 0 & 0 & 0 & e^{s\theta} & e^{-s\theta} & -1 & -1 \\
-e^{-\theta} & -e^{\theta} & 0 & 0 & 1 & 1 & e^{(t-1)\theta} & e^{-(t-1)\theta} \\
e^{(q-1)\theta} & e^{-(q-1)\theta} & -e^{-\theta} & -e^{\theta} & 0 & 0 & 1 & 1 \\
1 & 1 & e^{(r-1)\theta} & e^{-(r-1)\theta} & -e^{-\theta} & -e^{\theta} & 0 & 0 \\
0 & 0 & 1 & 1 & e^{(s-1)\theta} & e^{-(s-1)\theta} & -e^{-\theta} & -e^{\theta}
\end{pmatrix}.
$$

$$(3.5.5)$$

We write $\det(\mathbf{M}) = \Omega = \Omega(q,r,s,t,\theta)$. Since $(A,B,C,D,E,F,G,H)^T \neq 0$, we have $\Omega = 0$ and this defines $\theta$ implicitly as a function of $q,r,s,t$. We find that

$$
\begin{aligned}
\Omega =\ & 32\sinh^4\theta\{1 - \cosh n\theta\} \\
& + 32\sinh^3\theta\{\sinh(q+r)\theta + \sinh(r+s)\theta + \sinh(s+t)\theta + \sinh(t+q)\theta\} \\
& + 8\sinh^2\theta\{2\cosh n\theta + 4\sinh q\theta \sinh s\theta + 4\sinh r\theta \sinh t\theta \\
& \qquad - \cosh(q+r-s-t)\theta - \cosh(q+t-r-s)\theta\} \\
& - 16\sinh q\theta \sinh r\theta \sinh s\theta \sinh t\theta.
\end{aligned}
$$

$$(3.5.6)$$

It is now a matter of calculus to show that if $\theta$ (and hence $\mu_1$) is minimal, $X(q,r,s,t)$ is isomorphic to one of $X(m,m,m,m)$, $X(m,m,m,m+1)$, $X(m,m+1,m,m+1)$, $X(m,m,m+1,m+1)$, $X(m,m+1,m+1,m+1)$ for appropriate $m$. Of the two graphs here with $4m+2$ vertices the second is excluded because we can show as follows that it has larger index than the first. Fig. 3.5 shows a graph $X$ isomorphic to $X(m,m+1,m,m+1)$ with vertices labelled $1,2,\ldots,4m+2$ in cyclic order. Let $X'$ be the graph obtained from $X$ by adding edges $\{1,2m+1\}$, $\{2,2m+2\}$ and deleting edges $\{1,2\}$, $\{2m+1, 2m+2\}$. Then $X'$ is isomorphic to $X(m,m,m+1,m+1)$. Let $X$ have adjacency matrix $A$, index $\mu_1$ and principal eigenvector $\mathbf{x} = (x_1,\ldots,x_n)^T$, with $A',\mu_1',\mathbf{x}',x_i'$ the corresponding notation for $X'$. Then $\mu_1' - \mu_1 \geq \mathbf{x}^T(A' - A)\mathbf{x} = 2(x_1 - x_{2m+2})(x_{2m+1} - x_2)$. Now $x_1 = x_{2m+2}$ by symmetry and so $\mu_1' \geq \mu_1$; moreover, if $\mu_1' = \mu_1$ then $A'\mathbf{x} = \mu_1'\mathbf{x}$, whence $\mathbf{x} = \mathbf{x}'$. But $x_1' \neq x_{2m+2}'$ by Lemma 3.5.1 and so $\mu_1' > \mu_1$ as required.    □

It remains to deal with the graphs $Y(q,r,s,t)$. Let $G' = Y(q,r,s,t)$, $G = X(q,r,s,t)$, where $s \geq q$ and $r \geq t$ without loss of generality. Let $\mu_1' = \mu_1(G')$, $\mu_1 = \mu_1(G)$, and let $(x_1,x_2,\ldots,x_n)^T$ be the principal eigenvector of $G$. In the notation of Fig. 3.3, $G'$ is obtained from $G$ by replacing edges

$ac$, $bd$ with edges $ab$, $cd$ and so $\mu_1' \geq \mu_1 + 2(x_a - x_d)(x_b - x_c)$. By Lemma 3.5.1 we have $\mu_1' \geq \mu_1$; moreover, if $\mu_1' = \mu_1$ then $q = s$. Conversely, if $q = s$ we have $x_b = x_c$ and so by comparison with equations (3.5.1) and (3.5.2) the index of $Y(q, r, s, t)$ in this case is $2 \cosh \theta$ where $\theta$ is the largest solution of $\Omega(q, r, s, t, \theta) = 0$. Hence if $q = s$ then $\mu_1' = \mu_1$. We can now identify all the graphs in $\mathscr{G}_n$ with minimal index.

**3.5.2 Theorem** [RoYu] *Let $\mathscr{G}_n$ ($n \geq 4$) denote the class of tricyclic Hamiltonian graphs with $n$ vertices. A graph $G$ has minimal index among the graphs in $\mathscr{G}_n$ if and only if one of the following holds for some positive integer $m$:*

(a) $n \equiv 0 \bmod 4$ *and $G$ is isomorphic to one of* $X(m, m, m, m)$, $Y(m, m, m, m)$.

(b) $n \equiv 1 \bmod 4$ *and $G$ is isomorphic to one of* $X(m, m, m, m + 1)$, $Y(m, m, m, m + 1)$.

(c) $n \equiv 2 \bmod 4$ *and $G$ is isomorphic to one of* $X(m, m + 1, m, m + 1)$, $Y(m, m + 1, m, m + 1)$, $Y(m + 1, m, m + 1, m)$.

(d) $n \equiv 3 \bmod 4$ *and $G$ is isomorphic to one of* $X(m, m + 1, m + 1, m + 1)$, $Y(m + 1, m + 1, m + 1, m)$.

For our second illustration we turn to the class $\mathscr{S}(m)$ of all graphs with $m$ edges ($m > 0$). The graphs in $\mathscr{S}(m)$ with minimal index are clearly those with $m$ independent edges, while Theorem 3.3.2 describes those with maximal index in the special case that $m = \binom{d}{2}$. The picture is completed by the following result.

**3.5.3 Theorem** [Row4] *Let $m = \binom{d}{2} + t$ where $0 < t < d$, and let $\mathscr{S}(m)$ denote the class of all graphs with $m$ edges. A graph with maximal index in $\mathscr{S}(m)$ has a unique non-trivial component $G_m$, and $G_m$ is obtained from $K_d$ by adding a vertex of degree $t$.*

This result had already been conjectured by Brualdi and Hoffman for some ten years (see [Fri2] and p. 438 of [BeFVS]) before Friedland [Fri1] dealt with the case $t - d - 1$ in 1985. Subsequently Stanley [Sta] showed that the maximal index $f(m)$ is at most $\frac{1}{2}(-1 + \sqrt{1 + 8m})$, with equality precisely when $m = \binom{d}{2}$. Friedland [Fri3] used a refinement of Stanley's inequality to prove the result in the cases $t = 1$, $d - 3$, $d - 2$. The proof in the general case uses equation (3.3.1) to compare the indices of two graphs in $\mathscr{S}(m)$. In view of Lemma 3.3.1 we may restrict our attention to stepwise matrices; and in what follows the matrices in question are described solely by their upper triangular entries.

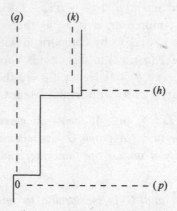

Fig. 3.6. Part of a stepwise matrix $(a_{ij})$ with $a_{hk} = 1$, $a_{pq} = 0$.

We need to show that for any $n > d$, the $n \times n$ matrix $A^*$ with entries

$$
\begin{array}{cc}
(2) & (d) \\
\end{array}
$$

$$
\begin{array}{ccccc}
1 & 1\,1\,\ldots\, 1 & 1 \\
  & 1\,1\,\ldots\, 1 & 1 \\
  & \vdots \,\ldots\, \vdots & \vdots \\
  & 1\,\ldots\, 1 & 1\ (t) \\
  & 1 & \\
  & \vdots & \\
  & 1 & (d-1) \\
\end{array}
$$

is the unique $n \times n$ stepwise adjacency matrix with spectral radius $f(m)$. This would follow by successive relocation of edges *provided* we could prove that $\mu_1(A') > \mu_1(A)$ whenever $A$ is an $n \times n$ stepwise adjacency matrix $(a_{ij})$ as shown in Fig. 3.6 with $h < p < q < k$, and $A'$ is obtained from $A$ by interchanging the $(h,k)$ and $(p,q)$ entries. This however is not true in general, but it *is* true whenever $p + q \geq h + k + 2$. This is a matter of arithmetic (see Lemma 2 of [Row4]): one uses the fact that $x_1 \geq x_2 \geq \cdots \geq x_n$ to prove that $\mathbf{x}^T(A' - A)\mathbf{x} > 0$ when $p + q \geq h + k + 2$. Accordingly it suffices to consider stepwise matrices $A = (a_{ij})$ such that

$$p + q \leq h + k + 1 \tag{3.5.7}$$

whenever $a_{hk} = 1$ and $a_{pq} = 0$ as in Fig. 3.6 with $h < p < q < k$. We let

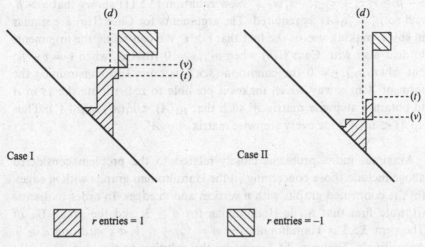

Fig. 3.7. The matrix $A^* - A$.

$\mathbf{x}^*$ denote the principal eigenvector of $A^*$ : then we have to show that the biquadratic form $\mathbf{x}^T(A^* - A)\mathbf{x}^*$ is positive whenever $A \neq A^*$.

There are three cases to consider, determined by the way in which the two matrices overlap, i.e. by the form of $A^* - A$: note that there exists $r > 0$ such that $A^* - A$ has $2r$ entries equal to $+1$ and $2r$ entries equal to $-1$. Referring to Fig. 3.7 the three cases are (I) $v \leq t$, (IIa) $v > t$, $r \geq t$, (IIb) $v > t > r$. In each case, $\mathbf{x}^T(A^* - A)\mathbf{x}^*$ is expressible as $\alpha - \beta$ where each of $\alpha, \beta$ is a sum of $r$ terms of the form $x_i x_j^* + x_i^* x_j$ $(i < j)$. Note that $x_1^* = \cdots = x_t^*$, $x_{t+1}^* = \cdots = x_d^*$ and $x_{d+2}^* = \cdots = x_n^* = 0$. The equation $A^* \mathbf{x}^* = \mu_1(A^*)\mathbf{x}^*$ is therefore equivalent to the following three equations, where we write $\mu_1^*$ for $\mu_1(A^*)$:

$$(\mu_1^* + 1)x_1^* = tx_1^* + (d - t)x_d^* + x_{d+1}^*, \tag{3.5.8}$$

$$(\mu_1^* + 1)x_d^* = tx_1^* + (d - t)x_d^*, \tag{3.5.9}$$

$$\mu_1^* x_{d+1}^* = tx_1^*. \tag{3.5.10}$$

On subtracting (3.5.10) and (3.5.9) from (3.5.8), we obtain

$$x_d^* + x_{d+1}^* = x_1^* + x_1^* t(\mu_1^* + 1)^{-1}. \tag{3.5.11}$$

In Case (I), $\alpha$ is the sum of $r$ terms $x_i x_j^* + x_i^* x_j$ for which $i < j \leq d + 1$, and so $\alpha \geq r(x_d^* + x_{d+1}^*)x_{d+1}$. On the other hand, $\beta$ is the sum of $r$ terms $x_i x_j^* + x_i^* x_j$ for which $j \geq d + 2$ and so $\beta \leq rx_1^* x_{d+2}$. It follows that

$\alpha - \beta \geq r(x_d^* + x_{d+1}^* - x_1^*)x_{d+1}$. Now equation (3.5.11) shows that $\alpha > \beta$, and so $\mu_1^* > \mu_1(A)$ as required. The arguments for Case (II$a$) are similar in spirit, making use of the fact that $r \geq t$. We can refine the arguments to deal also with Case (II$b$) when $a_{1,d+2} = 0$ (that is, when $r = v - t$); but when $a_{1,d+2} \neq 0$ the condition (3.5.7) is crucial in constraining the form of $A$ in a way which makes it possible to redistribute the 1s in $A$ to obtain a stepwise matrix $A'$ such that $\mu_1(A) < \mu_1(A') < \mu_1(A^*)$. Thus $\mu_1(A) < u_1(A^*)$ for every stepwise matrix $A \neq A^*$.

Maximal index problems closely related to the problem considered above include those concerning (i) the Hamiltonian graphs with $m$ edges, (ii) the connected graphs with $n$ vertices and $m$ edges. In order to discuss (i), note first that $K_d$ is Hamiltonian for $d \geq 3$, and the graph $G_m$ of Theorem 3.5.3 is Hamiltonian for $m = \binom{d}{2} + t$, $1 < t < d$ and $d \geq 3$. Accordingly Theorem 3.5.3 provides the solution to the maximal index problem for Hamiltonian graphs with $m$ edges unless $m = \binom{d}{2} + 1$. This case is dealt with in [Row12], where it is shown that for $d \geq 5$, the unique Hamiltonian graph with $\binom{d}{2} + 1$ edges and maximal index is the graph $H_d$ defined as follows: if $K_d^-$ denotes the graph obtained from $K_d$ by deleting an edge, then $H_d$ is obtained from $K_d^-$ by adding one new vertex adjacent to exactly two vertices of degree $d - 1$ in $K_d^-$. There are also unique Hamiltonian graphs with maximal index when $d = 3$ and $d = 4$, namely the 4-cycle and the unique maximal outerplanar graph with five vertices.

Finally we discuss the class $\mathscr{H}(n, m)$ of all *connected* graphs with $n$ vertices and $m$ edges, $n - 1 \leq m \leq \frac{1}{2}n(n - 1)$. The corresponding extremal index problems are trivial for $m > \frac{1}{2}n(n - 1) - 2$, and have been solved by a number of authors for $m = n - 1$, when the graphs in question are trees ([CoSi], [LoPe], [Sim2], [Wan]): among the trees with $n$ vertices, the path $P_n$ alone has smallest index and the star $K_{1,n-1}$ alone has largest index.

Accordingly, let $m = n + k$ where $k \geq 0$. The maximal index problem for $\mathscr{H}(n, n + k)$ has been solved by Brualdi and Solheid [BrSo] for $k = 0$ (unicyclic graphs), $k = 1$ (bicyclic graphs), $k = 2$ (tricyclic graphs) and by Bell [Bel1] for $k$ of the form $\binom{d-1}{2} - 1$ $(4 < d < n)$. Brualdi and Solheid showed that for any value of $e$, a graph in $\mathscr{H}(n, m)$ with maximal index again has a stepwise adjacency matrix. It then follows from connectedness that such a graph has a spanning star. In all known cases, a graph in $\mathscr{H}(n,m)$ with maximal index is of one of two types, $G_{n,k}$ and $H_{n,k}$, which we now describe in terms of adjacency matrices. For $k+1 = \binom{d-1}{2} + t, 0 \le t < d-1$, let $G_{n,k}$ be the graph with adjacency matrix of the form

$$
\begin{array}{cccc}
(1) & (d) & (n) \\
\end{array}
$$

$$
\begin{array}{ccccccccccc}
0 & 1 & 1 & \ldots\ldots & 1 & 1 & 1 & \ldots\ldots & 1 & (1) \\
 & 0 & 1 & \ldots\ldots & 1 & 1 & & & & \\
 & & & \ldots\ldots & \vdots & \vdots & & & & \\
 & & 0 & \ldots\ldots & 1 & 1 & & & (t+1) & \\
 & & & 0 & \ldots & 1 & & & & \\
 & & & & & \vdots & & & & \\
 & & & & 0 & 1 & & & & \\
 & & & & & 0 & & & (d) & .
\end{array}
$$

We can now see that the case $t = 0$, i.e. the case $k = \binom{d-1}{2} - 1$, corresponds to the special case $m = \binom{d}{2}$ of the maximal index problem for $\mathscr{S}(m)$ considered in Theorem 3.3.2.

For $k \le n-3$ let $H_{n,k}$ be the graph with adjacency matrix of the form

$$
\begin{array}{cccc}
(1) & (k) & (n) \\
\end{array}
$$

$$
\begin{array}{ccccccccccc}
0 & 1 & 1 & \ldots & 1 & 1 & 1 & 1 & 1 & \ldots & 1 & (1) \\
 & 1 & 1 & \ldots & 1 & 1 & 1 & 1 & 0 & \ldots & 0 & \\
 & & 0 & \ldots & 0 & 0 & 0 & 0 & & & & .
\end{array}
$$

In order to describe Bell's results, let

$$ g(d) = \frac{1}{2}d(d+5) + 7 + \frac{32}{d-4} + \frac{16}{(d-4)^2} \qquad (d > 4). $$

Bell extends the techniques of Section 3.3 to show that if $k = \binom{d-1}{2} - 1$ and $4 < d < n$ then $G_{n,k}$ is the unique graph in $\mathscr{H}(n, n+k)$ with maximal index when $n < g(d)$, and $H_{n,k}$ is the unique graph in $\mathscr{H}(n, n+k)$ with maximal index when $n > g(d)$. If $n = g(d)$ then $(d,n,m) \in$ $\{(5,60,69),(6,68,88),(8,80,85)\}$ and in these cases both $G_{n,k}$ and $H_{n,k}$ have maximal index.

For arbitrary $k \ge 3$, we have an asymptotic result [CvRo3]: there exists $N(k) > 0$ such that for all $n \geq N(k)$, $H_{n,k}$ is the unique graph in $\mathcal{H}(n, n + k)$ with maximal index.

Finally we remark that the graphs with maximal index in a class $\mathcal{G}$ are of course extremal graphs in the context of ordering $\mathcal{G}$ by index or lexicographically by spectrum; and for graphs with prescribed numbers of vertices and edges, ordering by $\mu_1$ is identical to ordering by $\mu_1 - \bar{d}$, where $\bar{d}$ denotes mean degree. Collatz and Sinogowitz [CoSi] observed that $\mu_1 \geq \bar{d}$ for any graph $G$, with equality if and only if $G$ is regular. This led them to propose $\mu_1 - \bar{d}$ as a measure of irregularity: for a discussion of $\mu_1 - \bar{d}$ in this context see [Bel2], [CvRo1] and Section 3 of [CvRo5].

# 4

# Graph angles

In this chapter we introduce invariants of eigenspaces known as *graph angles*. Angles were first mentioned explicitly in [Cve15], and their basic properties were described in [CvDo2] and [CvRo2]. Some notions related to graph angles were introduced independently in [DeoHS].

## 4.1 Motivation and definitions

The fact that a graph $G$ is reconstructible from its spectrum and a set of corresponding linearly independent eigenvectors points to the important role of eigenspaces in algebraic graph theory. In general a basis of eigenvectors is far from being an algebraic invariant, but eigenspaces themselves are invariant to within a permutation of the vertices of $G$. Explicitly, if $G, G'$ are cospectral graphs with adjacency matrices $A, A'$ respectively then $G$ and $G'$ are isomorphic if and only if there exists a permutation matrix $P$ such that $P \mathscr{E}_A(\mu) = \mathscr{E}_{A'}(\mu)$ for each eigenvalue $\mu$. Accordingly, in seeking algebraic invariants to extend spectral methods in graph theory we look to geometric attributes of eigenspaces having an algebraic formulation which is essentially independent of vertex labelling. One example is provided by the angles between eigenspaces and the all-1 vector $\mathbf{j}$; and another is the angle matrix defined below.

Suppose that $G$ has vertices $1, 2, \ldots, n$ and adjacency matrix $A$ with spectral decomposition $A = \mu_1 P_1 + \mu_2 P_2 + \cdots + \mu_m P_m$ where $\mu_1 > \mu_2 > \cdots > \mu_m$. Let $\{\mathbf{e}_1, \mathbf{e}_2, \ldots, \mathbf{e}_n\}$ be the standard orthonormal basis of $\mathbb{R}^n$ and let $\beta_{ij}$ be the angle between $\mathscr{E}(\mu_i)$ and $\mathbf{e}_j$. We write $\alpha_{ij} = \cos \beta_{ij}$ ($i = 1, 2, \ldots, m$; $j = 1, 2, \ldots, n$) : it is customary to abuse terminology and refer to the numbers $\alpha_{ij}$ as the *angles* of $G$. Note that since $P_i$ represents the orthogonal projection of $\mathbb{R}^n$ onto $\mathscr{E}(\mu_i)$ we have $\alpha_{ij} = \|P_i \mathbf{e}_j\|$. If the vertices of $G$ are labelled so that the columns of the $m \times n$ matrix $(\alpha_{ij})$

are ordered lexicographically then this matrix is an algebraic invariant, called the *angle matrix* of G. Its *i*-th row is called the *angle sequence* for the eigenvalue $\mu_i$. The *main angles* of G are the cosines of the angles between the eigenspaces of G and the all-1 vector **j**; these angles will be considered in Section 4.5.

The angles with $\mathscr{E}(\mu_i)$ can be computed from an orthonormal basis $\{\mathbf{x}_1, \mathbf{x}_2, \ldots, \mathbf{x}_k\}$ of $\mathscr{E}(\mu_i)$ as follows. If $\mathbf{x}_h = (x_{h1}, x_{h2}, \ldots, x_{hk})^T$ then, as we show in Section 4.2,

$$\cos \beta_{ij} = \left( \sum_{h=1}^{k} x_{hj}^2 \right)^{1/2}.$$

Note that the entries of a principal eigenvector are just the angles with the corresponding eigenspace $\mathscr{E}(\mu_1)$. We give three examples before proceeding.

**4.1.1 Example** The complete graph $K_n$ has the eigenvalues $n-1$ (with multiplicity 1) and $-1$ (with multiplicity $n-1$). The eigenspace of $n-1$ is spanned by **j**, whose orthogonal complement is the eigenspace of $-1$. It follows that the angle matrix of $K_n$ is the $2 \times n$ matrix

$$\begin{pmatrix} \frac{1}{\sqrt{n}} & \frac{1}{\sqrt{n}} & \cdots & \frac{1}{\sqrt{n}} \\ \sqrt{\frac{n-1}{n}} & \sqrt{\frac{n-1}{n}} & \cdots & \sqrt{\frac{n-1}{n}} \end{pmatrix}$$

and the main angles corresponding to $n-1, -1$ are $1, 0$ respectively. □

**4.1.2 Example** The graphs $C_4 \cup K_1$ and $K_{1,4}$ have the same spectrum, namely $2, 0, 0, 0, -2$, but different angle matrices. The graph $C_4 \cup K_1$ has angle matrix

$$\begin{pmatrix} \frac{1}{2} & \frac{1}{2} & \frac{1}{2} & \frac{1}{2} & 0 \\ \frac{1}{\sqrt{2}} & \frac{1}{\sqrt{2}} & \frac{1}{\sqrt{2}} & \frac{1}{\sqrt{2}} & 1 \\ \frac{1}{2} & \frac{1}{2} & \frac{1}{2} & \frac{1}{2} & 0 \end{pmatrix},$$

while $K_{1,4}$ has angle matrix

$$\begin{pmatrix} \frac{1}{\sqrt{2}} & \frac{1}{2\sqrt{2}} & \frac{1}{2\sqrt{2}} & \frac{1}{2\sqrt{2}} & \frac{1}{2\sqrt{2}} \\[2mm] 0 & \frac{\sqrt{3}}{2} & \frac{\sqrt{3}}{2} & \frac{\sqrt{3}}{2} & \frac{\sqrt{3}}{2} \\[2mm] \frac{1}{\sqrt{2}} & \frac{1}{2\sqrt{2}} & \frac{1}{2\sqrt{2}} & \frac{1}{2\sqrt{2}} & \frac{1}{2\sqrt{2}} \end{pmatrix}.$$

The main angles of $K_{1,4}$ are $2/\sqrt{5}, 1/\sqrt{5}, 0$ and $3/\sqrt{10}, 0, 1/\sqrt{10}$ corresponding to $2, 0, -2$, respectively. □

**4.1.3 Example** Suppose that we have a strongly regular graph with parameters $n, r, e, f$. One can compute the three eigenvalues $\mu_1 = r$, $\mu_2$ and $\mu_3$, and their multiplicities $k_1$, $k_2$ and $k_3$, in terms of $n, r, e$ and $f$ (see, for example, [CvDS], p. 195). Then it is an easy computation using results from Section 4.2 to verify that the angle matrix is $(\alpha_{ij})$, where $\alpha_{ij} = \sqrt{k_i/n}$ for $i = 1, 2, 3$ and $j = 1, ..., n$ (see also Section 4.3).

Thus any two strongly regular graphs with the same parameters have the same angle matrix. In particular this means that the angle matrix does not determine a graph up to isomorphism. □

The angles of all non-trivial connected graphs with up to five vertices are tabulated in Appendix B.

The idea of angles was first introduced formally in [Cve15] although there was some investigation of the concept earlier using different nomenclature. Angles are also used by Godsil and McKay [GoMK2] as an auxiliary tool in the graph reconstruction problem, and those authors show that angles are reconstructible from vertex-deleted subgraphs. It had been known before that the eigenvalues are reconstructible from the vertex-deleted subgraphs (see [Tut]). This means that both eigenvalues and eigenvectors can be used to recover information about a graph from its vertex-deleted subgraphs. (See also (4.2.8) and Section 5.4.)

In the Hückel molecular orbital theory of quantum chemistry, the carbon skeleton of a non-saturated hydrocarbon molecule is represented by a graph whose eigenvalues correspond to the energies of electrons in the molecule, and whose eigenvectors correspond to molecular orbitals (see [Mal] and [CvDS], Chapter 8). The *electron charge* $q_j$ on the vertex $j$, which is essentially the probability of finding an electron at the $j$-th atom, is expressible as a weighted sum of squares of angles: $q_j = \sum_{i=1}^{m} w_i \alpha_{ij}^2$. We return to this topic in Section 9.3.

Although the spectrum and the angle matrix do not constitute a complete set of graph invariants, they do in general provide more information about a graph than the spectrum alone. The calculations for small graphs in [Cve17] give a first indication of the extent to which angles enrich spectral techniques. These calculations show that many pairs of cospectral graphs are distinguishable by angles; for example the unique cospectral pair of graphs on five vertices (Example 4.1.2), the unique cospectral pair of connected graphs on six vertices, the unique cospectral pair of trees on eight vertices, and all three cospectral pairs of cubic graphs with at most 14 vertices. Further, in each of the following families there are no PINGs with the same angles:

- the connected graphs with seven vertices (where there are 29 cospectral pairs and one triplet)
- trees with at most eleven vertices (where, as well as a cospectral pair with eight vertices, there are five pairs with nine vertices, four pairs with ten vertices, and 26 pairs and two triplets with eleven vertices)
- unicyclic graphs with at most nine vertices (where there are five pairs with eight vertices and ten pairs and three triplets with nine vertices)
- bicyclic graphs on eight vertices (where there are eleven cospectral pairs).

Angles also serve to distinguish the graphs in a remarkable set of ten cospectral graphs with nine vertices and sixteen edges found by Godsil and McKay [GoMK1]. The same holds for their complements, which are themselves cospectral. No examples are known of a pair of cospectral graphs with the same angles and less than ten vertices.

Turning to regular graphs, we know that there are no pairs of cospectral regular graphs on less than ten vertices. There are however two pairs of cospectral regular graphs of degree 4 on ten vertices [GoMK1] and one pair is displayed in Fig. 4.1. Their complements form two pairs of cospectral regular graphs of degree 5. Graphs from these pairs are switching-equivalent (to see this, switch with respect to the set of black vertices in Fig. 4.1) and this is why they have the same angles. (Switching of graphs is defined in Section 1.1.) Generally, regular switching-equivalent graphs of the same degree are cospectral and have the same angles, as is shown in Section 4.3.

Because of their low degree cubic graphs cannot be switching-equivalent. (By Corollary 2.4.11 we have $r \geq n/4$ for switching-equivalent, regular graphs of degree $r$ on $n$ vertices.) Therefore it seems reasonable to investigate the extent to which the structure of cubic graphs is reflected

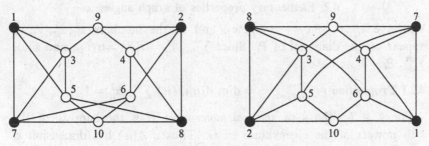

Fig. 4.1. A pair of cospectral regular graphs.

in eigenvalues and angles. It is not the case however that all cubic graphs are determined up to isomorphism by eigenvalues and angles. In [BuCv] two non-isomorphic cospectral cubic graphs on 20 vertices are given. They have the eigenvalues $3, 2, 1, -1, -2, -3$ with multiplicities $1, 4, 5, 5, 4, 1$, respectively. One of these graphs is known as the Desargues graph and the other was constructed in [BuCv]. Computations show that both graphs have the same angles.

The Desargues graph is transitive while its cospectral mate has twelve vertices of eccentricity 5 and eight vertices of eccentricity 4. However, both are walk-regular. (A graph is called *walk-regular* if for each $k \in \mathbb{N}$ the number of closed walks of length $k$ which start and terminate at a vertex $i$ is independent of $i$.) It was shown in [GoMK2] that a graph is walk-regular if and only if all vertices have the same angle sequences, a fact which follows from equation (4.2.5) below (cf. [CvRo2]). Motivated by the above example, and having in mind the fact that the sum of the squares of angles with a given eigenspace is equal to the dimension of the eigenspace (see Proposition 4.2.1), we can formulate the following theorem [Cve17].

**4.1.4 Theorem** *The angles of a walk-regular graph can be calculated from its eigenvalues.*

**4.1.5 Corollary** *Cospectral walk-regular graphs have the same angles.*

Although there exist non-isomorphic cubic graphs with the same angles and the same spectrum, we may still ask to what extent cubic graphs are characterized by their eigenvalues and angles, and we return to this question in Subsection 5.3.3.

## 4.2 Elementary properties of graph angles

We have $\alpha_{ij}^2 = \|P_i e_j\|^2 = e_j^T P_i e_j$, and so the numbers $\alpha_{i1}^2, \alpha_{i2}^2, \ldots, \alpha_{in}^2$ appear on the diagonal of $P_i$. Since $\sum_{j=1}^n \alpha_{ij}^2 = \text{tr}(P_i) = \text{tr}(E_i)$, and since $\sum_{i=1}^m P_i = I$ we have

**4.2.1 Proposition** *(i)* $\sum_{j=1}^n \alpha_{ij}^2 = \dim \mathscr{E}(\mu_i)$, *(ii)* $\sum_{i=1}^m \alpha_{ij}^2 = 1$.

For $k \in N$, the $k$-th *spectral moment* of $G$ is the sum $s_k$ of the $k$-th powers of the eigenvalues of $G$. Thus if $\mathscr{E}(\mu_i)$ has dimension $k_i$ $(i = 1, \ldots, m)$, then $s_k = \sum_{i=1}^m k_i \mu_i^k$ and we have the following:

**4.2.2 Corollary** *The $k$-th spectral moment $s_k$ is expressible as*

$$s_k = \sum_{i=1}^m \sum_{j=1}^n \alpha_{ij}^2 \mu_i^k.$$

**4.2.3 Remark** Let $\{e_1, e_2, \ldots, e_n\}$ be an orthonormal basis of $\mathbb{R}^n$. For any subspace $\mathscr{S}$ of $\mathbb{R}^n$, let $\beta_j$ be the smallest angle between $e_j$ and a vector from $\mathscr{S}$. The sequence $\beta_1, \beta_2, \ldots, \beta_n$ is independent of the choice of a basis in $\mathscr{S}$, but can be recovered from an arbitrary orthonormal basis as follows.

Let $\mathscr{S}$ have dimension $k$ and let $\{x_1, x_2, \ldots, x_k\}$ be an orthonormal basis of $\mathscr{S}$. Any unit vector $x \in \mathscr{S}$ can be represented in the form

$$x = c_1 x_1 + c_2 x_2 + \cdots + c_k x_k$$

where

$$c_1^2 + c_2^2 + \cdots + c_k^2 = 1. \tag{4.2.1}$$

For a fixed $y \in \mathbb{R}^n$ let $x^T y = \theta$ and $x_h^T y = \theta_h$ $(h = 1, 2, \ldots, k)$. Then

$$\theta = \sum_{h=1}^k c_h x_h^T y = \sum_{h=1}^k c_h \theta_h. \tag{4.2.2}$$

If $y$ is normalized, then $\theta$ is the cosine of the angle between $x$ and $y$. For the maximum value of (4.2.2) subject to condition (4.2.1) we have $\theta_{max} = (\sum_{h=1}^k \theta_h^2)^{1/2}$. Let $x_h = (x_{h1}, x_{h2}, \ldots, x_{hn})^T$ $(h = 1, 2, \ldots, k)$. For $y = e_j$ we have $\theta_h = e_j^T x_h = x_{hj}$ and so

$$\cos \beta_j = (\sum_{h=1}^k x_{hj}^2)^{1/2}. \tag{4.2.3}$$

Of course, in the case that $S = \mathscr{E}(\mu_i)$ we can obtain this result from the remarks in Section 1.1 as follows:

$$\alpha_{ij}^2 = \|P_i \mathbf{e}_j\|^2 = \mathbf{e}_j^T(\mathbf{x}_1\mathbf{x}_1^T + \cdots + \mathbf{x}_k\mathbf{x}_k^T)\mathbf{e}_j = \sum_{h=1}^{k}(\mathbf{e}_j^T\mathbf{x}_h)^2 = \sum_{h=1}^{k} x_{hj}^2.$$

$\square$

From the spectral decomposition of $A$ (see Section 1.1) we have

$$A^s = \sum_{i=1}^{m} \mu_i^s P_i, \qquad (4.2.4)$$

whence

$$a_{jj}^{(s)} = \sum_{i=1}^{n} \mu_i^s \alpha_{ij}^2. \qquad (4.2.5)$$

Recall that $a_{jj}^{(s)}$ is the number of closed walks of length $s$ which start and terminate at vertex $j$. In particular, $a_{jj}^{(1)} = \sum_{i=1}^{m} \alpha_{ij}^2 \mu_i = 0$, while $a_{jj}^{(2)}$ is equal to the degree $d_j$ of the vertex $j$. Hence the degree sequence of a graph can be obtained from the eigenvalues and the angle matrix. Columns of the angle matrix together with the eigenvalues can be viewed as *generalized vertex degrees* of the graph.

It also follows from (4.2.5) that for walk-regular graphs we have $\alpha_{i1} = \alpha_{i2} = \cdots = \alpha_{in}$ $(i = 1, 2, \ldots, m)$; together with Proposition 4.2.1(i) this proves the claim in Example 4.1.3.

Several other numerical invariants of a graph can also be calculated. The number of triangles $t(G)$ is just $\frac{1}{6}s_3$, while the sum $t(G) + t(\overline{G})$ can be determined from the vertex degrees. (To see this, count induced subgraphs on three vertices.) Hence the number of triples of three independent vertices (which is equal to $t(\overline{G})$) can be determined from the angle matrix and eigenvalues.

From the spectrum of a graph $G$ some information about the cycle structure of $G$ can be retrieved (see, for example, [CvDS], pp. 87-88, 95-97). Although the number of triangles can be found in this way, the number of quadrangles or cycles of greater length cannot be determined from the spectrum (cf. Example 4.1.2) except in special cases (e.g., in regular graphs [CvDS], p. 97). A knowledge of angles provides further information as the following results show.

**4.2.4 Proposition** *The degree $d_j$ of the vertex $j$, and the number $t_j$ of* *triangles containing the vertex j, are given by*

$$d_j = \sum_{i=1}^{m} \alpha_{ij}^2 \mu_i^2, \quad t_j = \frac{1}{2} \sum_{i=1}^{m} \alpha_{ij}^2 \mu_i^3.$$

*Proof* Take $s = 2$, $s = 3$ in (4.2.5).                                    $\square$

**4.2.5 Theorem** [CvRo2] *The number q of quadrangles in a graph is given by*

$$q = \frac{1}{8} \sum_{i=1}^{m} \sum_{j=1}^{n} \alpha_{ij}^2 \mu_i^3 (\mu_i^2 + 1 - 2 \sum_{h=1}^{m} \alpha_{hj}^2 \mu_h^2).$$

*Proof* We have $s_4 = 2e + 4f + 8q$ where $e$ is the number of edges and $f$ is the number of subgraphs isomorphic to $K_{1,2}$. Now $2e = \sum_{j=1}^{n} d_j$ and $f = \sum_{j=1}^{n} \binom{d_j}{2}$. Now we use Proposition 4.2.4 and Corollary 4.2.2 to express $d_j$ ($j = 1, 2, ..., n$) and $s_4$ in terms of eigenvalues and angles to obtain the result.                                    $\square$

**4.2.6 Theorem** [CvRo2] *The number p of pentagons in a graph is given by*

$$p = \frac{1}{10} \sum_{i=1}^{m} \sum_{j=1}^{n} \alpha_{ij}^2 \mu_i^3 \left( \mu_i^2 + 5 - 5 \sum_{h=1}^{m} \alpha_{hj}^2 \mu_h^2 \right).$$

*Proof* We have $s_5 = 10p + 10s + 30t$ where $t$ is the number of triangles and $s$ is the number of subgraphs consisting of a triangle with a pendant edge attached. Thus $s = \sum_{j=1}^{n} t_j(d_j - 2)$. Now we express $s_5, t, t_j$ and $d_j$ ($j = 1, 2, ..., n$) in terms of eigenvalues and angles to obtain the result. $\square$

Let $G$ be a graph with adjacency matrix $A$, and let $N_k(j) = a_{jj}^{(k)}$, the number of walks of length $k$ in $G$ originating and terminating at vertex $j$. Let $H_j(t)$ be the generating function $\sum_{k=0}^{\infty} N_k(j) t^k$. Using (4.2.5) we obtain

$$H_j(t) = \sum_{k=0}^{\infty} t^k \sum_{i=1}^{m} \alpha_{ij}^2 \mu_i^k = \sum_{i=1}^{m} \frac{\alpha_{ij}^2}{1 - \mu_i t}. \qquad (4.2.6)$$

From the spectral decomposition of $A$ we find that

$$(xI - A)^{-1} = \sum_{i=1}^{m} (x - \mu_i)^{-1} P_i. \qquad (4.2.7)$$

Now we use the fact that $(xI - A)^{-1} = \text{adj}(xI - A)/\det(xI - A)$, and equate diagonal entries in (4.2.7) to obtain

$$P_{G-j}(x) = P_G(x) \sum_{i=1}^{m} \frac{\alpha_{ij}^2}{x - \mu_i}. \tag{4.2.8}$$

Hence the characteristic polynomials of vertex-deleted subgraphs can be determined from the spectrum and the corresponding angles.

The relation between vertex-deleted subgraphs and angles was established in [GoMK3] by expressing in two ways a diagonal entry of the matrix generating function $\sum_{k=0}^{\infty} x^{-k} A^k$.

Again it is known ([CvDS], p.47, or [GoMK3]) that

$$H_j(t) = P_{G-j}(\frac{1}{t})/tP_G(\frac{1}{t}), \tag{4.2.9}$$

and so, using (4.2.9) and (4.2.6), we can obtain (4.2.8) in yet a third way.

**4.2.7 Definition** *If a graph invariant, or property, can be determined from the eigenvalues and angles of the graph, then the invariant, or property, is said to be EA-reconstructible.*

Most of the following observations regarding EA-reconstructibility are taken from [Cve14]. The first two lemmas follow from equations (4.2.8) and (4.2.9).

**4.2.8 Lemma** *The characteristic polynomials of vertex-deleted subgraphs of a graph are EA-reconstructible.*

**4.2.9 Lemma** *The function $H_i(t)$ is EA-reconstructible for each $i = 1, 2, ..., n$.*

**4.2.10 Lemma** *Vertices belonging to components whose index coincides with the index of the graph are EA-reconstuctible.*

*Proof* By the Perron-Frobenius theory of non-negative matrices (see Section 1.1), angles belonging to the index $\mu_1$ are different from zero precisely for those vertices described in the lemma. □

**4.2.11 Corollary** *The properties of a graph of being connected and of being disconnected are EA-reconstructible.*

**4.2.12 Proposition** *The properties of being a tree, of being a unicyclic graph, and of being a bicyclic graph are EA-reconstructible.*

*Proof* The number of vertices and the number of edges are clearly EA-reconstructible. By Corollary 4.2.11, the connectedness property is also EA-reconstructible.                                                      □

**4.2.13 Definition** *If $v$ is a vertex in a graph $G$, then the pair $(d, e)$, where $d$ is the degree of $v$ and $e$ is the sum of degrees of all neighbours of $v$ in $G$, is called the degree pair of $v$.*

**4.2.14 Proposition** *The sequence of vertex degree pairs is EA-reconstructible in trees.*

*Proof* By Lemma 4.2.9, the functions $H_i(t)$ are EA-reconstructible and hence so are the numbers $N_k(i)$. We have $N_2(i) = d_i$, the degree of $i$, and $N_4(i) = d_i^2 - d_i + e_i$ where $e_i$ is the sum of degrees of neighbours of $i$.  □

Next we study some metric properties of graphs, as discussed in [Cve18]. From the spectral decomposition of $A^s$ we have

$$a_{jk}^{(s)} = \sum_{i=1}^{m} \mu_i^s P_i \mathbf{e}_j \cdot P_i \mathbf{e}_k.$$

Since $|P_i \mathbf{e}_j \cdot P_i \mathbf{e}_k| \leq \|P_i \mathbf{e}_j\| \|P_i \mathbf{e}_k\|$ we obtain

$$a_{jk}^{(s)} \leq \sum_{i=1}^{m} |\mu_i|^s \alpha_{ij} \alpha_{ik}.$$

Let $d(j, k)$ be the distance between vertices $j$ and $k$ in $G$.

**4.2.15 Proposition** *If $g = \min\{s \; : \; \sum_{i=1}^{m} |\mu_i|^s \alpha_{ij} \alpha_{ik} \geq 1\}$, then $d(j, k) \geq g$.*

*Proof* By definition of $g$ we have $a_{jk}^{(s)} = 0$ for $s < g$. Hence there is no walk of length $s$ ($s < g$) between $j$ and $k$.                        □

**4.2.16 Corollary** *If $\sum_{i=1}^{m} |\mu_i| \alpha_{ij} \alpha_{ik} < 1$, then $j$ and $k$ are not adjacent.*

This corollary provides a necessary condition for two vertices to be adjacent, an alternative to the 'edge condition' formulated in Subsection 5.3.4. We also have the following result.

**4.2.17 Theorem** *Let $G$ be a graph on $n$ vertices with distinct eigenvalues $\mu_1, \mu_2, ..., \mu_m$ and let $\alpha_{ij}$ ($i = 1, 2, ..., m$; $j = 1, 2, ..., n$) be angles of $G$. Let*

$$d = \max_{j,k} \; \min\{s \; : \; \sum_{i=1}^{m} |\mu_i|^s \alpha_{ij} \alpha_{ik} \geq 1\}.$$

*Then the diameter of G is at least d.*

As we see from Theorem 1.3.14, bipartite graphs $G$ are characterized by the property that if $\lambda$ is an eigenvalue of $G$ then $-\lambda$ is an eigenvalue of $G$ with the same multiplicity. An analogous property is exhibited by the angle matrix of a bipartite graph [CvRo2].

**4.2.18 Proposition** *If $\lambda$ is an eigenvalue of a bipartite graph then the eigenvalues $\lambda$ and $-\lambda$ have the same angle sequence.*

*Proof* Consider a 2-colouring of the given bipartite graph. The eigenspace of $-\lambda$ is obtained from the eigenspace of $\lambda$ by reflection with respect to coordinates corresponding to vertices of one colour. The result now follows from equation (4.2.3). □

Finally we remark that if $\pi$ is an automorphism of $G$ then for each vertex $j$, $\alpha_{ij} = \alpha_{i\pi(j)}$ $(i = 1, 2, \ldots, m)$. For if $P(\pi)$ denotes the corresponding permutation matrix then each eigenspace is invariant under the transformation $\mathbf{x} \mapsto P(\pi)\mathbf{x}$ $(\mathbf{x} \in \mathbb{R}^n)$; and since this transformation is an isometry the angle between $\mathbf{e}_j$ and $\mathscr{E}(\mu_i)$ is the same as the angle between $P(\pi)\mathbf{e}_j$ and $P(\pi)\mathscr{E}(\mu_i)$, i.e. between $\mathbf{e}_{\pi(j)}$ and $\mathscr{E}(\mu_i)$. In particular, transitive graphs have constant angle sequences for any eigenvalue. However, angles corresponding to vertices from different orbits may still be equal.

## 4.3 Graph transformations and angles

For certain graph operations the spectra of the resulting graphs can be calculated from the spectra of the graphs on which the operation is performed: see [CvDS], Chapter 2. We describe some analogous results involving angles.

Consider first the union of two disjoint graphs $G_1, G_2$. If $G_1$ and $G_2$ have no eigenvalue in common then the angle sequence of any eigenvalue of $G_1$ should be extended by zeros for coordinates corresponding to vertices of $G_2$, and vice versa. The angle sequence of an eigenvalue common to $G_1$ and $G_2$ is a permutation of the concatenation of the corresponding angle sequences for $G_1$ and $G_2$. To be precise, we have the following:

**4.3.1 Proposition** [CvRo2] *Let $\lambda_1, \lambda_2, \ldots, \lambda_t$ be the distinct eigenvalues common to $G_1$ and $G_2$. For $i = 1, 2$ let $B_i$ be the matrix whose rows in order are the angle sequences of $G_i$ corresponding to $\lambda_1, \lambda_2, \ldots, \lambda_t$, and let $A_i$ be a matrix whose rows are the remaining angle sequences of $G_i$. The angle* *matrix of $G_1 \cup G_2$ is obtained from the matrix*

$$\begin{pmatrix} A_1 & O \\ B_1 & B_2 \\ O & A_2 \end{pmatrix}$$

*by first permuting the rows so that the corresponding eigenvalues of $G_1 \cup G_2$ are in decreasing order, and then ordering the columns lexicographically.*

The next observation concerns the complement $\overline{G}$ of a regular connected graph $G$.

**4.3.2 Proposition** [CvRo2] *Let $G$ be a regular connected graph on $n$ vertices of degree $r$, with distinct eigenvalues $\mu_1 = r, \mu_2, ..., \mu_m$ in decreasing order. For $i = 2, ..., m - 1$ the angle sequence for the eigenvalue $\mu_i$ of $G$ is a permutation of the angle sequence for the eigenvalue $-\mu_i - 1$ of $\overline{G}$. If $\mu_m \neq r - n$ then the angle sequence for the eigenvalue $\mu_m$ of $G$ is a permutation of the angle sequence for $-\mu_m - 1$ of $\overline{G}$, while the eigenvalue $r$ of $G$ and the eigenvalue $n - r - 1$ of $\overline{G}$ have the same angle sequence, namely $\frac{1}{\sqrt{n}}, ..., \frac{1}{\sqrt{n}}$.*

*Proof* The graph $\overline{G}$ has eigenvalues $n - 1 - r, -\mu_2 - 1, ..., -\mu_m - 1$ and these $m$ values are distinct unless $n - 1 - r$ coincides with $-\mu_m - 1$. For $i = 2, ..., m - 1$, the eigenvalue $\mu_i$ of $G$ always determines the same eigenspace as the eigenvalue $-\mu_i - 1$ of $\overline{G}$. When $\mu_m \neq r - n$, the same is true of $\mu_m$ and $-\mu_m - 1$, while for each of $G$ and $\overline{G}$ the one remaining eigenspace is spanned by the all-1 vector $\mathbf{j}$. $\square$

**4.3.3 Corollary** *Let $G$ be a regular graph. If both $G$ and $\overline{G}$ are connected then their angle sequences coincide (to within order).*

*Proof* When $\overline{G}$ is connected its largest eigenvalue $n - r - 1$ has multiplicity 1 (see [CvDS], p. 18) and so $n - r - 1$ does not coincide with $-\mu_m - 1$. $\square$

**4.3.4 Theorem** [Cve16] *Let $G$ be a regular connected graph with $n$ vertices of degree $r$. If both $G$ and $\overline{G}$ are connected, then*

$$\frac{P_{G-i}(x)}{P_G(x)} + \frac{P_{\overline{G}-i}(-x-1)}{P_{\overline{G}}(-x-1)} = \frac{1}{(x-r)(x+n-r)}. \tag{4.3.1}$$

*Proof* By (4.2.8) and Corollary 4.3.3 we have

$$P_{G-i}(x) = P_G(x) \left( \frac{\alpha_{1i}^2}{x-r} + \sum_{j=2}^{m} \frac{\alpha_{ji}^2}{x - \mu_j} \right),$$

$$P_{\overline{G}-i}(x) = P_{\overline{G}}(x) \left( \frac{\alpha_{1i}^2}{x-(n-1-r)} + \sum_{j=2}^{m} \frac{\alpha_{ji}^2}{x+\mu_i+1} \right).$$

Since $\alpha_{1i} = 1/\sqrt{n}$ we readily obtain (4.3.1). □

**4.3.5 Remark** Let $H_G^i(t) = H_i(t) = \sum_{k=0}^{\infty} N_k(i)t^k$, the generating function for the numbers $N_k(i)$ of *i-i* walks of length $k$ in $G$. Using (4.2.6), formula (4.3.1) can be written in the form

$$\frac{1}{x}H_G^i\left(\frac{1}{x}\right) - \frac{1}{x+1}H_G^i\left(-\frac{1}{x+1}\right) = \frac{1}{(x-r)(x+n-r)}.$$

□

We note also that the characteristic polynomials of certain subgraphs and supergraphs of $G$ can be expressed in terms of $P_G(x)$ together with the eigenvalues and angles of $G$.

We have already seen that

$$P_{G-j}(x) = P_G(x) \sum_{i=1}^{m} \frac{\alpha_{ij}^2}{x-\mu_i}. \tag{4.3.2}$$

**4.3.6 Proposition** (cf. [CvDo2]) *Let $G_j$ be the graph obtained from $G$ by adding a pendant edge at vertex $j$. Then*

$$P_{G_j}(x) = P_G(x) \left( x - \sum_{i=1}^{m} \frac{\alpha_{ij}^2}{x-\mu_i} \right).$$

*Proof* This follows from (4.3.2) and the determinantal expansion

$$P_{G_j}(x) = xP_G(x) - P_{G-j}(x). \tag{4.3.3}$$

□

Propositions 4.3.7 to 4.3.10 are taken from [CvRo2].

**4.3.7 Proposition** *Let $G^j$ be the multigraph obtained from $G$ by adding a loop at vertex $j$. Then*

$$P_{G^j}(x) = P_G(x) \left( 1 - \sum_{i=1}^{m} \frac{\alpha_{ij}^2}{x-\mu_i} \right).$$

*Proof* A simple transformation of the determinant representing $P_{G^j}(x)$ yields $P_{G^j}(x) = P_G(x) - P_{G-j}(x)$, and the result follows from (4.3.2). $\square$

**4.3.8 Proposition** *Let $G_j^n$ be the graph obtained from $G$ by adding a path of length $n$ at vertex $j$, and let $r_{jn}$ be the index of $G_j^n$. Suppose that $r_{jn} \to \lambda_j > 2$ as $n \to \infty$. Then $\lambda_j$ is the largest positive solution of the equation*

$$\frac{1}{2}(x + \sqrt{x^2 - 4}) - \sum_{i=1}^{m} \frac{\alpha_{ij}}{x - \mu_i} = 0.$$

*Proof* We use an argument of Hoffman [Hof6]. Let us write $\phi_n(x)$ for the characteristic polynomial of $G_j^n$ ($n \geq 0$). Then from equation (4.3.3) we have

$$\phi_{n+1}(x) = x\phi_n(x) - \phi_{n-1}(x),$$

a recurrence relation which holds for all $n \geq 0$ if we define $\phi_{-1}(x)$ as $P_{G-j}(x)$. If we solve this recurrence relation, and let $n \to \infty$, then we find that $\lambda_j$ is the largest positive solution of the equation

$$\frac{1}{2}(x + \sqrt{x^2 - 4})P_G(x) - P_{G-j}(x) = 0.$$

Moreover $P_G(\lambda_j) \neq 0$ because $\{r_{jn}\}$ is an increasing sequence and $r_{j1} > \mu_1$ ([CvDS], Theorem 0.7). The result therefore follows immediately from (4.3.2). $\square$

The next two results are concerned with constructions from a pair of graphs $G$ and $H$. We retain the notation above for the eigenvalues and angles of $G$ and we let $v_h$ ($h = 1, ..., p$) and $\gamma_{hk}$ ($h = 1, ..., p; k = 1, ..., q$) be the eigenvalues and angles of $H$.

**4.3.9 Proposition** *Let $G$ and $H$ be (disjoint) graphs and let $F$ be the graph obtained from $G \cup H$ by introducing an edge between vertex $j$ of $G$ and vertex $k$ of $H$. Then*

$$P_F(x) = P_G(x)P_H(x)\left\{1 - \sum_{i=1}^{m}\sum_{h=1}^{p} \frac{\alpha_{ij}^2 \gamma_{hk}^2}{(x - \mu_i)(x - v_h)}\right\}.$$

*Proof* The result follows from (4.3.2) and the well-known formula (see, e.g., [CvDS], p. 59) $P_F(x) = P_G(x)P_H(x) - P_{G-j}(x)P_{H-k}(x)$. $\square$

**4.3.10 Proposition** *Let F be the coalescence of graphs G and H at vertices j and k respectively (i.e. F is obtained by identifying vertex j of G with vertex k of H). Then*

$$P_F(x) = \frac{1}{x} P_G(x) P_H(x) \left\{ 1 - \left( 1 - x \sum_{i=1}^{m} \frac{\alpha_{ij}^2}{x - \mu_i} \right) \left( 1 - x \sum_{h=1}^{p} \frac{\gamma_{hk}^2}{x - \nu_h} \right) \right\}.$$

*Proof* The result follows from (4.3.2) and the well-known formula (e.g. [CvDS], p. 159)

$$P_F(x) = P_G(x)P_{H-k}(x) + P_{G-j}(x)P_H(x) - xP_{G-j}(x)P_{H-k}(x). \qquad (4.3.4)$$

$\square$

Before we formulate and prove a general formula for the characteristic polynomial of a graph obtained from the graph $G$ by adding a vertex we rewrite some of the results already described. If we delete a vertex $u$ then formula (4.3.2) can be written in the form

$$P_{G-u}(x) = P_G(x) \sum_{i=1}^{m} \frac{\|P_i e_u\|^2}{x - \mu_i}, \qquad (4.3.5)$$

where $P_1, \ldots, P_m$ are the projection matrices from formula (1.1.1). When adding a vertex we must specify the set of vertices of $G$ adjacent to the new vertex. Three special cases appear in the literature: the first is given by formula (4.3.3) which may be written in the form

$$P_{G_u}(x) = P_G(x) \left( x - \sum_{i=1}^{m} \frac{\|P_i e_u\|^2}{x - \mu_i} \right). \qquad (4.3.6)$$

Secondly, if we add a vertex adjacent to all vertices of $G$ then the resulting graph is the join $K_1 \triangledown G$ and we have (e.g. [CvRo2], Theorem 5)

$$P_{K_1 \triangledown G}(x) = P_G(x) \left( x - \sum_{i=1}^{m} \frac{\|P_i \mathbf{j}\|^2}{x - \mu_i} \right). \qquad (4.3.7)$$

where $\mathbf{j}$ is the all-1 vector. (See also Proposition 4.5.5 and the comments thereafter.)

Thirdly, if $u$ and $v$ are non-adjacent vertices and $G(u,v)$ is obtained from $G$ by adding a so-called *bridging vertex* [LoSo] between $u$ and $v$ then [Row14]:

$$P_{G(u,v)}(x) = P_G(x) \left( x - \sum_{i=1}^{m} \frac{\|P_i e_u + P_i e_v\|^2}{x - \mu_i} \right). \qquad (4.3.8)$$

In the next theorem we establish a general formula for the characteristic polynomial of a graph $G^*$ obtained from $G$ by adding a new vertex with any prescribed set of neighbours in $V(G)$.

**4.3.11 Theorem** [Row9] *Let $G$ be a finite graph whose adjacency matrix $A$ has spectral decomposition $A = \sum_{i=1}^{m} \mu_i P_i$, and let $G^*$ be the graph obtained from $G$ by adding one new vertex whose neighbours are the vertices in the non-empty set $S$. Then*

$$P_{G^*}(x) = P_G(x) \left( x - \sum_{i=1}^{m} \frac{\rho_i^2}{x - \mu_i} \right),$$

*where*

$$\rho_i = \left\| \sum_{k \in S} P_i e_k \right\|.$$

**Proof** Let $\mathbf{r} = \sum_{k \in S} \mathbf{e}_k$, so that $G^*$ has adjacency matrix $\begin{pmatrix} 0 & \mathbf{r}^T \\ \mathbf{r} & A \end{pmatrix}$. We order the eigenvalues $\mu_1, ..., \mu_m$ so that $P_i \mathbf{r} \neq 0$ for $i \in \{1, ..., s\}$ and $P_i \mathbf{r} = 0$ for $i \in \{s+1, ..., m\}$. (Note that $s \geq 1$ because $\sum_{i=1}^{m} P_i \mathbf{r} = \mathbf{r} \neq \mathbf{0}$.) Let $M(x)$ be the matrix $\sum_{i=1}^{s} (x - \mu_i)^{-1} P_i$. We have $\mathbf{r}^T M(x) \mathbf{r} = \sum_{i=1}^{s} (x - \mu_i)^{-1} \mathbf{r}^T P_i \mathbf{r} = \sum_{i=1}^{s} (x - \mu_i)^{-1} \rho_i^2$, and so the equation $x - \mathbf{r}^T M(x) \mathbf{r} = 0$ has $s+1$ different solutions in $\mathbb{R} \setminus \{\mu_1, ..., \mu_s\}$. We show first that these $s + 1$ numbers are eigenvalues of $G^*$.

Indeed if $\lambda = \mathbf{r}^T M(\lambda) \mathbf{r}$ then $\begin{pmatrix} 1 \\ M(\lambda)\mathbf{r} \end{pmatrix}$ is a corresponding eigenvector because

$$\begin{pmatrix} 0 & \mathbf{r}^T \\ \mathbf{r} & A \end{pmatrix} \begin{pmatrix} 1 \\ M(\lambda)\mathbf{r} \end{pmatrix} = \begin{pmatrix} \mathbf{r}^T M(\lambda)\mathbf{r} \\ \mathbf{r} + AM(\lambda)\mathbf{r} \end{pmatrix}$$

$$= \begin{pmatrix} \lambda \\ \mathbf{r} - (\lambda I - A)M(\lambda)\mathbf{r} + \lambda M(\lambda)\mathbf{r} \end{pmatrix}$$

and

$$(\lambda I - A)M(\lambda)\mathbf{r} = \left( \sum_{i=1}^{m} (\lambda - \mu_i) P_i \right) \left( \sum_{i=1}^{s} \frac{P_i}{\lambda - \mu_i} \right) \mathbf{r} = \sum_{i=1}^{s} P_i \mathbf{r} = \mathbf{r}.$$

Secondly we observe that $\begin{pmatrix} 0 & \mathbf{r}^T \\ \mathbf{r} & A \end{pmatrix} \begin{pmatrix} 0 \\ \mathbf{x} \end{pmatrix} = \mu \begin{pmatrix} 0 \\ \mathbf{x} \end{pmatrix}$ if and only if $\mathbf{r}^T \mathbf{x} = 0$ and $A\mathbf{x} = \mu\mathbf{x}$. For each $i \in \{1, 2, ..., m\}$, $\mu_i$ arises in this way as an eigenvalue of $G^*$ with multiplicity (possibly zero) equal to $\dim(\mathscr{E}(\mu_i) \cap \mathscr{W})$, where $\mathscr{W} = \langle \mathbf{r} \rangle^{\perp}$. Let $k_i$ be the multiplicity of $\mu_i$ as an eigenvalue of $G$.

Then $\dim(\mathscr{E}(\mu_i) \cap \mathscr{W}) = k_i - 1 + \varepsilon_i$ where $\varepsilon_i = 0$ if $\mathscr{E}(\mu_i) \nsubseteq \mathscr{W}$ and $\varepsilon_i = 1$ if $\mathscr{E}(\mu_i) \subseteq \mathscr{W}$. Now $\mathscr{E}(\mu_i) \subseteq \mathscr{W}$ if and only if $\mathbf{r}$ is orthogonal to $\mathscr{E}(\mu_i)$; equivalently $P_i \mathbf{r} = 0$; equivalently $i \in \{s + 1, ..., m\}$. Thus the number of such eigenvalues $\mu$ of $G^*$, counted according to their multiplicities, is given by $\sum_{i=1}^{m} \dim(\mathscr{E}(\mu_i) \cap \mathscr{W}) = \sum_{i=1}^{m}(k_i - 1) + (m - s) = n - s$.

We have now obtained $s + 1$ eigenvalues algebraically and a further $n - s$ eigenvalues geometrically. Hence we have all $n + 1$ eigenvalues of $G^*$; moreover they are the roots of the monic polynomial $P_G(x)(x - \sum_{i=1}^{s}(x - \mu_i)^{-1}\rho_i^2)$ because the multiplicity of $\mu_i$ as a root of this polynomial is $k_i - 1$ for $i = 1, 2, ..., s$. This polynomial is therefore the characteristic polynomial of $G^*$; and it may be written as in the statement of the theorem because $\rho_i = 0$ for $i \in \{s + 1, ..., m\}$. $\qquad\square$

It is easy to see that equations (4.3.6), (4.3.7), (4.3.8) are special cases of Theorem 4.3.11.

Next we study graph operations in which the vertex set of the resulting graph is the Cartesian product of the vertex sets of graphs on which the operation is performed.

Let $A$ and $B$ be real symmetric square matrices of orders $m, n$ respectively. Let $\lambda$ be an eigenvalue of $A$, $\mu$ an eigenvalue of $B$, and let $\{\mathbf{u}_1, \mathbf{u}_2, ..., \mathbf{u}_p\}$, $\{\mathbf{v}_1, \mathbf{v}_2, ..., \mathbf{v}_q\}$ be orthonormal bases of the corresponding eigenspaces. Let $\mathbf{u}_i = (u_{1i}, u_{2i}, ..., u_{mi})^T$ $(i = 1, 2, ..., p)$ and $\mathbf{v}_k = (v_{1k}, v_{2k}, ..., v_{nk})^T$ $(k = 1, 2, ..., q)$.

As is well known, the eigenvalues of the Kronecker product $A \otimes B$ are all possible products $\lambda_r \mu_s$ where $\lambda_r$ is an eigenvalue of $A$ and $\mu_s$ is an eigenvalue of $B$. Moreover, in the case that $\lambda\mu$ does not coincide with any other product $\lambda_r \mu_s$, the eigenspace of $\lambda\mu$ has an orthonormal basis consisting of the vectors $\mathbf{u}_i \otimes \mathbf{v}_k$ $(i = 1, 2, ..., p; \ k = 1, 2, ..., q)$.

For $h \in \{1, 2, ..., m\}$ and $j \in \{1, 2, ..., n\}$, the angle $\gamma_{hj}$ with the eigenspace of $\lambda_\mu$ corresponding to the ordered pair $(h, j)$ is given by

$$\gamma_{hj} = \left\{ \sum_{i=1}^{p} \sum_{k=1}^{q} (u_{hi}v_{jk})^2 \right\}^{1/2} = \left( \sum_{i=1}^{p} u_{hi}^2 \right)^{1/2} \left( \sum_{k=1}^{q} v_{jk}^2 \right)^{1/2} = \alpha_h \beta_j,$$

where $\alpha_h$ is the angle corresponding to the eigenvalue $\lambda$ of $A$ and to the coordinate $h$ and $\beta_j$ is related analogously to $\mu, B$ and $j$.

In particular, this means that for NEPS (see Section 2.3), the angles of the resulting graphs are, roughly speaking, the products of the corresponding angles of the graphs on which the operation is performed. If we extend the foregoing argument to deal with the case of coincident products $\lambda_r \mu_s$ then we obtain the following.

**4.3.12 Proposition** [CvRo2] *Let $A, B$ be real symmetric square matrices of orders $m, n$ respectively. Let $\lambda$ be an eigenvalue of $A$ and let $\mu$ be an eigenvalue of $B$. Let $\lambda_{i_1}, ..., \lambda_{i_t}$ and $\mu_{i_1}, ..., \mu_{i_t}$ be all distinct eigenvalues of $A, B$ respectively such that $\lambda\mu = \lambda_{i_1}\mu_{i_1} = \cdots = \lambda_{i_t}\mu_{i_t}$. Let $\alpha_{i_k h}$ be the angle corresponding to the eigenvalue $\lambda_{i_k}$ and the coordinate $h$ ($h = 1, 2, ..., m; k = 1, 2, ..., t$) and let $\beta_{i_k j}$ ($j = 1, 2, ..., n; k = 1, 2, ..., t$) be the angle corresponding to the eigenvalue $\mu_{i_k}$ and the coordinate $j$. Then the angle $\gamma_{hj}$ corresponding to the eigenvalue $\lambda\mu$ of $A \otimes B$ and the coordinate pair $(h, j)$ satisfies*

$$\gamma_{hj}^2 = \alpha_{i_1 h}^2 \beta_{i_1 j}^2 + \alpha_{i_2 h}^2 \beta_{i_2 j}^2 + \cdots + \alpha_{i_t h}^2 \beta_{i_t j}^2.$$

Theorem 4.5 of [GoMK2], which is given without a proof, shows that the authors were aware of the above facts at least for some special cases of NEPS.

The following lemma is a direct consequence of the distributivity of the Kronecker product over matrix addition.

**4.3.13 Lemma** *Given matrices $A_1, ..., A_k$ (in particular, the adjacency matrices of graphs $G_1, ..., G_k$, respectively) with spectral decompositions*

$$\sum_{s_1=1}^{m_1} \mu_{s_1}^{(1)} P_{s_1}^{(1)}, \ ..., \ \sum_{s_k=1}^{m_k} \mu_{s_k}^{(k)} P_{s_k}^{(k)},$$

*we have*

$$A_1 \otimes \cdots \otimes A_k = \sum_{(s_1,...,s_k)} \mu_{s_1}^{(1)} \cdots \mu_{s_k}^{(k)} (P_{s_1}^{(1)} \otimes \cdots \otimes P_{s_k}^{(k)}). \tag{4.3.9}$$

**4.3.14 Remark** If some of the products $\mu_{s_1}^{(1)}...\mu_{s_k}^{(k)}$ ($1 \le s_i \le m_i; i = 1, ..., k$) coincide, then (4.3.9) can be written in the form

$$A_1 \otimes \cdots \otimes A_k = \sum_r \mu_r P_r, \tag{4.3.10}$$

where $\mu_1 > \cdots > \mu_t$ are the different product of the form $\mu_s^{(1)} \cdots \mu_s^{(k)}$ ($= \mu_r$ for some $r$), and $P_r = \sum P_{s_1}^{(1)} \otimes \cdots \otimes P_{s_k}^{(k)}$ where the sum is extended over all $k$-tuples $(s_1, ..., s_k)$ such that $\mu_{s_1}^{(k)} \cdots \mu_{s_k}^{(k)} = \mu_r$. Then (4.3.10) is in fact the spectral decomposition of $A_1 \otimes \cdots \otimes A_k$.

To prove (4.3.10) it is sufficient to prove that $P_i P_j = \delta_{ij} I$, and this follows from the fact that

$$\left(\sum X_{i_1} \otimes \cdots \otimes X_{i_k}\right)\left(\sum Y_{j_1} \otimes \cdots \otimes Y_{j_k}\right) = \sum (X_{i_1} Y_{j_1}) \otimes \cdots \otimes (X_{i_k} Y_{j_k}).$$

$\square$

Remark 4.3.14 gives rise to the following definition.

**4.3.15 Definition** *A NEPS of graphs is called coincidence-free if no two eigenvalues obtained by Theorem 2.3.4 coincide.*

In Proposition 4.3.12 virtually the same result as in Remark 4.3.14 was deduced for the Kronecker product of two matrices, but expressed in terms of graph angles.

In view of the fact that diagonal entries of the projection matrices $P_i$ are squares of angles, we see that formula (4.3.10) yields Proposition 4.3.12 for $k = 2$. Both Remark 4.3.14 and Proposition 4.3.12 can be generalized to hold for any NEPS but we avoid a more general statement in order to avoid technical difficulties which would not contribute much to the essence of the matter.

**4.3.16 Remark** We see that in a coincidence-free NEPS the angles are the products of the corresponding angles on which the operation is performed. Moreover we see that the angles of a NEPS are unaffected by any change in the basis of the NEPS which does not affect the coincidences between eigenvalues. $\square$

Now we turn to the operation of Seidel switching. Recall from Section 1.1 that if $G_1$, $G_2$ are switching-equivalent graphs with Seidel matrices $S_1$, $S_2$ respectively, then $S_2 = D^{-1}S_2 D$ where $D$ is a diagonal matrix with diagonal entries $\pm 1$. The matrix $D$ represents a reflection $R$ of $\mathbb{R}^n$ with the property that if $\mathcal{U}$ is an eigenspace of $S_1$ then $R(\mathcal{U})$ is an eigenspace of $S_2$.

Now let $G$ be a regular graph with $n$ vertices of degree $r$. If $G$ has adjacency matrix $A$ and Seidel matrix $S$ then $S = J - 2A - I$ and $\mathscr{E}_A(r) = \mathscr{E}_S(n - 2r - 1)$; moreover $\mathscr{E}_A(\mu) = \mathscr{E}_S(-2\mu - 1)$ for all eigenvalues $\mu$ of $A$ other than $r$. Thus $\mathcal{U}$ is an eigenspace of $A$ if and only if it is an eigenspace of $S$. Since $\mathcal{U}$ and $R(\mathcal{U})$ have the same angle sequence we conclude:

**4.3.17 Theorem** *Cospectral, regular, switching-equivalent graphs have the same angle matrix.*

Let $G$ be a regular graph of degree $r$ and let $G'$ be obtained from $G$ by switching with respect to $U$. The above considerations provide a necessary condition in terms of eigenvectors for $G'$ to be regular (in which case $U$ induces a regular subgraph of $G$). For if **u** is any eigenvector of $G$ not corresponding to $r$ then in these circumstances both **u** and $R(\mathbf{u})$ are orthogonal to the all-1 vector **j**; and it follows that the components of **u** corresponding to $U$ have zero sum. This is analogous to a result of H. Sachs which provides an eigenvector characterization of regular factors of a regular graph (see Theorem 2.6.8).

**4.3.18 Remark** Let $\mathscr{S}_1$ and $\mathscr{S}_2$ be subspaces of $\mathbb{R}^n$ of the same dimension and suppose that $\mathscr{S}_1$ and $\mathscr{S}_2$ have the same sequence of angles with the coordinate vectors of $\mathbb{R}^n$. Then it is not always true that there exists a reflection $R$ such that $\mathscr{S}_2 = R(\mathscr{S}_1)$, for consider the following example. Let

$$\mathbf{a}_1 = \left(\frac{1}{2}, \frac{1}{2}, \frac{1}{2}, \frac{1}{2}\right)^T , \quad \mathbf{a}_2 = \left(\frac{1}{2}, \frac{1}{2}, -\frac{1}{2}, -\frac{1}{2}\right)^T ;$$

$$\mathbf{b}_1 = \left(\frac{1}{2}, \frac{1}{2}, \frac{1}{2}, \frac{1}{2}\right)^T , \quad \mathbf{b}_2 = \left(\frac{1}{2}, -\frac{1}{2}, \frac{1}{2}, -\frac{1}{2}\right)^T .$$

Then the vector $\mathbf{c} = (1,1,0,0)^T$ is contained in the subspace of $\mathbb{R}^4$ generated by $\mathbf{a}_1, \mathbf{a}_2$. Neither **c** nor any of its reflected images is contained in the subspace generated by $\mathbf{b}_1$ and $\mathbf{b}_2$, although both eigenspaces have the same angle sequence.

This example shows that switching of graphs (i.e. reflection of subspaces) may not be the only transformation which preserves angles (and which also preserves eigenvalues in the regular case). It seems that there is a rich variety of such transformations of eigenspaces (and of decompositions into eigenspaces), and this is the reason that we come across large families of cospectral graphs with the same angle sequences.      □

We have seen in Example 4.1.3 that a similar conclusion holds for strongly regular graphs. Here $\alpha_{ij} = \sqrt{k_i/n}$ for all vertices $j$, and by (4.3.2), this implies that all vertex-deleted subgraphs of a strongly regular graph are mutually cospectral. Although there are strongly regular graphs which are cospectral but not switching-equivalent, all have the same angle matrix.

Strongly regular graphs are not necessarily transitive; indeed any finite group is isomorphic to the automorphism group of some strongly regular graph. Accordingly there are many examples of graphs in which non-similar vertices have the same angles.

We can also show that in graphs with just four distinct eigenvalues, angles are the same for all vertices. Note that these graphs include the graphs related to symmetric balanced incomplete block designs ([CvDS], pp. 166-167).

In conclusion, angles provide little information concerning strongly regular graphs or other graphs with a small number of distinct eigenvalues.

## 4.4 Angles and components

It was observed in [Cve14] (see Lemma 4.2.10) that one can reconstruct from the eigenvalues and angles of a graph $G$ the set of those vertices of $G$ which belong to components whose index coincides with the index of $G$. However, as noted in [CvRo2] and [Cve18], we can say much more in this direction.

A partition of the vertex set of $G$ is called *admissible* if no edge of $G$ connects vertices from different parts; and subgraphs induced by parts of an admissible partition are called *partial graphs*. (Thus a partial graph is a union of components, and the components are induced by the parts of the finest admissible partition.) The spectra and angles of these partial graphs are called the *partial spectra* and *partial angles* corresponding to the original partition.

**4.4.1 Lemma** *Given the eigenvalues, angles and an admisssible partition of the graph $G$, the corresponding partial spectra and partial angles of $G$ are determined uniquely.*

*Proof* We know from formula (4.2.5) that the $(j, j)$-entry $a_{jj}^{(k)}$ of $A^k$ is $\sum_{i=1}^{m} \alpha_{ij}^2 \mu_i^k$ $(j = 1, 2, ..., n)$. Let $j \in \tilde{V}$, where $\tilde{V}$ is the set of vertices of a partial graph $\tilde{G}$: then $a_{jj}^{(k)}$ is the number of $j$-$j$ walks of length $k$ in $G$ and hence also in $\tilde{G}$. The spectral moments of $\tilde{G}$ are therefore $\sum_{j \in \tilde{V}} a_{jj}^{(k)}$ $(k \in \mathbb{N})$, and these determine the spectrum of $\tilde{G}$. Moreover $a_{jj}^{(k)} = \sum_{i=1}^{t} \tilde{\alpha}_{ij} \tilde{\mu}_i^k$ $(k \in \tilde{V})$ where $\tilde{\mu}_1, ..., \tilde{\mu}_t$ are the distinct eigenvalues of $\tilde{G}$ and $\tilde{\alpha}_{ij}$ is the angle of $\tilde{G}$ corresponding to $\tilde{\mu}_i$ and $j$. These equations now determine $\tilde{\alpha}_{1j}, ..., \tilde{\alpha}_{tj}$ $(j \in \tilde{V})$. □

**4.4.2 Remark** The union of all partial spectra corresponding to a given admissible partition of $G$ is of course the spectrum of $G$. Moreover, by Proposition 4.3.1, the angle matrix of each partial graph is (to within ordering of columns) a submatrix of the angle matrix of $G$. These provide two necessary conditions for a given partition of $G$ to be admissible, since the procedures of Lemma 4.4.1 may be applied to any partition.     □

**4.4.3 Theorem** [CvRo2] *Given the eigenvalues and angles of a graph G, there is a uniquely determined admisssible partition of G such that*

*(i) in each partial graph all components have the same index, and*
*(ii) any two partial graphs have different indices.*

*Proof* By Lemma 4.2.10, vertices belonging to components whose index coincides with the index $\mu_1$ of $G$ can be identified from the angle sequence for $\mu_1$: they are the vertices $j$ for which the angle $\alpha_{1j}$ is non-zero. The bipartition of $G$ in which one part consists of these vertices is an admissible partition, and so we can apply Lemma 4.4.1 to determine the corresponding partial spectra and partial angles. In particular, we obtain the eigenvalues and angles of the subgraph induced by the other part of the bipartition. We can now apply the above arguments to this subgraph and repeat the procedure until we obtain the partition described in the theorem.     □

Two admissible partitions of $G$ are called *EA-equivalent* if their partial spectra and partial angles coincide to within a permutation of parts.

Having determined the partition of vertices described in Theorem 4.4.3 we can then try to find the components of $G$. Consider a partial graph $H$ of $G$ determined by Theorem 4.4.3: the number of its components is equal to the multiplicity $k$ of its index. We can now use Remark 4.4.2 to check all partitions of $H$ having $k$ parts. Many of them will not be admissible and if all those which survive this test are EA-equivalent then we can say that we have determined the components of $H$. Otherwise for a given admissible partition we have to investigate the existence of graphs with corresponding components having the specified eigenvalues and angles.

**4.4.4 Remark** [Cve18] Given eigenvalues and angles of a graph $G$, we can:

   (1) establish whether or not $G$ is connected (see Corollary 4.2.11);
   (2) separate vertices from partial graphs whose components have the same largest eigenvalue (see Theorem 4.4.3);

(3) determine the number of components (this also follows from Theorem 4.4.3);

(4) determine the sizes of regular components;

(5) separate regular components of different sizes.

To demonstrate (4) and (5) we assume, in view of (2), that we have already separated components having the same largest eigenvalue. Consider a group of components having the largest eigenvalue $r$ ($r$ being an integer). In looking for components which are regular graphs of degree $r$ we consider the eigenvalue angle sequence belonging to $r$. Coordinates corresponding to vertices belonging to a regular component on $s$ vertices are equal to $1/\sqrt{s}$. We can recognize the case when all components are regular by looking at vertex degrees (which are EA-reconstructible by Proposition 4.2.4). In this case we readily establish sizes of components. If components are of different sizes we can identify vertices in each component. □

However, components themselves may not be uniquely determined, as the following example shows.

**4.4.5 Example** [Cve18] Recall that $G \triangledown H$ denotes the graph obtained from graphs $G$ and $H$ by joining each vertex of $G$ to each vertex of $H$. The graphs $G_1 = (C_6 \triangledown C_6) \cup (2C_3 \triangledown 2C_3)$ and $G_2 = (C_6 \triangledown 2C_3) \cup (C_6 \triangledown 2C_3)$ are cospectral. They are regular of degree 8 and can be obtained one from the other by (Seidel) switching with respect to vertices from a copy of $C_6$ and a copy of $2C_3$. Since switching in regular graphs does not change angles (see Theorem 4.3.17), the graphs $G_1$ and $G_2$ have the same angles. However, these graphs have non-isomorphic components with different spectra. □

## 4.5 Main angles

We next look at some graph invariants that can be determined by means of the angles between the eigenspaces and the main direction. Recall that the *main direction* is that of the *main vector* $\mathbf{j} = (1, 1, ..., 1)^T$.

Let $N_k$ be the number of walks of length $k$ in an $n$-vertex graph $G$. Recall from Section 2.2 that

$$N_k = \sum_{i=1}^m D_i \mu_i^k, \qquad (4.5.1)$$

where $D_i = n\beta_i^2$ and $\beta_i$ is the cosine of the angle between the eigenspace of $\mu_i$ and the main direction. Hence we have

$$N_k = n \sum_{i=1}^{m} \beta_i^2 \mu_i^k. \tag{4.5.2}$$

Since $N_0 = n$, we also have

$$\sum_{i=1}^{m} \beta_i^2 = 1. \tag{4.5.3}$$

We shall call $\beta_i$ the *main angle* corresponding to $\mu_i$. Recall that an eigenvalue $\mu_i$ is a *main* eigenvalue if $D_i \neq 0$, equivalently if its main angle is non-zero. In [DeoHS], cospectral graphs are said to be *comain* if for each eigenvalue $\lambda$, the othogonal projections of $\mathbf{j}$ onto $\mathscr{E}(\lambda)$ have the same length in each graph. Thus cospectral graphs are comain if and only if they have the same main eigenvalues and the same main angles.

Main eigenvalues play an important role in many problems in the theory of graph spectra. For example, if we know the eigenvalues and the main angles of a graph, we can write the generating function

$$H_G(t) = \sum_{k=0}^{\infty} N_k t^k \tag{4.5.4}$$

in the form

$$H_G(t) = n \sum_{i=1}^{m} (\beta_i^2 / (1 - t\mu_i)). \tag{4.5.5}$$

Secondly we note a connection between the main angles and the spectral decomposition of the adjacency matrix of the graph (see Section 1.1). Since $\beta_i = \frac{1}{\sqrt{n}} \|P_i \mathbf{j}\|$ and $\mathbf{j}^T P_i \mathbf{j} = \operatorname{sum} P_i$, the sum of the entries of the matrix $P_i$, we have

**4.5.1 Proposition** $\beta_i = \sqrt{\frac{1}{n} \operatorname{sum} P_i}$ $(i = 1, ..., m)$.

Hence several graph characteristics not computable from eigenvalues alone can be determined if the main angles are known.

Regular graphs are precisely those for which the largest eigenvalue is the only main eigenvalue. It is an open problem to characterize graphs with precisely $k$ main eigenvalues when $k \geq 2$.

Main angles cannot be used to distinguish between cospectral regular graphs, but they do play an important role for non-regular graphs. Nevertheless, it still may happen that two non-isomorphic, non-regular cospectral graphs have the same main angles. By Proposition 4.5.2, such graphs have cospectral complements, and some examples have been noted in the literature [GoMK1]. From (4.5.7) it follows that they have the same Seidel spectrum. In attempting to distinguish such graphs, we may consider first the angle matrix, but even this may not be enough to distinguish them. For if $G$ and $H$ are regular cospectral graphs with the same angle sequence, then $G \cup K_1$ and $H \cup K_1$ are two non-regular graphs with the same angle matrix and the same main angles. (To see this, use Proposition 4.3.1 and formula (4.5.2).)

Main angles have also been used in ordering graphs, as for example in [CvPe2], where graphs were first ordered lexicographically by spectral moments.

Main angles also play a role in graph transformations, as we now show. We begin by recalling two results from [CvDo2], p. 167.

**4.5.2 Proposition** *The complement $\overline{G}$ of $G$ has characteristic polynomial*

$$P_{\overline{G}}(x) = (-1)^n P_G(-x - 1)\left(1 - n\sum_{i=1}^{m}\frac{\beta_i^2}{x + 1 + \mu_i}\right). \tag{4.5.6}$$

Recall that the *Seidel adjacency matrix* of $G$ is obtained from the $(0, 1)$-adjacency matrix $A$ of $G$ by replacing each off-diagonal zero with 1 and each 1 with $-1$. The spectrum of this $(0, -1, 1)$-adjacency matrix is called the *Seidel spectrum* of $G$.

**4.5.3 Proposition** *The characteristic polynomial $S_G(x)$ of the Seidel adjacency matrix of $G$ is given by*

$$S_G(x) = (-2)^n P_G(-\tfrac{1}{2}(x + 1))\left(1 - n\sum_{i=1}^{m}\frac{\beta_i^2}{x + 1 + 2\mu_i}\right). \tag{4.5.7}$$

In view of formula (4.5.5), we have ([CvDS], pp. 45 and 50)

$$P_{\overline{G}}(x) = (-1)^n P_G(-1 - x)(1 - (x + 1)^{-1}H_G(-1/(1 + x))),$$

$$S_G(x) = (-1)^n 2^n P_G(-(x + 1)/2)(1 - (x + 1)^{-1}H_G(-2/(x + 1))).$$

We see that, given the eigenvalues of $G$, knowledge of the main angles of $G$ is equivalent to the knowledge of either the eigenvalues of $\overline{G}$ or the Seidel spectrum of $G$.

We now give a generalization of Propositions 4.5.2 and 4.5.3 which shows that the eigenvalues and main angles of $G$ determine a generic characteristic polynomial of $G$. Let $a, b$ be distinct real numbers. Let $M = (m_{ij})$ where $m_{ii} = 0$ ($i = 1, ..., n$) and for $i \neq j$, $m_{ij} = a$ if $i$ is adjacent to $j$, $m_{ij} = b$ otherwise. Let $M_G(x) = \det(xI - M)$.

**4.5.4 Theorem** [CvRo2] *We have*

$$M_G(x) = (a - b)^n P_G \left( \frac{x + b}{a - b} \right) \left( 1 - bn \sum_{i=1}^{m} \frac{\beta_i^2}{x + b - \mu_i(a - b)} \right).$$

*Proof* The sum of all entries of a matrix $X$ is denoted by sum $X$. Thus if $N_k$ is the number of walks of length $k$ in $G$ then $N_k = $ sum $A^k$. Let $H_G(t)$ denote the generating function $\sum_{k=0}^{\infty} N_k t^k$. Then

$$H_G(t) = \text{sum} \sum_{k=0}^{\infty} A^k t^k = \text{sum}\,(I - tA)^{-1} = \text{sum adj}\,(I - tA)/\det(I - tA).$$

In what follows we shall also use the fact that if $X$ is a square matrix and $J$ is the all-1 matrix of the same size then $\det(X + tJ) = \det X + t$ sum adj $X$. Indeed,

$$\det(xI - M) = \det(xI + bI - (a - b)A - bJ)$$

$$= \det(xI + bI - (a - b)A) - b \text{ sum adj}\,(xI + bI - (a - b)A).$$

In this last term the first summand is $\det \left\{ (a - b)\left( \frac{x+b}{a-b}I - A \right) \right\}$, i.e. $(a - b)^n P_G \left( \frac{x+b}{a-b} \right)$. The second summand is:

$$\text{sum adj} \left( (x + b) \left( I - \frac{a - b}{x + b} A \right) \right)$$

$$= (x + b)^{n-1} \text{sum adj} \left( I - \frac{a - b}{x + b} A \right)$$

$$= (x + b)^{n-1} H_G \left( \frac{a - b}{x + b} \right) \det \left( I - \frac{a - b}{x + b} A \right)$$

$$= (x + b)^{-1} H_G \left( \frac{a - b}{x + b} \right) \cdot \det\{(x + b)I - (a - b)A\}$$

$$= (x + b)^{-1} (a - b)^n H_G \left( \frac{a - b}{x + b} \right) P_G \left( \frac{x + b}{a - b} \right).$$

Now the result follows from (4.5.5). $\qquad\qquad\qquad\qquad\qquad\square$

**4.5.5 Proposition** [CvRo2] *Let $G \triangledown H$ denote the join of the (disjoint) graphs $G$ and $H$. Suppose that $G$ has distinct eigenvalues $\mu_1, \ldots, \mu_m$ and corresponding main angles $\beta_1, \ldots, \beta_m$, while $H$ has distinct eigenvalues $\nu_1, \ldots, \nu_p$ and corresponding main angles $\delta_1, \ldots, \delta_p$. Then*

$$P_{G \triangledown H}(x) = P_G(x) P_H(x) \left\{ 1 - nq \sum_{i=1}^{m} \sum_{h=1}^{p} \frac{\beta_i^2 \delta_h^2}{(x - \mu_i)(x - \nu_h)} \right\}$$

where $n$ is the number of vertices of $G$ and $q$ is the number of vertices of $H$.

*Proof* The result follows from Proposition 4.5.2 and the following formula (see, e.g., [CvDS], p. 57):

$$P_{G\triangledown H}(x) = (-1)^q P_G(x)P_{\overline{H}}(-x-1) + (-1)^n P_H(x)P_{\overline{G}}(-x-1)$$

$$-(-1)^{n+q}P_{\overline{G}}(-x-1)P_{\overline{H}}(-x-1).$$

$\square$

Several authors have discussed what additional algebraic invariants might be introduced to extend spectral techniques beyond consideration of eigenvalues alone. The following attempt is described in [PrDe] (cf. also [DeoHS] and [CvDGT], Chapter 3).

For cospectral graphs $G$ and $H$ it may happen that the graphs $G \triangledown K_1$ and $H \triangledown K_1$ have different spectra, and this property can of course be used to detect non-isomorphism of the graphs $G$ and $H$. It was also noted in [PrDe] that if $G \triangledown K_1$ and $H \triangledown K_1$ are also cospectral then for each $h \in \mathbb{N}$, $G \triangledown K_h$ and $H \triangledown K_h$ are also cospectral. Thus there is no advantage in introducing more than one vertex joined to all other vertices.

From Proposition 4.5.5 we see that the additional information in the form of the spectrum of $G \triangledown K_1$ is equivalent to a knowledge of the main angles of $G$. Indeed, given any graph $H$, knowledge of $P_G(x)$ and $P_{G\triangledown H}(x)$ enables us to compute the main angles of $G$. Conversely $P_{G\triangledown H}(x)$ can be determined from $P_G(x)$ and the main angles of $G$.

The usefulness of main angles has also been demonstrated implicitly in [Wil]. One of the results is that $n/(n - \mu_1\beta_1^{-2})$ is a lower bound for the chromatic number of $G$. Since $\beta_1 \leq 1$ this result improves the lower bound $n/(n - \mu_1)$ derived previously by Cvetković ([CvDS], p. 92).

Other related results exist where the adjacency matrix concept itself is extended. Johnson and Newman [JoNe] used the $(0, \lambda)$-adjacency matrix (where a variable $\lambda$ is used instead of 1 to indicate adjacency). Hence the characteristic polynomial is a two-variable function $P_G(x, \lambda)$. The authors found that cospectral graphs exist only under special circumstances:

**4.5.6 Theorem** [JoNe] *Let $G$ and $H$ be graphs whose $(0, \lambda)$-adjacency matrices are $A(G)$ and $A(H)$. Then the following are equivalent:*

(i) $P_G(x, \lambda) = P_H(x, \lambda)$,

(ii) *the complements $\overline{G}$ and $\overline{H}$ are cospectral,*

(iii) *there are two values of $\lambda$ different from 1 such that $P_G(x, \lambda) = P_H(x, \lambda)$,*

(iv) *$A(G) = UA(H)U^T$ where $U$ is orthogonal and has all row and column sums equal to 1.*

Now we consider main angles in a NEPS of graphs (see Definition 2.3.1).

It is easy to see that $\text{sum}(X \otimes Y) = \text{sum}\,X\,\text{sum}\,Y$. In a coincidence-free NEPS we find from Lemma 4.3.13 and Proposition 4.5.1 that the main angle of the eigenvalue $\mu_{s_1}^{(1)} \cdots \mu_{s_k}^{(k)}$ is equal to

$$\sqrt{\frac{1}{n}\,\text{sum}\,(P_{s_1}^{(1)} \otimes \cdots \otimes P_{s_k}^{(k)})} = \sqrt{\frac{1}{n_1}\,\text{sum}\,P_{s_1}^{(1)} \cdots \frac{1}{n_k}\,\text{sum}\,P_{s_k}^{(k)}} = \beta_{s_1}^{(1)} \cdots \beta_{s_k}^{(k)},$$

where $n_1, ..., n_k$ are numbers of vertices of graphs $G_1, ..., G_k$. Hence we have proved the following proposition.

**4.5.7 Proposition** [CvSi2] *In a coincidence-free NEPS main angles are products of the corresponding main angles of the graphs on which the operation is performed.*

**4.5.8 Remark** In the case of coincident eigenvalues we have for main angles the same effect as that described for angles in Proposition 4.3.12. Note that the main angles of a NEPS are also independent of any change to the basis of the NEPS which does not affect the coincidences between eigenvalues. □

If the graph $G$ has adjacency matrix $A$ and Seidel matrix $S$ then $S = J - 2A - I$, and this relation enables us to describe a connection between the eigenvalues of $A$ and those of $S$. As usual, let $\mathbf{j}$ denote the all-1 vector and let $\mathscr{W} = \langle \mathbf{j} \rangle^{\perp}$. Since $J\mathbf{v} = \mathbf{0}$ for all $\mathbf{v} \in \mathscr{W}$ we have $\mathscr{E}_A(\lambda) \cap \mathscr{W} = \mathscr{E}_S(-2\lambda - 1) \cap \mathscr{W}$; moreover, if this subspace has dimension $d$ then each of $\mathscr{E}_A(\lambda)$, $\mathscr{E}_S(-2\lambda - 1)$ has dimension $d$ or $d + 1$ because $\mathscr{W}$ has codimension 1.

Now suppose that $G, G'$ are switching-equivalent graphs with adjacency matrices $A, A'$ and Seidel matrices $S, S'$ respectvely. Let $\dim \mathscr{E}_A(\lambda) = k$, $\dim \mathscr{E}_{A'}(\lambda) = k'$. Since $S$ and $S'$ are similar, $\mathscr{E}_S(-2\lambda - 1)$, $\mathscr{E}_{S'}(-2\lambda - 1)$ have the same dimension, say $h$. By the foregoing remarks we have $|h - k| \leq 1$ and $|h - k'| \leq 1$, whence $|k - k'| \leq 2$. In words: the multiplicity of $\lambda$ as an eigenvalue of $G$ differs by at most 2 from the multiplicity of $\lambda$ as an eigenvalue of $G'$ [Cve10]. If $|k - k'| = 2$ and without loss of generality, $k \geq k'$, then necessarily $\dim(\mathscr{E}_A(\lambda) \cap \mathscr{W}) = k - 1$, $\mathscr{E}_A(\lambda) \not\subseteq \mathscr{W}$, $\mathscr{E}_S(-2\lambda - 1) \subseteq \mathscr{W}$, $\mathscr{E}_{S'}(-2\lambda - 1) \not\subseteq \mathscr{W}$ and $\mathscr{E}_{A'}(\lambda) \subseteq \mathscr{W}$. The above remarks show that if $k = k' + 2$ then necessarily $\lambda$ is a main eigenvalue of $G$ but not of $G'$. In the following example we have $\lambda = 0$, $k = 3$, $k' = 1$.

**4.5.9 Example** Let $G = \overline{K}_3$, $G' = K_{1,2}$ (so that $G'$ is obtained from $G$ by switching with respect to a single vertex). Here $0$ is a main eigenvalue of $G$ but not of $G'$. □

The relation between $\mathscr{E}_A(\lambda)$ and $\mathscr{E}_S(-2\lambda - 1)$ is similar to the relation between $\mathscr{E}_A(\lambda)$ and $\mathscr{E}_{J-A-I}(\lambda)$. Now $J - I - A$ is the adjacency matrix of $\overline{G}$, and so the multiplicity of $\lambda$ as an eigenvalue of $G$ differs by 1 at most from the multiplicity of $-1 - \lambda$ as an eigenvalue of $\overline{G}$. We conclude the section with some remarks in this context concerning the case $\lambda = 0$.

Let $\mathscr{S}$ be the set of all graphs $G$ such that $-1$ is an eigenvalue of $\overline{G}$ and $0$ is not an eigenvalue of $G$. It follows from Proposition 4.5.2 that $G$ belongs to $\mathscr{S}$ if and only if

$$P_G(0) \neq 0 \quad \text{and} \quad \sum_{i=1}^{m} \frac{n\beta_i^2}{\mu_i} = 1. \tag{4.5.8}$$

We deduce the following result.

**4.5.10 Theorem** [BeRo3] *A graph $G$ belongs to $\mathscr{S}$ if and only if the adjacency matrix of $G$ has an inverse whose entries sum to 1.*

*Proof* When $A$ is invertible (equivalently, $P_G(0) \neq 0$), we have $A^{-1} = \sum_{i=1}^{m} \mu_i^{-1} P_i$ and $\sum_{i=1}^{m} \mu_i^{-1} n\beta_i^2 = \sum_{i=1}^{m} \mu_i^{-1} \mathbf{j}^T P_i \mathbf{j} = \mathbf{j}^T A^{-1} \mathbf{j}$. Thus condition (4.5.8) holds if and only if $A$ is invertible and $\mathbf{j}^T A^{-1} \mathbf{j} = 1$. □

Condition (4.5.8) also gives us a way of deriving the following examples. By Proposition 4.5.5, the characteristic polynomial of $K_1 \nabla G$ is given by

$$P_{K_1 \nabla G}(x) = P_G(x) \left( x - \sum_{i=1}^{m} \frac{n\beta_i^2}{x - \mu_i} \right). \tag{4.5.9}$$

If $G$ belongs to $\mathscr{S}$ then, from (4.5.8) and (4.5.9), $P_{K_1 \nabla G}(0) = P_G(0) \neq 0$, while $\overline{K_1 \nabla G} \ (= K_1 \cup \overline{G})$ has $-1$ as an eigenvalue; thus $K_1 \nabla G \in \mathscr{S}$. Since $K_m \nabla G = K_1 \nabla (K_{m-1} \nabla G)$, repetition of this process shows that $K_m \nabla G \in \mathscr{S}$ for all $m \geq 1$. We may take $\overline{G} = T_6$ (Fig. 3.2).

The construction of $K_1 \nabla G$ may be placed in the context of bordered matrices as follows. If $A$ is a symmetric matrix with an inverse whose entries sum to 1 then the same is true of the matrix $A'$, where $A' = \begin{pmatrix} 0 & \mathbf{j}^T \\ \mathbf{j} & A \end{pmatrix}$. In this situation the inverse of $A'$ is $\begin{pmatrix} -1 & \mathbf{b}^T \\ \mathbf{b} & A^{-1} - \mathbf{b}\mathbf{b}^T \end{pmatrix}$, where $\mathbf{b} = A^{-1}\mathbf{j}$.

# 5

# Angle techniques

Graph angles were introduced in Chapter 4. In this chapter we use them to study problems such as the construction of cospectral graphs, the reconstruction of a graph with given eigenvalues and angles, and the Ulam reconstruction conjecture.

## 5.1 Angles and cospectral graphs

A well-known theorem of Schwenk [Sch1] states that almost all trees have a cospectral mate. We describe this result in more detail.

**5.1.1 Definition** *A branch of a tree at a vertex v is a maximal subtree containing v as an endvertex. The union of one or more branches at v is called a limb at v.*

Considered in its own right, a limb at the vertex $v$ is a rooted tree, with $v$ as its root.

Schwenk [Sch1] proved that the proportion of trees on $n$ vertices which avoid a specified limb tends to zero as $n$ tends to infinity. Moreover, the number of trees on $n$ vertices which do not contain a specified limb depends only on the number of edges of the limb.

**5.1.2 Definition** *Vertices u and v in cospectral (not necessarily non-isomorphic) graphs G and H are said to be cospectral if $P_{G-u}(x) = P_{H-v}(x)$.*

Schwenk observed that vertices $u$ and $v$ in the tree $T$ of Fig. 5.1 are cospectral but not similar under the automorphism group of $T$. By a well-known formula for the characteristic polynomial of the coalescence of two rooted graphs, (4.3.4), we deduce that graphs $G_1$ and $G_2$ of Fig.5.1 are cospectral for any rooted graph $G$ (for details, see Proposition 5.1.3).

105

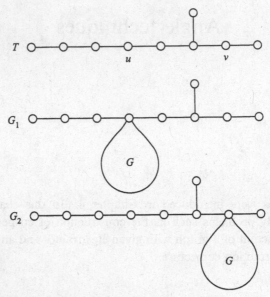

Fig. 5.1. Schwenk's construction

Now, Schwenk's argument was that almost all trees are of the form $G_1$ and hence have a non-isomorphic cospectral mate $G_2$.

It is important to note that the graphs $G_1$ and $G_2$ in Fig. 5.1 have different sequences of vertex degree pairs (see Definition 4.2.13); hence by Proposition 4.2.14 they have different angles. Accordingly all of the cospectral graphs constructed by Schwenk can be distinguished by angles.

In Subsection 5.3.1 we shall see to what extent trees are better characterized if not only the eigenvalues but also the angles are known. The results are of a constructive character in that an algorithm for constructing the trees is described.

Most of the propositions in this section stem from the paper [Row14], which contains a mathematical explanation of some facts concerning graph angles recently noted in the chemical literature, but first we provide details of the above procedure for constructing cospectral graphs from a graph with cospectral vertices.

**5.1.3 Proposition** *Let $H$, $K$ be graphs and let $w$ be a fixed vertex of $K$. For any vertex $u$ of $H$, let $G^u$ be the graph obtained from $H \cup K$ by identifying $u$ and $w$. If $u$, $v$ are cospectral vertices of $H$ then the graphs $G^u$, $G^v$ are cospectral.*

*Proof* The graph $G^u$ is the coalescence of two rooted graphs and so by (4.3.4) its characteristic polynomial is

$$P_{H-u}(x)P_K(x) + P_H(x)P_{K-w}(x) - xP_{H-u}(x)P_{K-w}(x). \qquad (5.1.1)$$

It follows that if $P_{H-u}(x) = P_{H-v}(x)$ then $G^u$, $G^v$ are cospectral. $\square$

Next we have the following characterization of cospectral vertices which follows from formula (4.2.8).

**5.1.4 Proposition** *Vertices $u$, $v$ of a graph are cospectral if and only if the angles at $u$ coincide with the angles at $v$, that is, $\alpha_{iu} = \alpha_{iv}$ ($i = 1, \ldots, m$).*

If $\mu_i$ is a simple eigenvalue then to within sign $\alpha_{iu}$ is the $u$-th entry of a unit vector which spans $\mathscr{E}(\mu_i)$; hence the remark in [HeEl2], p. 101, that cospectral vertices 'must have identical absolute values of eigenvectors in every non-degenerate eigenlevel'. In the general case (with $i$ fixed), let $\{x_1, \ldots, x_d\}$ be an orthonormal basis for $\mathscr{E}(\mu_i)$, with $x_k = (x_{1k}, x_{2k}, \ldots, x_{nk})^T$ ($k = 1, 2, \ldots, d$). Then $P_i = x_1 x_1^T + \cdots + x_d x_d^T$ and so $\alpha_{ij}^2 = x_{j1}^2 + \cdots + x_{jd}^2$: in the language of [LoSo], p. 25, 'the sum-over-degenerate-eigenvalues of squares of coefficients at isospectral points must be equal'.

The next result concerns the construction in [LoSo] of infinite families of graphs with cospectral vertices. Let $H$ be a graph with cospectral vertices $u$, $v$. A third vertex $t$ of $H$ is said to be an *unrestricted substitution vertex* with respect to $u$, $v$ if for any graph $K$, the vertices $u$, $v$ remain cospectral in the graph obtained from $H \dot\cup K$ by identifying $t$ with a vertex of $K$. Now we have

**5.1.5 Proposition** *Let $u$, $v$ be cospectral vertices of the graph $H$ and let $t$ be a third vertex of $H$. Then $t$ is an unrestricted substitution vertex with respect to $u$, $v$ if and only if $u$, $v$ are cospectral vertices of $H - t$.*

*Proof* Let $G$ be the graph obtained from $H \cup K$ by identifying $t$ with a vertex $w$ of $K$. By (5.1.1), we have $P_{G-u}(x) = P_{H-t-u}(x)P_K(x) + P_{H-u}(x)P_{K-w}(x) - xP_{H-t-u}(x)P_{K-w}(x)$, together with a similar expression for $P_{G-v}(x)$. Since $P_{H-u}(x) = P_{H-v}(x)$, it follows that $u$, $v$ are cospectral in $G$ if and only if

$$\left(P_{H-t-u}(x) - P_{H-t-v}(x)\right) \cdot \left(P_K(x) - xP_{K-w}(x)\right) = 0. \qquad (5.1.2)$$

Hence $P_{G-u}(x) = P_{G-v}(x)$ for all choices of $K$ if and only if $P_{H-t-u}(x) = P_{H-t-v}(x)$. $\square$

**5.1.6 Remark** It follows from (5.1.2) that if $P_{G-u}(x) = P_{G-v}(x)$ for just one choice of $K$ in which $w$ is not isolated then $t$ is an unrestricted substitution vertex; for if $P_K(x) = xP_{K-w}(x)$ then $w$ is an isolated vertex. To see this we apply (4.2.8) to $K$: taking $\mu_h = 0$ we have $\alpha_{hw} = 1$ and $\alpha_{iw} = 0$ $(i \neq h)$, whence $\mathbf{e}_w = P_h\mathbf{e}_w \in \mathscr{E}(\mu_h)$ and $A\mathbf{e}_w = \mathbf{0}$.  $\square$

Some authors have considered cospectral pairs of vertices, defined as follows. Let $G[u, v]$ be the graph obtained from $G$ by adding a vertex adjacent to the two vertices $u$ and $v$. (Thus $G[u, v] = G(u, v)$ when $u, v$ are non-adjacent.) The pairs $\{u, v\}$ and $\{u', v'\}$ are said to be *cospectral pairs* if the graphs $G[u, v]$ and $G[u', v']$ have the same spectrum. By Theorem 4.3.11, we have

$$P_{G[u,v]}(x) = P_G(x)\left(x - \sum_{i=1}^{m} \frac{\| P_i\mathbf{e}_u + P_i\mathbf{e}_v \|^2}{x - \mu_i}\right), \qquad (5.1.3)$$

and so we obtain the following result.

**5.1.7 Proposition** *Let* $\{u, v\}$, $\{u', v'\}$ *be pairs of vertices in the graph* $G$. *Then* $\{u, v\}$, $\{u', v'\}$ *are cospectral pairs in* $G$ *if and only if*

$$\| P_i\mathbf{e}_u + P_i\mathbf{e}_v \| = \| P_i\mathbf{e}_{u'} + P_i\mathbf{e}_{v'} \| \quad (i = 1, 2, ..., m).$$

For pairs of non-adjacent vertices, this is the 'absolute value sum rule for isospectral pairs', conjectured in [DAGT], and justified in [HeEl2] by means of second-order approximations in perturbation theory.

For future reference we note some related results from [Row11]: the spectra of the graphs $G + uv$ $(u \not\sim v)$, $G - uv$ $(u \sim v)$ and $G - u - v$ (among others) are determined by the spectrum of $G$ and the lengths $\| P_i\mathbf{e}_u \|$, $\| P_i\mathbf{e}_v \|$, $\| P_i\mathbf{e}_u + P_i\mathbf{e}_v \|$ $(i = 1, 2, ..., m)$. We outline the proof for $G + uv$ and $G - u - v$ in the case that $u, v$ are non-adjacent. Here $\theta_{uv}(xI - A)$ denotes the $(u, v)$-entry of $\text{adj}(xI - A)$.

First one can show by expanding determinants that

$$P_{G[u,v]}(x) = P_{G+uv}(x) + (x-1)P_G(x) - P_{G-u}(x) - P_{G-v}(x) + P_{G-u-v}(x) \quad (5.1.4)$$

and

$$\theta_{uv}(xI - A) = \frac{1}{2}\{P_G(x) - P_{G-u-v}(x) + P_{G+uv}(x)\}. \qquad (5.1.5)$$

(Equations (5.1.4) and (5.1.5) may be seen in the context of the deletion-contraction algorithm for characteristic polynomials established in [Row3]; in particular, (5.1.4) can be obtained by applying the algorithm to $G + uv$.)

It follows from (5.1.3) and (5.1.4) that $P_{G+uv}(x) + P_{G-u-v}(x)$ is known in terms of $P_G(x)$ and the lengths $\| P_i e_u \|$, $\| P_i e_v \|$, $\| P_i e_u + P_i e_v \|$ $(i = 1, 2, \ldots, m)$. By (4.2.7) and (5.1.5) the same is true of $P_{G+uv}(x) - P_{G-u-v}(x)$, and hence of both $P_{G+uv}(x)$ and $P_{G-u-v}(x)$. Note that knowledge of $\| P_i e_u \|$, $\| P_i e_v \|$, $\| P_i e_u + P_i e_v \|$ is equivalent to knowledge of the principal submatrix of $P_i$ determined by $u$ and $v$.

## 5.2 Ordering unicyclic graphs

In this section we follow [CvRo1] in discussing the ordering of unicyclic graphs lexicographically by spectrum (the $\Lambda$-ordering) and lexicographically by spectral moments (the $S$-ordering). Some results of Simić [Sim2] relevant to $\Lambda$-ordering have already been mentioned in Section 3.2. Both orderings have been used in producing graph catalogues ([BuČCS], [CvPe2]), and intuitively both are natural orderings.

The paper [CvPe2], however, suggests that $S$-order might be preferable precisely because of certain phenomena exhibited by unicyclic graphs. One such phenomenon is described here in the context of adding a pendant edge to a graph $G$. The effect on the index $\mu_1$ suggests a role for the angles of $G$ corresponding to $\mu_1$.

First let us define $\mathcal{U}_{e,f}$ $(e \geq 3, f \geq 1)$ as the family of unicyclic graphs in which the unique cycle is $C_e$ and the total number of vertices is $e + f$.

Secondly let $\mathcal{U}^g_{e,f}$ $(g > 0)$ consist of those graphs in $\mathcal{U}_{e,f}$ in which exactly $g$ trees are attached to $C_e$.

Thirdly let $E_{e,f}$ denote the graph in $\mathcal{U}^1_{e,f}$ obtained by coalescence of a vertex in $C_e$ with an endvertex of $P_{f+1}$.

In Section 3.2 we have already noted a result of [LiFe] concerning the index of a graph $G^{k,m-k}$ obtained from the rooted graph $G$ by attaching paths of lengths $k, m - k$ at the root: for fixed $m$, the index of $G^{k,m-k}$ is strictly increasing for $0 \leq k \leq \lfloor \frac{1}{2}m \rfloor$ (see also Theorem 6.2.2). If we apply this result to unicyclic graphs, relocating edges of trees as necessary, we obtain the following theorem.

**5.2.1 Theorem** *Graphs with minimal index in $\mathcal{U}^g_{e,f}$ are graphs in which all trees attached to the cycle are paths attached by endvertices. In particular, the graph $E_{e,f}$ alone has least index in $\mathcal{U}^1_{e,f}$.*

Li and Feng [LiFe] conjectured that $E_{e,f}$ is of minimal index in $\mathcal{U}_{e,f}$, and verified that this is indeed the case for $3 \leq e \leq 6$. In order to describe a counterexample let $G_{m,n,r,s}$ $(n \geq m \geq 1, m+n \geq 3, r \geq 1, s \geq 1)$ denote the graph obtained from $C_{m+n}$ by attaching paths $P_{r+1}$, $P_{s+1}$ by endvertices

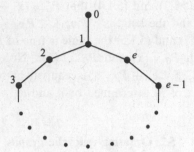

Fig. 5.2. The graph $E_{e,1}$.

at vertices distance $m$ apart in $C_{m+n}$. It was pointed out in [Cve20] that $\mu_1(G_{5,14,1,1}) \approx 2.0880$, while $\mu_1(E_{19,2}) \approx 2.0945$. Further counterexamples were subsequently described in [Cao]. The tables in [CvPe2], however, show that $E_{3,3}$ is the first graph in the $S$-ordering of $\mathcal{U}_{3,3}$.

Now the graphs $E_{19,2}$ and $G_{5,14,1,1}$, which represent a counterexample to the conjecture, can both be obtained by adding a pendant edge to the graph $E_{19,1}$. More generally, consider the graph $E_{e,1}$ which is represented in Fig. 5.2. Adding a pendant edge to vertex 0 we obtain the graph $E_{e,2}$, while adding it to the vertex $k+1$ ($k = 1, 2, ..., \lfloor e/2 \rfloor$) results in the graph $G_{k,e-k,1,1}$. The increase of the index of the graph after adding a pendant edge depends, of course, on the vertex where it is attached.

Let $G_i$ be the graph obtained from $G$ by attaching a pendant edge at vertex $i$ of $G$. Using Proposition 4.3.6 we see that the largest eigenvalue of $G_i$ is the largest root of the equation

$$x - \sum_{p=1}^{m} \frac{\alpha_{pi}^2}{x - \mu_p} = 0.$$

If the increase in the largest eigenvalue, when adding the pendant edge, is small we may neglect all but the first term in the above sum. The equation obtained is then $x^2 - \mu_1 x - \alpha_{1i}^2 = 0$: consider the solution

$$\lambda = \frac{\mu_1}{2} \left( 1 + \sqrt{1 + \frac{4\alpha_{1i}^2}{\mu_1^2}} \right).$$

If $\alpha_{1i}$ is sufficiently small with respect to $\mu_1$ we may use the approximation $\sqrt{1+x} \approx 1 + \frac{x}{2}$ ($x \to 0$) to obtain $\lambda \approx \mu_1 + \alpha_{1i}^2 \mu_1^{-1}$ as an estimate for the index of $G_i$. Since $\mu_1$ is a simple eigenvalue, $\alpha_{1i}$ is the $i$-th coordinate of the principal eigenvector of $G$. Hence, the coordinates of that eigenvector roughly determine the increase of the index when adding pendant edges. One should expect that the approximation obtained is good enough in graphs with sufficiently large number of vertices.

Let $(x_0, x_n, ..., x_e)^T$ be the principal eigenvector of $\mu_1$ in the graph $E_{e,1}$ (see Fig. 5.2). It is a technical task to verify that for some $a$ $(a > 0)$ we have $x_0 = \frac{a}{\mu_1}$ and

$$x_k = \frac{a}{1 + \alpha^e}(\alpha^{k-1} + \alpha^{e+1-k}) \quad (k = 1, 2, ..., e)$$

where $\alpha = (\mu_1 + \sqrt{\mu_1^2 - 4})/2$. Now suppose, for example, that $e = 2s$. Then $x_{s+1} = 2ae^s/(1 + \alpha^{2s})$. Since $2 < \mu_1 < 3$ we have $x_0 > \frac{a}{3}$, while $\lim_{e \to \infty} x_{s+1} = 0$. Hence, $x_s$ is smaller than $x_0$ for a sufficiently large length $e$ of the cycle in $E_{e,1}$. Therefore, in that case one should expect that the largest eigenvalue is smaller in $G_{k,e-k,1,1}$ $(k = \lfloor e/2 \rfloor)$ than in $E_{e,2}$. This explains why the above conjecture should fail for large cycles. Computer experiments show that for $e < 12$, $E_{e2}$ has a smaller index than $G_{k,e-k,1,1}$ $(k = \lfloor e/2 \rfloor)$; for $e = 12$ the indices are equal in both graphs (although the graphs are not cospectral); and for $e > 12$ we have the opposite situation.

See Section 6.3 for a deeper explanation of these effects.

## 5.3 Constructing graphs with given eigenvalues and angles

In this section we treat the problem of constructing all graphs with prescribed eigenvalues and angles. The results for various families of graphs are classified within different subsections.

### 5.3.1 Constructing trees

Recall from Proposition 4.2.12 that given the eigenvalues and angles of a graph $G$ we can tell whether or not $G$ is a tree. Here we present an algorithm for constructing all trees with given eigenvalues and angles. The algorithm is based on the following result, known as the *Reconstruction Lemma*.

**5.3.1 Lemma** [Cve14] *Given a limb $R$ of a tree $T$ at a vertex $i$ which is adjacent to a unique vertex of $T$ not in $R$, that vertex is among the vertices $j$ for which $P_{T-j}(x) = g_i^R(x)$, where*

$$g_i^R(x) = \frac{P_R(x)}{P_{R-i}(x)^2}\{P_R(x)P_{T-i}(x) - P_{R-i}(x)P_T(x)\}. \quad (5.3.1)$$

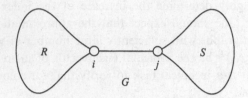

Fig. 5.3. Construction of a tree.

*Proof* Let $S$ denote the maximal limb of $T$ at $j$ not containing $i$, as shown in Fig. 5.3. We have (cf. e.g., [CvDS], p. 59)

$$P_T(x) = P_R(x)P_S(x) - P_{R-i}(x)P_{S-j}(x). \qquad (5.3.2)$$

Obviously, $P_{T-i}(x) = P_{R-i}(x)P_S(x)$ and $P_{T-j} = P_R(x)P_{S-j}(x)$. By eliminating $P_S(x)$ and $P_{S-j}(x)$ we obtain (5.3.1). $\qquad\square$

By specifying that $R$ consists only of vertex $i$, so that $P_R(x) = x$ and $P_{R-i}(x) = 1$, we obtain the following corollary.

**5.3.2 Corollary** *If $i$ is a vertex of degree 1 in a tree $T$, then the neighbour of the vertex $i$ is among those vertices $j$ such that $P_{T-j}(x) = f_i(x)$, where*

$$f_i(x) = x^2 P_{T-i}(x) - x P_T(x).$$

Now we describe the reconstruction algorithm. Although this is the name given to the algorithm in [Cve14] it is important to realize that in general trees are not EA-reconstructible in the sense of Definition 4.2.7. Indeed we shall see that almost all trees have non-isomorphic mates with the same eigenvalues and angles.

*Reconstruction Algorithm* [Cve14]. Let $T$ be a tree with prescribed eigenvalues and angles. First we use Proposition 4.2.4 to find the degree of vertices in $T$, and then we begin to construct possible edges as follows. For each vertex $i$ of degree 1 we choose a neighbour $j$ from the set $A_i = \{j \in V(T) : P_{T-j}(x) = f_i(x)\}$ (cf. Corollary 5.3.2). The number of times an individual vertex $j$ is chosen as a neighbour of an endvertex is bounded above by the degree of $j$. Now let $T'$ be the graph obtained from $T$ by deleting all endvertices. A vertex of degree 1 in $T'$ is necessarily one of the vertices $j$ chosen above and in this case we may apply Lemma 5.3.1 to the limb $R$ at $j$ consisting of all pendant edges at $j$. The neighbour of $j$ in $T'$ lies in the set $B_j^R = \{k \in V(T') : P_{T-k}(x) = g_j^R(x)\}$, and for each such $j$ we choose a neighbour $k \in B_j^R$. Continuing in this way we may construct a tree by successive construction of limbs provided that

(i) at each stage there are vertices $j$ of degree 1 in the subtree $T''$ which remains to be constructed, and (ii) the corresponding sets $B_j^R$ are non-empty. If $T''$ is non-trivial and one or other of these requirements is not met, then the algorithm proceeds with a different choice of neighbours at the previous stage. If $T''$ is trivial then a tree $T$ has been constructed and the algorithm is repeated with a new choice of neighbour. Using such a backtracking algorithm one constructs a collection of trees which includes all those with the given eigenvalues and angles. Finally one excludes those which do not have the specified eigenvalues. □

Let us discuss the question of how big a step has been made in characterizing the structure of trees when angles are introduced. The difference is that now we can construct all of the trees in question, while without angles that seems not to be possible in a reasonable way. This is related to the fact that we know exactly which features are responsible for the existence of non-isomorphic trees with the same eigenvalues and angles. Indeed, in the notation of Fig. 5.3, non-isomorphic trees can arise as follows.

(1) The limb $R$ may be replaced with a cospectral limb $R'$ such that
   $P_{R-i}(x) = P_{R'-i}(x)$.
(2) The choice of neighbours $j$ with given $P_{T-j}(x)$ may not be unique.

In view of (1), we may use four copies of the tree $T$ from Fig. 5.1 to construct the trees $T_1$, $T_2$ shown in Fig. 5.4, where $H$ denotes a rooted tree. For any choice of $H$, the trees $T_1$ and $T_2$ are non-isomorphic and have the same eigenvalues and angles. Corresponding vertices (i.e. vertices for which the vertex-deleted subgraphs in $T_1$ and $T_2$ are cospectral) are denoted by the same numbers for some specific vertices.

The construction illustrated in Fig. 5.4 also shows that almost all trees are not characterized by eigenvalues and angles. It also shows (e.g. by reference to the vertices labelled 2 in $T_1$ and $T_2$) that eigenvalues and angles do not determine degree sequences of vertices. (The *degree sequence* of a vertex $v$ consists of the degrees of the neighbours of $v$, in non-increasing order.)

If, in applying the reconstruction algorithm, we know the degree sequences in $T$ (in particular if we know the vertex-deleted subgraphs of $T$) then the choice in (2) above is limited to the extent that the trees in question are determined up to cospectral limbs with a constant degree sequence of the root. We do not know of an example of non-isomorphic cospectral trees $G_1, G_2$ for which there exists a bijection

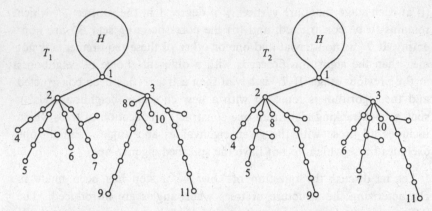

Fig. 5.4. Non-isomorphic trees with the same eigenvalues and the same angles.

$\theta : V(G_1) \to V(G_2)$ such that for each $v \in V(G_1)$, the vertices $v$ and $\theta(v)$ are cospectral with the same degree sequence.

It is well known that a tree is in fact reconstructible from its vertex-deleted subgraphs, but it is not known whether the characteristic polynomial of a tree is reconstructible from the characteristic polynomials of vertex-deleted subgraphs (see [GuCv]). The latter is true of many trees and if it proves to be true for all trees then the reconstruction algorithm can be used to construct all trees for which only the characteristic polynomials of vertex-deleted subgraphs are specified.

### 5.3.2 *Constructing unicyclic and bicyclic graphs*

It was shown in Subsection 5.3.1 that for trees, knowledge of the angles in addition to the spectrum is 'almost' sufficient for us to reconstruct the graphs in question. Although there exist non-isomorphic trees with the same eigenvalues and angles, and even although this happens almost always, we are still in a position to construct easily all trees with given eigenvalues and angles. Starting with endvertices and working towards the centre, we can 'calculate' the remaining neighbour of each particular vertex by the use of the Reconstruction Lemma given in Subsection 5.3.1. Here, following [Cve18], we describe the extent to which these ideas carry over to unicyclic and bicyclic graphs. We noted in Section 4.2 that the properties of being unicyclic and of being bicyclic are EA-reconstructible.

It was noted in [Cve14] that the reconstruction algorithm for trees can be extended to unicyclic graphs. Such a graph $G$ consists of a cycle with

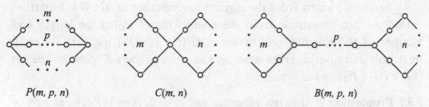

Fig. 5.5. Some bicyclic graphs.

trees attached, and we can construct the various possibilities for these trees until the subgraph not yet constructed is regular of degree 2. If this subgraph has vertex set $V$ of size $k$ then in general we must consider all $\frac{1}{2}k!$ possibilities for the cycle with vertex set $V$. One question which arises is whether $V$ is determined uniquely, and to address this question we first recall from Theorem 1.3.14 that $G$ is bipartite if and only if its characteristic polynomial has the property

$$P_G(-x) = (-1)^n P_G(x) \tag{5.3.3}$$

where $n$ is the number of vertices in $G$. If $P_G(x)$ satisfies (5.3.3) then we say that $P_G(x)$ is *symmetric*. Now the unicyclic graph $G$ is bipartite if and only if its cycle is even, and so we know from the spectrum whether $|V|$ is even or odd. If $|V|$ is odd, $V$ is determined uniquely as $\{i \in V(G) : P_{G-i}(x) \text{ is symmetric}\}$. If $|V|$ is even then we resort to *ad hoc* arguments. For example, we can make use of the bridge condition (see Theorem 5.3.9), the constant term of the characteristic polynomial [Cve18], and, for each vertex-deleted subgraph, the following result on the multiplicity of 0 as an eigenvalue: if $H$ is bipartite on sets of size $n_1, n_2$ then the multiplicity of 0 as an eigenvalue of $H$ is at least $|n_1 - n_2|$ (see [CvDS], p. 233).

Now we turn to bicyclic graphs. Those without vertices of degree 1 are of types $P(m, p, n)$, $C(m, n)$ and $B(m, p, n)$ illustrated in Fig. 5.5, where $m, n, p$ refer to the numbers of edges. In general, a bicyclic graph $G$ consists of a *central* part $X$ of type $P(m, p, n)$, $C(m, n)$ or $B(m, p, n)$, and a number of trees (*tails*) attached to the central part. In this situation let $V = V(X)$.

Given the eigenvalues and angles of $G$, the various possibilities for these trees can be constructed as before. Now the question of uniqueness of $V$ is even less tractable and we discuss the more modest problem of identifying the type of $X$ once its vertex set $V$ has been constructed. Note that in this situation the degrees of vertices in $X$ are known.

As before we know from the eigenvalues whether or not $G$ is bipartite; indeed we can determine from the coefficients of $P_G(x)$ the length and number of the shortest odd cycles in $G$ (see [CvDS], p. 87). We shall treat only the non-bipartite case, and in what follows $K$ denotes the set $\{i \in V(G) : P_{G-i}(x) \text{ is symmetric}\}$.

**5.3.3 Proposition** *If $G$ is not bipartite and $K = \emptyset$, then the central part of $G$ is isomorphic to $B(m, p, n)$ for some uniquely determined integers $m$, $p$, $n$ with $m, n$ odd.*

*Proof* Since $G$ is not bipartite, $G$ contains odd cycles. Since $K = \emptyset$, $G$ contains no vertex whose deletion destroys all odd cycles. Hence $G$ has two disjoint odd cycles. The set $U$ of vertices belonging to odd cycles consists of those vertices $u$ for which $G - u$ contains only one cycle. (If the odd cycles have different lengths then we can specify the vertex set of each cycle.) The lengths $m, n$ of the two odd cycles can be determined from the coefficients of the polynomials $P_{G-u}(x)$ $(u \in U)$, while $p = |X| - m - n + 1$. $\qquad\qquad\square$

The following proposition has a straightforward proof.

**5.3.4 Proposition** *Suppose that $G$ is not bipartite and $K \neq \emptyset$. There exist positive integers $m, n, p$ such that the central part $X$ of $G$ is isomorphic to:*

  (i) $C(m, n)$ *if $X$ contains a vertex of degree 4;*
 (ii) $P(m, p, n)$ *if $X$ contains two vertices of degree 3 and both belong to $K$;*
(iii) $B(m, p, n)$ *if $X$ contains two vertices of degree 3 and exactly one belongs to $K$.*

In case (i), if $|K| = 1$, then $m$ and $n$ are odd and the cycle vertex sets can be determined from the characteristic polynomials of vertex-deleted subgraphs as before; and if $|K| > 1$, then $K$ is just the odd cycle vertex set.

In case (ii), the set $K$ is just the set of vertices lying in both odd cycles. In case (iii), $K$ is the odd cycle vertex set.

The case in which $G$ is bipartite is more difficult. For some observations on this case see [Cve18].

### 5.3.3 *Tree-like cubic graphs*

We have seen in Subsection 5.3.2 that the Reconstruction Lemma from Subsection 5.3.1 can also be applied to unicyclic and bicyclic graphs.

More generally, we can reconstruct in a similar way all graphs consisting of a central part, which is known to be characterized by eigenvalues (and angles), and trees attached to the central part. In this subsection, we demonstrate further possibilities for the application of the Reconstruction Lemma and develop other reconstruction tools based on angles. Following some general observations we shall treat a class of cubic graphs described in [Cve17].

From eigenvalues and the angle sequence of a particular vertex $i$ we can calculate the number $N_k(i)$ of closed walks of any length $k$ originating and terminating at $i$ (see (4.2.9)). Recall that $N_2(i)$ is the degree of $i$ and $N_3(i)$ is twice the number $t_i$ of triangles passing through $i$. Further, $t_i$ is equal to the number of edges in the subgraph induced by the neighbours of $i$. The numbers $N_k(i)$ ($k = 4, 5, ...$) contain some information on the further neighbourhood of vertex $i$ but in general we cannot deduce very much about the actual graph structure. In regular graphs of degree $r$ we have

$$q_i = (N_4(i) - r^2)/2, \tag{5.3.4}$$

where $q_i$ is the number of quadrangles passing through $i$. However, in general the number of pentagons passing through $i$ is not uniquely determined in this way, even in the regular case. (On the other hand, the number of shortest cycles passing through $i$ can be calculated in the regular case.) We omit the details.

In this subsection we consider cubic graphs, where, because of the low degree, the first and second neighbourhoods of a vertex can be described fairly well; in particular, the subgraph induced by the three neighbours of a vertex is determined uniquely by $t_i$. Using $q_i$ and $N_5(i)$ we can say something about the second neighbours of vertex $i$, and in some cases even more. The possibilities are displayed in Fig. 5.6.

The ordered pair $(t_i, q_i)$ is called the *type* of vertex $i$. As Fig. 5.6 illustrates, the type of a vertex limits the possibilites for its second neighbourhood, and sometimes determines it uniquely.

We shall confine ourselves to a special class of cubic graphs. A connected cubic graph is called *tree-like* if it consists of blocks of the three types illustrated in Fig. 5.7. Blocks of the type (a) in Fig. 5.7 are called *terminal* blocks while those of the type (b) or (c) are *internal* blocks. Vertices such as $u$ and $v$ in a terminal block (see Fig. 5.7) are called *terminal* vertices. Blocks of type (a) and (b) are called *non-trivial* while those of type (c) are *trivial*. The substitution of trivial blocks for non-trivial ones results in a tree, and if this tree is a path then the

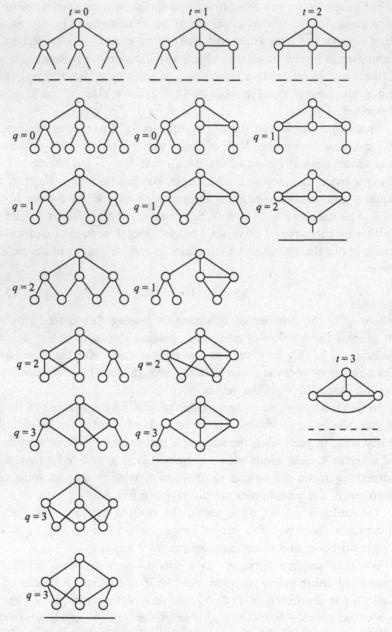

Fig. 5.6. Neighbourhoods in a cubic graph.

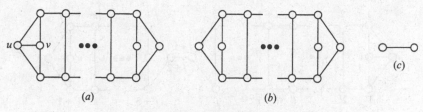

Fig. 5.7. Three types of block.

graph is called a *path-like* cubic graph. We show that such graphs are EA-reconstructible.

**5.3.5 Theorem** [Cve17] *A path-like cubic graph is characterized up to an isomorphism by eigenvalues and angles.*

*Proof* Given the eigenvalues and angles of a path-like cubic graph we can identify a terminal vertex $u$ from the condition $N_3(u) = 4$. Now we can gradually reconstruct the whole graph starting from $u$ (see Fig. 5.8).

Suppose that we have established the graph structure between vertices $u$ and $i, i'$. Note that vertices $i$ and $i'$ are at distance $i$ from $u$. The value of $N_{2i+2}(u)$ tells us whether $i+1$ and $i'+1$ are identical or distinct. In the first case $N_{2i+2}$ is greater by 2 than in the second case. The two possible values are easily determined because most of the walks to be counted are walks in the subgraph already reconstructed. By considering $N_{2i+3}(u)$ we can determine in a similar way whether $i+1$ and $i'+1$ are adjacent or not.

If $i+1$ and $i'+1$ are identical we have the situation illustrated with vertices $j, j', j+1$ in Fig. 5.8. Vertex $j+1$ is an endvertex of a block and we consider vertices $j+1, j+2, j+3, j'+3$. By considering $N_{2j+7}(u)$ we can decide whether $j+3$ and $j'+3$ are adjacent or not. Finally, we come to vertices $k+1, k'+1$ which are not adjacent, while only two vertices $u', v'$ remain. Their adjacencies must be as given in Fig. 5.8 and the whole graph is reconstructed. Hence we can construct in unique fashion a path-like cubic graph with given eigenvalues and angles, or establish that no such graph exists.     □

Note that in the proof of Theorem 5.3.5 we have used only the angle sequence of the vertex $u$, and we have not identified other vertices in the sense that we have not specified which angle sequence belongs to which vertex.

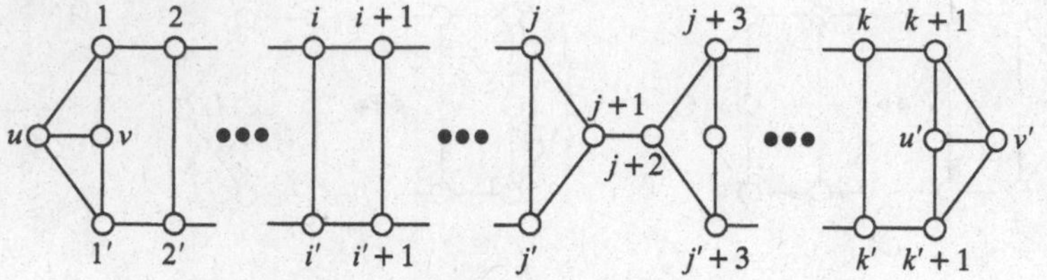

Fig. 5.8. Reconstructing a path-like cubic graph.

Now we combine Theorem 5.3.5 and results from Subsection 5.3.1 in order to study tree-like cubic graphs. First we need some more definitions.

Consider a tree-like cubic graph $G$. Vertices in which three blocks meet are called *centres*. A maximal connected subgraph of $G$ not containing any centre as a cutvertex is called a *branch*. Branches of $G$ contain one or two centres. Branches with one centre are called *terminal* while those with two centres are called *internal* branches. A terminal branch contains a terminal block (which includes two terminal vertices) and a centre which is called the *end* of the branch. In internal branches, *ends* are the two centres of $G$ belonging to the branch. A branch consisting of a single edge is called *trivial*. Thus non-trivial branches contain non-trivial blocks.

A *limb* of a tree-like cubic graph is defined recursively as follows:

(i) a terminal branch is a limb;
(ii) two limbs meeting at a vertex form a limb;
(iii) limbs are obtained by a finite number of applications of (i) and (ii).

Using the procedure from the proof of Theorem 5.3.5 we can construct all terminal branches. Now we have the problem of identifying the ends of these branches, that is, determining the angle sequences of the ends.

Suppose that the vertex $i$ is the end of the last non-trivial block in the branch, and $j$ is the end of the branch. Since the edge $ij$ is a bridge of $G$, if we know the angle sequence of $i$ we can calculate the angle sequence of $j$ by the Reconstruction Lemma.

We can restrict the set of possible candidates for vertex $i$. First, we know the type of this vertex (type $(1,0)$ in Fig. 5.6). Secondly, the polynomial $P_i(x)$ should contain as a factor the characteristic polynomial of that part of the branch containing the vertices constructed before vertex $i$. Thirdly, in applying the Reconstruction Lemma to calculate $P_j(x)$, there must be at least one vertex to which $P_j(x)$ is associated.

If there is no vertex satisfying these three requirements, the graph is not tree-like. If there is more than one such vertex we have to examine all possibilities in further constructions. But for each terminal branch we have a set of candidates for the endvertex of the branch.

In a tree-like cubic graph we must have two terminal branches with a common end which is a centre. By the Reconstruction Lemma we can find the third neighbour of $v$. The type of this neighbour determines whether it is (a) the end of a non-trivial block or (b) another centre.

In case (a) we have a putative non-trivial branch which can be constructed as in Theorem 5.3.5 using the angle sequence of $v$. Here we are constructing a limb of $G$. In case (b) we check whether some other branches have their ends at the new centre. If so, we can continue the construction as in case (a), while otherwise we turn our attention to other branches (or limbs). Gradually we can construct the whole graph, and then we check whether it has the given eigenvalues and angles.

Let us denote the algorithm described above as *Algorithm C*. We can now formulate the following theorem.

**5.3.6 Theorem** [Cve17] *Let the eigenvalues and angles of a graph be specified. Using Algorithm C we can construct all tree-like cubic graphs with the given eigenalues and angles or establish that no such graph exists.*

### 5.3.4 Fuzzy images of graphs

In this subsection we follow [Cve16] in developing additional techniques for the construction of graphs with given eigenvalues and angles. It turns out that for arbitrary graphs it is possible at least to construct a 'fuzzy image' of a graph which contains some information on graph structure.

If the spectrum of a graph $G$ is known, then knowledge of the angles of $G$ is equivalent to knowledge of the spectra of all vertex-deleted subgraphs $G - i$ (see (4.2.8)). In what follows we shall speak in terms of characteristic polynomials of $G$ and vertex-deleted subgraphs $G - i$, rather than in terms of eigenvalues and angles of $G$.

Let $\mathscr{C}$ be the set of cycles of $G$ containing the edge $uv$ and let $\mathscr{P}$ be the set of all paths in $G$ connecting the vertices $u$ and $v$.

We start with the following formulas (see, for example, [CvDGT], p. 144, with an obvious misprint concerning (5.3.7)). For ease of presentation, we write $P_G$ for $P_G(x)$.

$$P_G = P_{G-uv} - P_{G-u-v} - 2\sum_{C \in \mathscr{C}} P_{G-C}, \qquad (5.3.5)$$

$$P_{G-u}P_{G-v} - P_G P_{G-u-v} = \left(\sum_{P \in \mathscr{P}} P_{G-P}\right)^2, \qquad (5.3.6)$$

$$4(P_{G-u}P_{G-v} - P_G P_{G-u-v}) = (P_G - P_{G-uv} - P_{G-u-v})^2. \qquad (5.3.7)$$

Formula (5.3.5) is due to Schwenk (see for example [CvDS], p. 78). Formula (5.3.6) holds also for the case when $u$ and $v$ are not adjacent. Formula (5.3.7) follows from (5.3.5) and (5.3.6). Formula (5.3.6) has been obtained from a classical formula on determinants due to Jacobi using a more recent graph-theoretic interpretation of cofactors by Coates [Coa] (see [Cve16] for further references on these formulas).

Our first result is obtained by solving (5.3.7) as a quadratic equation for $P_G$, the choice of sign being determined by (5.3.5).

**5.3.7 Theorem** [Cve16] *Let $u$ and $v$ be adjacent vertices in a graph G. Then*

$$P_G = P_{G-uv} - P_{G-u-v} - 2\sqrt{P_{G-u}P_{G-v} - P_{G-uv}P_{G-u-v}} \qquad (5.3.8)$$

*where the square root is interpreted as a polynomial with a positive coefficient in the highest term.*

Formula (5.3.8) provides a recursive tool for calculating the characteristic polynomial of the graph. See [Row3] for another recursion formula of this type. We now consider the case that the square root in (5.3.8) is equal to zero, that is,

$$P_{G-u}P_{G-v} - P_{G-uv}P_{G-u-v} = 0. \qquad (5.3.9)$$

In this case (5.3.8) reduces to

$$P_G = P_{G-uv} - P_{G-u-v}. \qquad (5.3.10)$$

**5.3.8 Theorem** [Cve16] *Let $u$ and $v$ be adjacent vertices in a graph G. Formula (5.3.10) holds if and only if the edge $uv$ is a bridge of G.*

*Proof* It is well known (see, for example, [CvDS], p. 59) that formula (5.3.10) holds if the edge $uv$ is a bridge. To prove the converse, compare (5.3.10) with (5.3.5). It follows that $\sum_{C \in \mathscr{C}} P_{G-C} = 0$, i.e. the set of cycles $\mathscr{C}$ containing the edge $uv$ is empty. Hence, $uv$ is a bridge.      □

The next result involves only the characteristic polynomials of $G, G-u$ and $G-v$.

**5.3.9 Theorem** [Cve16] *Let $uv$ be a bridge of a graph G. Then the expression $P_G^2 + 4P_{G-u}P_{G-v}$ is a square.*

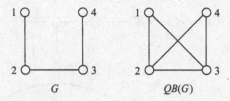

Fig. 5.9. A graph $G$ and the corresponding quasi-bridge graph $QB(G)$.

*Proof* Using (5.3.10) to substitute for $P_G$ in (5.3.7) we obtain

$$P_{G-u-v}^2 + P_G P_{G-u-v} - P_{G-u} P_{G-v} = 0.$$

The discriminant of this quadratic equation in $P_{G-u-v}$ should be a square.

□

**5.3.10 Remark** Theorem 5.3.9 can also be proved in the following way. Let $G$ be the graph obtained from graphs $R$ and $S$ by introducing an edge between vertex $u$ of $R$ and vertex $v$ of $S$ (see Fig. 5.3, where we have $i$ and $j$ instead of $u$ and $v$). Let $A$, $B$, $A_u$, $B_v$ be characteristic polynomials of $R$, $S$, $R - u$, $S - v$ respectively. We have $P_G = AB - A_u B_v$ (cf. [CvDS], p. 59). Since $P_{G-u} = A_u B$ and $P_{G-v} = AB_v$, we readily obtain $P_G^2 + 4P_{G-u}P_{G-v} = (AB + A_u B_v)^2$.             □

Theorem 5.3.9 provides a necessary condition for two vertices $u$ and $v$ to be joined by a bridge. Let us call this condition the *bridge condition*. The following example shows that the bridge condition is not sufficient for the existence of the bridge.

**5.3.11 Example** Consider the graph $G$ in Fig. 5.9. There is an automorphism of $G$ taking 2 to 3. Hence, $P_{G-2} = P_{G-3}$ and the bridge condition is fulfilled for $u = 1$, $v = 3$ in the same way as it is for $u = 1$, $v = 2$; but 1 and 3 are not connected with a bridge.             □

It seems useful to introduce the following notion.

**5.3.12 Definition** *The quasi-bridge graph $QB(G)$ of the graph $G$ is the graph with the same vertices as $G$, two vertices being adjacent if and only if they fulfil the bridge condition.*

**5.3.13 Example** In Figure 5.9 the quasi-bridge graph $QB(G)$ is shown alongside the labelled graph $G$.             □

It is clear that if $G$ is a tree then $G$ is a spanning subgraph of $QB(G)$. This fact can be used to formulate a procedure for determining all

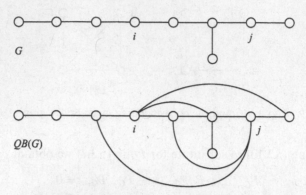

Fig. 5.10. The graph of Example 5.3.14.

trees with prescribed eigenvalues and angles. Since vertex degrees are
EA-reconstructible (see Proposition 4.2.4), the procedure is to construct
$QB(G)$ and then take as candidates the spanning trees of $QB(G)$ in which
vertices have the correct degrees.

**5.3.14 Example** It can happen that $P_{G-u} = P_{G-v}$ when there is no
automorphisim of $G$ taking $u$ to $v$. A tree with this property is shown
in Fig. 5.10 (cf. [CvDS], p.159). The quasi-bridge graph $QB(G)$ contains
four 'quasi-bridges'.                                                 □

In all examples investigated to date, if two vertices are adjacent in
$QB(G)$ then they are adjacent in some graph which has the same eigenvalues and angles as $G$.

We remark in passing that the procedure for constructing tree-like
cubic graphs with given eigenvalues and angles, described in Subsection
5.3.3, can be improved by the use of the bridge condition at an early
stage.

Next we derive a necessary condition for two vertices to be adjacent.

**5.3.15 Theorem** [Cve16] *Let $G$ be a graph with $n$ vertices and $m$ edges,
and let $uv$ be an edge of $G$. Then there exists a polynomial $q(x)$ of degree
at most $n - 3$ such that*

$$(x^n - (m-1)x^{n-2} + q(x))P_G(x) + P_{G-u}(x)P_{G-v}(x) \qquad (5.3.11)$$

*is a square.*

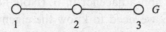

Fig. 5.11. The graph of Exmple 5.3.16.

*Proof* Formula (5.3.7) can be rewritten in the form

$$P_{G-u-v}^2 + 2P_{G-u-v}(P_G + P_{G-uv}) + (P_G - P_{G-uv})^2 - 4P_{G-u}P_{G-v} = 0.$$

The discriminant of this quadratic equation in $P_{G-u-v}$ is

$$(P_G + P_{G-uv})^2 - (P_G - P_{G-uv})^2 + 4P_{G-u}P_{G-v} = 4(P_G P_{G-uv} + P_{G-u}P_{G-v}).$$

Hence $P_G P_{G-uv} + P_{G-u}P_{G-v}$ should be a square. We have $P_{G-uv}(x) = x^n - (m-1)x^{n-2} + q(x)$, where $q(x)$ is a polynomial of degree at most $n-3$.  □

In Theorem 5.3.15, the necessary condition for two vertices to be adjacent, given by (5.3.11) is called the *edge condition*. It can be further elaborated by studying properties of $q(x)$.

We do not have much experience with numerical examples concerning Theorem 5.3.15, it seems to be difficult to use the edge condition practically.

**5.3.16 Example** For the graph $G$ in Fig. 5.11 we have $P_G(x) = x^3 - 2x$, $P_{G-1}(x) = P_{G-3}(x) = x^2 - 1$ and $P_{G-2}(x) = x^2$. For vertices 1 and 3 the edge condition is not satisfied since $q(x) = a$, where $a$ is a constant and (5.3.11) reduces to $(x^3 - 2x)(x^3 - x + a) + (x^2 - 1)^2$ which is not the square of any polynomial.  □

Strongly regular graphs show that the edge condition can be satisfied by non-adjacent vertices because in such graphs all vertex-deleted subgraphs have the same characteristic polynomial. (Since the edge condition is satisfied for adjacent vertices it follows in this case that it is satisfied also for non-adjacent pairs of vertices.)

Analogous to the quasi-bridge graph $QB(G)$ of a graph $G$ is the *quasi-graph* $Q(G)$ defined as follows.

**5.3.17 Definition** *The quasi-graph $Q(G)$ of a graph $G$ has the same vertex set as $G$, two vertices being adjacent if they fulfil the edge condition.*

Any graph is a spanning subgraph of its quasi-graph. Quasi-graphs of strongly regular graphs are complete. Just as the edge condition in $G$ is a necessary condition for adjacency in $G$, so the edge condition in $\overline{G}$ is a necessary condition for non-adjacency in $G$. (Note however that to apply the edge condition in $\overline{G}$ we need to know the eigenvalues and angles of $\overline{G}$.)

Any two distinct vertices of $G$ are adjacent either in $Q(G)$ or in $Q(\overline{G})$. If they are adjacent in one and not adjacent in the other then their status in $G$ coincides with that in $Q(G)$. Remaining pairs of vertices may or may not be adjacent in $G$. These considerations give use to the following definition.

**5.3.18 Definition** *The fuzzy image $FI(G)$ has the same vertex set as $G$ and two kinds of edges, solid and fuzzy. Vertices $u$ and $v$ of $FI(G)$ are*

(1) *non-adjacent if they are non-adjacent in $Q(G)$ and adjacent in $Q(\overline{G})$;*
(2) *joined by a solid edge if they are adjacent in $Q(G)$ and non-adjacent in $Q(\overline{G})$;*
(3) *joined by a fuzzy edge if they are adjacent in both $Q(G)$ and $Q(\overline{G})$.*

We can construct $Q(\overline{G})$, and hence $FI(G)$, if we know the eigenvalues and angles of $\overline{G}$. If $G$ is regular and both $G$ and $\overline{G}$ are connected then this information is already known from the eigenvalues and angles of $G$ (see Section 4.3). Accordingly the fuzzy image graph is a further aid to constructing a regular graph with given eigenvalues and angles. Note however that if the eigenvalues are those of a strongly regular graph then all pairs of vertices in the fuzzy image are joined by a fuzzy edge.

### 5.4 The Ulam graph reconstruction problem

In this section we present, following [CvRo4], several facts supporting the idea that graphs obtained from certain strongly regular graphs by deleting a vertex are good candidates for counterexamples to the following conjecture of Ulam.

**Reconstruction Conjecture** *Any graph with at least three vertices can be reconstructed uniquely from the collection of its vertex-deleted subgraphs.*

Much attention has been paid in the literature to this conjecture: for a survey of results see [BoHe], [Nas]. The analogous conjecture for digraphs is not valid [Sto1].

A counterexample to the reconstruction conjecture would consist of a pair of non-isomorphic graphs which have the same collection of vertex-deleted subgraphs. The possibility that a counterexample might necessarily consist of cospectral graphs was suggested in [GuCv]. This was verified when Tutte [Tut] proved that the spectrum of a graph is reconstructible from the collection of its vertex-deleted subgraphs. Since it also follows from the results of Tutte that the Seidel spectrum of a graph is likewise reconstructible, it has been conjectured in [CvDS] that a counterexample consists of switching-equivalent graphs. In any case it appears that the theory of graph spectra must play a central role in the reconstruction problem (cf. [Cve12]), an assertion further supported by the results of [Sto2]. See [Sch4] and [CvDGT], Section 3.5, for further details describing connections between spectra and the graph reconstruction problem.

It was suggested in [Cve19] that a counterexample might be sought among graphs with a small number of eigenvalues. This is substantiated by the results presented in this section which are motivated by the following observations. If a counterexample exists then the two graphs are very similar (since they have the same collection of vertex-deleted subgraphs), and this indicates that it would be hard to establish that they are non-isomorphic. Now it is known that, in general, strongly regular graphs with the same parameters present a difficult case for graph-isomorphism algorithms. On the other hand, regular graphs are reconstructible [BoHe] and so we might consider first some non-regular graphs which are closely related to strongly regular graphs. Strongly regular graphs have just three distinct eigenvalues and the modifications of them which we consider here turn out to have at most four distinct eigenvalues. The cases of one or two distinct eigenvalues correspond to complete graphs and their complements and so do not arise in our considerations (but perhaps one should note that $K_2$ and $\overline{K}_2$ are not reconstructible).

Further information on the nature of a counterexample may be obtained by considering graph angles. In view of equation (4.2.8), knowledge of the angles of $G$ is equivalent to knowledge of the spectra of all vertex-deleted subgraphs $G - u$, provided that the spectrum of $G$ is known. Thus angles are reconstructible from the collection of vertex-deleted subgraphs [GoMK3].

Analogous to the situation for angles, knowledge of the main angles of the graph $G$ is equivalent to that knowledge of the spectrum of the complement $\overline{G}$ of $G$, provided the spectrum of $G$ is known (see Proposition 4.5.2). Since $P_{\overline{G}}(x)$ is reconstructible from the collection of vertex-deleted subgraphs [Tut], so are the main angles of $G$. Thus graphs forming a counterexample to the reconstruction conjecture should have the same angles and the same main angles [CvDo2].

An observation, first made in [GoMK2], that all vertex-deleted subgraphs of a strongly regular graph have the same characteristic polynomial (see also [CvDo2]) is a key to further considerations. By (4.2.8) such graphs have at most four distinct eigenvalues.

Let $G$ be a strongly regular graph. If $G - u$ and $G - v$ are non-isomorphic, we have a pair of non-isomorphic cospectral graphs which are close to strongly regular graphs. Since $\overline{G}$ is also a strongly regular graph and since $\overline{G - u} = \overline{G} - u$ we conclude that $G - u$ and $G - v$ have the same main angles. We shall prove that they also have the same angles.

**5.4.1 Proposition** *Given a strongly regular graph* $G$, $P_{G-u-v}(x)$ *depends only on whether or not* $u$ *and* $v$ *are adjacent in* $G$.

*Proof* Let $\mu_1 > \mu_2 > \mu_3$ be the distinct eigenvalues of $G$. By two applications of the interlacing theorem (see, for example, [CvDS], p. 19) the graph $G - u - v$ has distinct eigenvalues $v_1, ..., v_5$ with

$$\mu_1 > v_2 > \mu_2 = v_2 \geq v_3 \geq v_4 \geq v_5 = \mu_3.$$

If $q_1, ..., q_5$ are the multiplicities of $v_1, ..., v_5$, respectively, the quantities $s_k = \sum_{i=1}^{5} q_i v_i^k$ $(k = 0, 1, 2, ...)$ are the spectral moments of $G - u - v$. We have $q_1 = q_3 = q_4 = 1$ and $q_2 = p_2 - 2$, $q_5 = p_3 - 2$, where $k_1 = 1$, $k_2$, $k_3$ are multiplicities of $\mu_1$, $\mu_2$, $\mu_3$ respectively. To calculate $v_1$, $v_3$ and $v_4$ we can use the equations

$$s_1 = 0, \quad s_2 = 2m,$$

$$\left. \frac{nt}{3} - \frac{s_3}{6} = \begin{cases} 2t - e & \text{if } u \text{ and } v \text{ are adjacent,} \\ 2t & \text{otherwise,} \end{cases} \right\} \tag{5.4.1}$$

where $n$ is the number of vertices of $G$, $m$ is the number of edges of $G - u - v$, $t$ is the number of triangles in $G$ containing a given vertex, and $e$ is the number of common neighbours of any two adjacent vertices in $G$. The first two equations (5.4.1) are immediate, and we deduce that

$$m = \begin{cases} \frac{nr}{2} - (2r - 1) & \text{if } u \text{ and } v \text{ are adjacent,} \\ \frac{nr}{2} - 2r & \text{otherwise,} \end{cases}$$

where $r$ is the degree of $G$. The third equation in (5.4.1) follows from the fact that the number of triangles containing two adjacent vertices is $e$, while $s_3$ is the number of closed walks of length 3. From (5.4.1) we derive a unique cubic whose roots are $v_1$, $v_3$ and $v_4$. $\square$

From Proposition 5.4.1 and equation (4.2.8) it follows that always $G - u$

and $G - v$ have the same angles when $G$ is strongly regular. It is clear that $G - u$ and $G - v$ can provide a counterexample to the reconstruction conjecture only when $G$ is not transitive: this rules out the Petersen graph and the Chang graphs, but imposes no serious constraint on $G$ because it is known that almost every strongly regular graph has a trivial automorphism group.

Let $n$, $r$, $e$, $f$ be the parameters of a strongly regular graph $G$ ($n$ the number of vertices; $r$ the degree; $e$, $f$ the number of common neighbours of any two adjacent, non-adjacent vertices, respectively). A graph $G - u$ has $r$ vertices of degree $r - 1$ and $n - 1 - r$ vertices of degree $r$. Vertices of degree $r - 1$ induce a regular subgraph of degree $e$ and those of degree $r$ induce a regular subgraph of degree $r - f$. Each vertex from the first set is adjacent to exactly $r - 1 - e$ vertices from the second set and each vertex from the second set is adjacent to exactly $f$ vertices from the first set. Hence in all graphs $G - u$ not only degrees of vertices are the same but also degrees of the neighbours of vertices: this is another necessary condition for a counterexample to the reconstruction conjecture [Nas]. And in a counterexample with just two different degrees, these degrees necessarily differ by 1 (cf. [BoHe], p. 254).

Let us also note that, due to the structural features described above, each graph $G - u$ has a divisor (for the definition see Section 2.4) with adjacency matrix $\begin{pmatrix} e & r-1-e \\ f & r-f \end{pmatrix}$. Let $N$ be the set of neighbours of a vertex $u$ in $G$, and let $G(u)$ be the graph obtained from $G - u$ by switching with respect to $N$.

**5.4.2 Proposition** *If $G$ is a strongly regular graph then all the graphs $G(u)$ are cospectral.*

*Proof* Let $G$ have parameters $n$, $r$, $e$, $f$, and suppose that $G^*$ is obtained from $G$ by switching with respect to the neighbours of an arbitrary vertex $u$. Let $A$, $A^*$, $A_u$ denote the adjacency matrices of $G$, $G^*$, $G(u)$ respectively; and let $S$, $S^*$ be the Seidel $(0, -1, 1)$-adjacency matrices of $G$, $G^*$ respectively.

Let $\lambda_1$, $\lambda_2$, with multiplicities $m_1$, $m_2$ respectively, be the eigenvalues of $G$ different from $r$. Thus for $i = 1, 2$, the eigenspace of $\lambda_i$ is an $m_i$-dimensional subspace of $\langle j \rangle^\perp$. Since $S = J - I - 2A$, $-1 - 2\lambda_i$ is an eigenvalue of $S$, and hence of $S^*$, with multiplicity $m_i$ ($i = 1, 2$). Therefore, $S^*$ has (at least) $m_i - 1$ linearly independent eigenvectors in $\langle j \rangle^\perp$ with corresponding eigenvalue $-1 - 2\lambda_i$. Since $S^* = J - I - 2A^*$, $A^*$ has (at least) $m_i - 1$ linearly independent eigenvectors in $\langle j \rangle^\perp$ with corresponding eigenvalue $\lambda_i$ ($i = 1, 2$). The same is true for $A_u$ (in $n - 1$ dimensions) since $u$ is an isolated vertex of $G^*$.

Now $A_u$ also has two eigenvalues which arise as eigenvalues of the divisor of $G(u)$ with adjacency matrix $\begin{pmatrix} e & n - 2r + e \\ r - f & r - f \end{pmatrix}$. If $r = f$ then $G$ is a regular complete multipartite graph, hence transitive, and our result is clearly true in this case. Accordingly we suppose that $r > f$. Then $G(u)$ is connected and one eigenvalue of the divisor, $\lambda_3$ say, is necessarily the index of $G(u)$, a simple eigenvalue with a non-negative eigenvector ([CvDS], p. 132). We now know that the characteristic polynomial of $G(u)$ is divisible by $(x - \lambda_1)^{m_1-1}(x - \lambda_2)^{m_2-1}(\lambda - \lambda_3)$. But $\text{tr}(A_u) = 0$ and so the one linear factor remaining is $x - \lambda_4$ where $(m_1 - 1)\lambda_1 + (m_2 - 1)\lambda_2 + \lambda_3 + \lambda_4 = 0$.                    $\square$

In the situation of Proposition 5.4.2, all of the graphs $\overline{G(u)}$ are cospectral since $\overline{G(u)} = \overline{G}(u)$ and $\overline{G}$ is strongly regular. Hence all $G(u)$ have the same main angles and so we have another way of producing 'good candidates' for counterexamples to the reconstruction conjecture. We can also start with non-isomorphic but cospectral strongly regular graphs $G$ and $H$ and consider either $G - u$ and $H - v$ or $G(u)$ and $H(v)$.

The case when $G - u$ and $G(u)$ are cospectral is especially interesting. The graph $G - u$ has a divisor with adjacency matrix $\begin{pmatrix} e & r - 1 - e \\ f & r - f \end{pmatrix}$, while $G(u)$ has a divisor with adjacency matrix $\begin{pmatrix} e & n - 2r + e \\ r - f & r - f \end{pmatrix}$. In the case that $n = 2r + 1$ and $r - f = f = e + 1$ these matrices have the same characteristic polynomial and so do $G - u$ and $G(u)$. The strongly regular graphs constructed in [Pau] have parameters $n = 25$, $r = 12$, $e = 5$, $f = 6$ and hence fulfil this requirement and those which are intransitive fulfil all the requirements noted so far. It is interesting to note also that some of these graphs on 25 vertices are self-complementary. According to [Cla] if $G$ is a regular self-complementary graph then there is at least one vertex $u$ such that $G - u$ is self-complementary; however, if $G - u$ and $G - v$ are both self-complementary, they are isomorphic. (One could perhaps consider two self-complementary graphs of the form $G - u$, $H - v$. Other interesting pairs are $G - u$, $G(u)$ and $G - u$, $\overline{G - u}$, where $G - u$ is not self-complementary.)

The self-complementary case shows a marked similarity to counterexamples to the reconstruction conjecture for digraphs in [Sto1]. These counterexamples consist of self-complementary (i.e. self-converse) tournaments in which players (vertices) have exactly two distinct scores (outdegrees). The similarity of the situation is marked by the facts that (i) tournaments are algebraically similar to self-complementary graphs and (ii) a special class of tournaments, called doubly regular tournaments, is similar to the class of strongly regular graphs (see Section 3.11 of [CvDGT]). What is different is that in the counterexamples of [Sto1] the tournaments are not cospectral [Sto2].

We have pointed to several ways of constructing pairs of graphs which are 'good candidates' for a counterexample to the reconstruction conjecture. Generally speaking, the reconstructibility of these graphs does not follow easily from known results.

In the case that $G$ is one of the 25-vertex graphs from [Pau], there is no counterexample consisting of vertex-deleted subgraphs $G-u$, $G-v$. This result of W. L. Kocay, proved using his program for testing isomorphism of graphs, was communicated by M. Doob in 1987.

# 6

# Graph perturbations

In this chapter we discuss how the eigenvalues of a graph are affected by small changes in the structure of the graph. In particular we investigate the role of angles and principal eigenvectors in this context. Techniques include both an analytic and an algebraic theory of matrix perturbations applied to the adjacency matrix of a graph.

## 6.1 Introduction

The theory of graph perturbations is concerned primarily with changes in eigenvalues which result from local modifications of a graph such as the addition or deletion of a vertex or edge. For a variety of such modifications, the corresponding characteristic polynomials are discussed in Section 4.3, and in these cases the eigenvalues of the perturbed graph are determined as implicit functions of algebraic and geometric invariants of the original graph. One of the problems investigated in this chapter is that of expressing the perturbed eigenvalues as series in these invariants. For this purpose we apply (in Section 6.3) the classical analytic theory of matrix perturbations to an adjacency matrix: here it is necessary to impose conditions which ensure convergence of the resulting power series, but when these conditions are satisfied we can find simultaneously an eigenvalue and eigenvector of the perturbed graph to any degree of accuracy. Clearly this is related to the second problem of estimating changes in eigenvalues, where we might hope to deal with a wider range of perturbations. For example, if $G_1, G_2$ are different perturbations of the same graph $G$ then we may be able to estimate $\mu_1(G_1), \mu_1(G_2)$ sufficiently closely to deduce that $\mu_1(G_1) > \mu_1(G_2)$, even if we do not have $\mu_1(G_1), \mu_1(G_2)$ as explicit functions of the relevant invariants (cf. Example 6.4.8 below). Here we can make use of the theory underlying

132

so-called intermediate eigenvalue problems of the second type (Section 6.4). A third problem concerns merely a comparison of original and perturbed eigenvalues, where for a given perturbation $G'$ of $G$ we seek the sign of $\mu_1(G') - \mu_1(G)$. In all cases — series approximations, estimates, comparisons — our discussion is focussed on the largest eigenvalue (or *index*) of a connected graph, reflecting the emphasis in the literature. Except for immediate consequences of the Perron-Frobenius theory, some of the earliest results concerned the change in index arising from the subdivision of an edge (see Section 3.2) and the relocation of a pendant edge (see Theorem 6.2.2). The influence of the index $\mu_1$ on graph structure is discussed in the survey article [CvRo5]: suffice it to say here that the maximum degree, chromatic number, clique number and extent of branching in a connected graph are all related to $\mu_1$. Moreover, as we saw in Theorem 2.2.5, the number of $i$-$j$ walks of length $k$ in a non-bipartite graph has the asymptotic approximation $\mu_1^k x_i x_j$ as $k \to \infty$, where $x_i$ is the $i$-th component of the principal eigenvector, a vector which plays a crucial role in most of our deliberations.

In Section 6.2 we present several results concerning estimates and comparisons which may be proved without recourse to the more elaborate techniques of Sections 6.3 and 6.4.

## 6.2 First observations

We begin by comparing the index of a connected graph $G$ with the index of a perturbation $G'$ of $G$. Any proper subgraph of $G$ has smaller index than $G$ (see Chapter 1) and so $\mu_1(G') < \mu_1(G)$ whenever $G'$ is obtained from $G$ by deleting an edge or vertex. By the same token, $\mu_1(G') > \mu_1(G)$ whenever $G'$ is obtained from $G$ by adding an edge or a non-isolated vertex. We saw in Chapter 3 that $\mu_1(G') < \mu_1(G)$ whenever $G'$ is obtained form $G$ by splitting a vertex, while Remark 3.2.2 and Theorem 3.2.3 describe the relation between $\mu_1(G')$ and $\mu_1(G)$ when $G'$ is obtained from $G$ by subdividing an edge. The first two results of this section provide comparisons of indices which are based on a comparison of characteristic polynomials.

We shall need the following observation.

**6.2.1 Lemma** (cf. [LiFe]) *If $H$ is a spanning subgraph of the graph $G$ then*

$$P_G(x) \le P_H(x) \quad \text{for all} \quad x \ge \mu_1(G).$$

*If further $\mu_1(H) < \mu_1(G)$ (in particular, if $G$ is connected and $H$ is a* *spanning subgraph of G) then*

$$P_G(x) < P_H(x) \quad \text{for all} \quad x \geq \mu_1(G).$$

*Proof* Let $G$ have vertex set $\{1, 2, \ldots, n\}$ and adjacency matrix $A$. We prove the first observation by induction on $n$. The result is immediate for $n = 1$. Accordingly suppose that $n > 1$ and the result holds for graphs with $n - 1$ vertices.

On differentiating $\det(xI - A)$ as a sum of $n!$ products we find that

$$P'_G(x) = \sum_{j=1}^{n} P_{G-j}(x).$$

We have a similar expression for $P'_H(x)$ and so

$$P'_H(x) - P'_G(x) = \sum_{j=1}^{n} \{P_{H-j}(x) - P_{G-j}(x)\}.$$

For each $j$, $H - j$ is a spanning subgraph of $G - j$ and so, by the induction hypothesis,

$$P_{G-j}(x) \leq P_{H-j}(x) \quad \text{for all} \quad x \geq \mu_1(G - j).$$

Since $\mu_1(G) \geq \mu_1(G - j)$ for each $j$, it follows that

$$P'_H(x) - P'_G(x) \geq 0 \quad \text{for all} \quad x \geq \mu_1(G).$$

Since $\mu_1(G) \geq \mu_1(H)$ and $P_H(x) \geq 0$ for all $x \geq \mu_1(H)$ the function $P_H(x) - P_G(x)$ is non-negative at $x = \mu_1(G)$, and hence non-negative for all $n \geq \mu_1(G)$.

For the second assertion, observe that if $\mu_1(G) > \mu_1(H)$ then the function $P_H(x) - P_G(x)$ is positive at $x = \mu_1(G)$, and hence positive for all $x \geq \mu_1(G)$.                                                                    □

**6.2.2 Theorem** [LiFe] *Let $u$ be a vertex of the non-trivial connected graph $G$, and (for non-negative integers $k, \ell$) let $G(k, \ell)$ denote the graph obtained from $G$ by adding pendant paths of lengths $k$ and $\ell$ at $u$. If $k \geq \ell \geq 1$ then $\mu_1(G(k, \ell)) > \mu_1(G(k + 1, \ell - 1))$.*

*Proof* By (4.3.3) we have

$$P_{G(k,\ell)}(x) = xP_{G(k,\ell-1)}(x) - P_{G(k,\ell-2)}(x)$$

when $\ell \geq 2$, and

$$P_{G(k+1,\ell-1)}(x) = xP_{G(k,\ell-1)}(x) - P_{G(k-1,\ell-1)}(x).$$

It follows that for $k \geq \ell \geq 1$ we have

$$P_{G(k,\ell)}(x) - P_{G(k+1,\ell-1)}(x) = P_{G(k-\ell+1,1)}(x) - P_{G(k-\ell+2,0)}(x).$$

Similarly, $P_{G(k-\ell+2,0)}(x) = xP_{G(k-\ell+1,0)}(x) - P_{G(k-\ell,0)}(x)$ and $P_{G(k-\ell+1,1)}(x) = xP_{G(k-\ell+1,0)}(x) - P_H(x)$, where $H$ is the graph $G(k-\ell+1,0) - u$. Thus

$$P_{G(k,\ell)}(x) - P_{G(k+1,\ell-1)}(x) = P_{G(k-\ell,0)}(x) - P_H(x).$$

Now $H$ is a proper spanning subgraph of $G(k-\ell,0)$ and so by Lemma 6.2.1 we have

$$P_{G(k,\ell)}(x) - P_{G(k+1,\ell-1)}(x) > 0 \quad \text{for all} \quad x \geq \mu_1(G(k-\ell,0)).$$

Since $G(k-\ell,0)$ is a proper subgraph of $G(k,\ell)$ we have $\mu_1(G(k-1,0)) < \mu_1(G(k,\ell))$. Hence $P_{G(k+1,\ell-1)}(x)$ is negative at $\mu_1(G(k,\ell))$ and the result follows. $\qquad\square$

The graph $G(k+1,\ell-1)$ of Theorem 6.2.2 may be obtained from $G(k,\ell)$ by relocating a pendant edge. The next theorem concerns the relocation of a bridge with a prescribed vertex. In Section 6.4 we shall see what can be said in the general case of a graph perturbed by the relocation of an arbitrary edge.

We write $HvwK$ for the graph obtained from disjoint graphs $H, K$ by adding an edge joining the vertex $v$ of $H$ to the vertex $w$ of $K$. Further, $H_v$ denotes the graph obtained from $H$ by adding a pendant edge at vertex $v$.

**6.2.3 Theorem** [ZhZZ] *If $P_{H_u}(x) < P_{H_v}(x)$ for all $x > \mu_1(H_v)$ then $\mu_1(HuwK) > \mu_1(HvwK)$ for all vertices $w$ of $K$.*

*Proof* As noted by Heilbronner [Hei2] we have

$$P_{HuwK}(x) = P_H(x)P_K(x) - P_{H-u}(x)P_{K-w}(x),$$

a result proved using an appropriate Laplacian determinantal expansion (see Theorem 2.12 of [CvDS]).

Similarly, or by equation (4.3.3), we have $P_{H_u}(x) = xP_H(x) - P_{H-u}(x)$, and so

$$P_{HuwK}(x) = P_H(x)P_K(x) - P_{K-w}(x)\{xP_H(x) - P_{H_u}(x)\}.$$

On subtracting the analogous expression for $P_{HvwK}(x)$ we obtain

$$P_{HuwK}(x) - P_{HvwK}(x) = P_{K-w}(x)\{P_{H_u}(x) - P_{H_v}(x)\}.$$

If now $x > \mu_1(HvwK)$ then $x > \mu_1(K - w)$ and so $P_{K-w}(x) > 0$. By hypothesis, $P_{H_u}(x) - P_{H_v}(x) < 0$, and the result follows. $\qquad\square$

There have been several estimates for the increase in index resulting from the attachment of a pendant edge, as in Section 3 of [BeRo1], Section 3 of [CvRo1] and Section 6 of [CvRo2]. These estimates are best seen in the context of Theorem 6.2.4 below. We shall see in Sections 6.3 and 6.4 that if $\mu_1 - \mu_2$ is large enough, reasonable upper bounds can be obtained in terms of $\mu_1, \mu_2$ and the angles $\alpha_{iu}$ of $G$ at the vertex $u$ of attachment. Here we observe that if $\mu_1(\mu_1 - \mu_2) > 1$ then an upper bound for $\mu_1(G_u)$ in terms of these parameters can be obtained by inspecting the characteristic polynomial of $G_u$.

**6.2.4 Theorem** [Row6] *Let $G$ be a connected graph and let $G_u$ denote the graph obtained from $G$ by adding a pendant edge at vertex $u$. If $\mu_1(\mu_1 - \mu_2) > 1$ then $\mu_1(G_u) < \mu_1(G) + \epsilon_u$ where $\epsilon_u = \dfrac{\alpha_{1u}^2}{\mu_1 - (\mu_1 - \mu_2)^{-1}}$.*

*Proof* Since the eigenvalues of $G$ interlace those of $G_u$ (by Theorem 2.4.1) it suffices to prove that $P_{G_u}(\mu_1 + \epsilon_u) > 0$. From Proposition 4.3.6 we have

$$P_{G_u}(\mu_1 + \epsilon_u) = P_G(\mu_1 + \epsilon_u)\{\epsilon_u + (\mu_1 - \mu_2)^{-1} - \sum_{i=2}^{m} \frac{\alpha_{iu}^2}{\mu_1 - \mu_i + \epsilon_u}\}.$$

Now

$$\sum_{i=2}^{m} \frac{\alpha_{iu}^2}{\mu_1 - \mu_i + \epsilon_\mu} \leq \sum_{i=2}^{m} \frac{\alpha_{iu}^2}{\mu_1 - \mu_2} = \frac{1 - \alpha_{1u}^2}{\mu_1 - \mu_2},$$

and so $P_{G_u}(\mu_1 + \epsilon_u) \geq P_G(\mu_1 + \epsilon_u)\{\epsilon_u + \alpha_{1u}^2(\mu_1 - \mu_2)^{-1}\} > 0$. $\qquad\square$

**6.2.5 Example** [Row6] If $G$ is the skeleton of an icosahedron then $m = 4$ and $\mu_1 = 5, \mu_2 = \sqrt{5}, \mu_3 = -1, \mu_4 = -\sqrt{5}$. In particular, $\mu_1(\mu_1 - \mu_2) > 1$. For any vertex $u$, we have $\alpha_{1u}^2 = \frac{1}{12}$ and $\epsilon_u \approx 0.01797$. Thus Theorem 6.2.4 gives an upper bound of 5.01797 for $\mu_1(G_u)$, compared with an actual value of 5.01728 (to five decimal places). $\qquad\square$

## 6.3 An analytical theory of perturbations

The various theorems from matrix theory (such as Theorem 2.4.1) which serve as basic tools in algebraic graph theory are listed in Section 0.3 of the monograph [CvDS]. Here we show how to make use of the classical theory of matrix perturbations as described for example in Sections 11.5, 11.6 of [LaTi]. In general one considers a perturbation of the matrix $A$

of the form $A(\zeta)$ where $\zeta$ is a complex variable and $A(0) = A$. Here we shall restrict ourselves to the case of a linear perturbation $A(\zeta) = A + \zeta B$, where $A$ and $B$ are fixed symmetric matrices. For our purposes, $A$ and $A + B$ will be taken to be the adjacency matrices of a connected graph $G$ and a perturbation $G'$ of $G$ respectively. The basic idea is to express eigenvalues and eigenvectors of $A(\zeta)$ as convergent power series in $\zeta$. For this to be of any use in our context we need to impose conditions which ensure a radius of convergence greater than 1, so that we may obtain information about the eigenvalues of $G'$ by setting $\zeta = 1$. For the purposes of exposition we consider changes in only the largest eigenvalue resulting from a perturbation of a connected graph $G$ (cf. [Row5]). The underlying theory is essentially the same for any simple eigenvalue. For corresponding results on multiple eigenvalues the reader is referred to Section 11.7 of [LaTi].

Suppose then that

$$(A + \zeta B)\mathbf{x}(\zeta) = \mu_1(\zeta)\mathbf{x}(\zeta) \tag{6.3.1}$$

where $\mathbf{x}(\zeta) \neq \mathbf{0}, \mathbf{x}(0) = \mathbf{x}$ (the principal eigenvector of $G$) and $\mu_1(0) = \mu_1$. We show first that if there exist convergent power series

$$\mu_1(\zeta) = \mu_1 + c_1\zeta + c_2\zeta^2 + \cdots, \tag{6.3.2}$$

$$\mathbf{x}(\zeta) = \mathbf{x} + \mathbf{x}_1\zeta + \mathbf{x}_2\zeta^2 + \cdots \tag{6.3.3}$$

such that

$$\mathbf{x}^T\mathbf{x}_r = 0 \quad \text{for each} \quad r \in I\!N \tag{6.3.4}$$

then the perturbation coefficients $c_r$ $(r \in I\!N)$ and the vectors $\mathbf{x}_r$ $(r \in I\!N)$ are determined by $\mathbf{x}, B$ and the spectral decomposition of $A$. Since $I\!R^n = \langle \mathbf{x} \rangle \oplus \langle \mathbf{x} \rangle^\perp$ and $\langle \mathbf{x} \rangle^\perp = \mathscr{E}(\mu_2) \oplus \cdots \oplus \mathscr{E}(\mu_m)$, condition (6.3.4) may be seen as normalizing $\mathbf{x}(\zeta)$ and constraining $\mathbf{x}(\zeta)$ to lie outside $\mathscr{E}(\mu_2) \oplus \cdots \oplus \mathscr{E}(\mu_m)$.

Let $E = \sum_{i=2}^m (\mu_1 - \mu_i)^{-1} P_i$, so that $(\mu_1 I - A)E = E(\mu_1 I - A) = I - P_1$ and $E\mathbf{x} = \mathbf{0}$. If we use equations (6.3.2) and (6.3.3) to substitute for $\mu_1(\zeta)$ and $\mathbf{x}(\zeta)$ in equation (6.3.1), and equate coefficients of $\zeta^r$, then we obtain

$$\left. \begin{array}{l} (\mu_1 I - A)\mathbf{x}_1 = B\mathbf{x} - c_1\mathbf{x}_1, \\ (\mu_1 I - A)\mathbf{x}_r = B\mathbf{x}_{r-1} - (c_1\mathbf{x}_{r-1} + \cdots + c_{r-1}\mathbf{x}_1) - c_r\mathbf{x} \quad (r \geq 2). \end{array} \right\} \tag{6.3.5}$$

On taking the scalar product with $\mathbf{x}$, we obtain

$$\left. \begin{array}{l} c_1 = \mathbf{x}^T B\mathbf{x}, \\ c_r = \mathbf{x}^T B\mathbf{x}_{r-1} \quad (r \geq 2). \end{array} \right\} \tag{6.3.6}$$

On multiplication by $E$, we obtain from equation (6.3.5):

$$\left.\begin{array}{l} \mathbf{x}_1 = EB\mathbf{x}, \\ \mathbf{x}_r = EB\mathbf{x}_{r-1} - E(c_1\mathbf{x}_{r-1} + \cdots + c_{r-1}\mathbf{x}_1) \ \ (r \geq 2). \end{array}\right\} \quad (6.3.7)$$

Equations (6.3.6) and (6.3.7) suffice to determine $c_r$ and $\mathbf{x}_r$ recursively in terms of $E, B$ and $\mathbf{x}$.

Let us now use equations (6.3.6) and (6.3.7) to *define* $c_r, \mathbf{x}_r$ and power series $\mu_1(\zeta) = \mu_1 + \sum_{r=1}^{\infty} c_r \zeta^r$, $\mathbf{x}(\zeta) = \mathbf{x} + \sum_{r=1}^{\infty} \mathbf{x}_r \zeta^r$. Then within the circle of convergence of these series, equations (6.3.5) and (6.3.1) hold. To see this, note that $(\mu_1 I - A)\mathbf{x}_1 = (I - P_1)B\mathbf{x} = B\mathbf{x} - \mathbf{x}\mathbf{x}^T B\mathbf{x} = B\mathbf{x} - c_1\mathbf{x}$; moreover $P_1\mathbf{x}_r = \mathbf{0}$ for all $r \in \mathbb{N}$ by induction on $r$, and so for $r \geq 2$, $(\mu_1 I - A)\mathbf{x}_r = (I - P_1)B\mathbf{x}_{r-1} - (c_1\mathbf{x}_{r-1} + \cdots + c_{r-1}\mathbf{x}_1)$; finally $P_1 B\mathbf{x}_{r-1} = \mathbf{x}\mathbf{x}^T B\mathbf{x}_{r-1} = c_r\mathbf{x}$.

The next thing to do is to estimate $R$, a common radius of convergence of the series $\mu_1(\zeta), \mathbf{x}(\zeta)$. A general result, Theorem 11.5.1 of [LaTi], asserts that $R > 0$, but here we seek a lower bound for $R$ in terms of $\|B\|$ and $\|E\|$, where $\|B\| = \sup\{\|B\mathbf{v}\| : \|\mathbf{v}\| = 1\}$ and $\|E\| = \sup\{\|E\mathbf{v}\| : \|\mathbf{v}\| = 1\} = (\mu_1 - \mu_2)^{-1}$. We use an argument from Section 2 of [BeRo2] to show how $R$ may be estimated when $\|B\| \leq 1$. We can then apply the theory in the case that $B$ corresponds to the addition of an edge.

Suppose that $\|B\| \leq 1$, and let $b = \|B\mathbf{x}\|$, $e = (\mu_1 - \mu_2)^{-1}$, $\mathbf{x}_0 = \mathbf{x}$. Define $V(\zeta) = \sum_{r=0}^{\infty} v_r \zeta^r$, with radius of convergence $R^*$, where

$$\left.\begin{array}{l} v_0 = 1, \ v_1 = eb, \\ v_r = e(v_0 v_{r-1} + v_1 v_{r-2} + \cdots + v_{r-1} v_0) \ \ (r \geq 2). \end{array}\right\} \quad (6.3.8)$$

By induction on $r$ we have $\|\mathbf{x}_r\| \leq v_r$ $(r \geq 0)$ and $|c_r| \leq v_{r-1}$ $(r \geq 1)$; and so $R \geq R^*$. Now from equations (6.3.8) we have

$$V(\zeta) = e\zeta V(\zeta)^2 + 1 + \zeta(eb - e), \quad (6.3.9)$$

and on solving this quadratic for $e\zeta V(\zeta)$, we find that $R^* \geq 1/2e(1 + \sqrt{b})$. It follows that if $\|B\| \leq 1$ and $\mu_1 - \mu_2 > 2(1 + \sqrt{\|B\mathbf{x}\|})$ then $R^* > 1$ and so $\mu_1(\zeta), \mathbf{x}(\zeta)$ converge at $\zeta = 1$. In particular, $\mu_1(1)$ is an eigenvalue of $A + B$, and the following lemma shows that it is the largest one.

**6.3.1 Lemma** *Let $G$ be a connected graph and let $u, v$ be non-adjacent vertices of $G$. Let $A, A+B$ be the adjacency matrices of $G, G+uv$ respectively. For each real $\zeta \in (0, R)$, $\mu_1(\zeta)$ is the largest eigenvalue of $A + \zeta B$.*

*Proof* The eigenvalues of $A + \zeta B$ are continuous functions of $\zeta$. Hence if $\mu_1(\zeta)$ does not remain the largest eigenvalue of $A + \zeta B$ as $\zeta$ increases from 0 to $R$ then for some $\xi \in (0, R), \mu_1(\xi)$ coincides with a second eigenvalue which is equal largest. But $A + \xi B$ is irreducible because $A$ is irreducible and all entries of $A$ and $\xi B$ are non-negative. Hence the largest eigenvalue of $A + \xi B$ is simple, a contradiction. $\qquad \square$

We noted in Section 5.1 that the spectrum of $G + uv$ is determined (implicitly) by the spectrum of $G$, the angles $\alpha_{iu}, \alpha_{iv}$ $(i = 1, 2, \ldots, m)$ and the numbers $p_{uv}^{[i]}$ $(i = 1, 2, \ldots, m)$, where $p_{uv}^{[i]}$ denotes the $(u, v)$-entry of $P_i$. As far as the index of $G + uv$ is concerned, we can now prove the following result.

**6.3.2 Theorem** [Row5] *Let $u, v$ be non-adjacent vertices of the connected graph $G$. If $\mu_1 - \mu_2 > 2(1 + \sqrt[4]{\alpha_{1u}^2 + \alpha_{1v}^2})$, in particular if $\mu_1 - \mu_2 > 4$, then $\mu_1(G + uv) = \mu_1 + \sum_{r=1}^{\infty} c_r$ where the $c_r$ are recursively defined functions of $\mu_i, \alpha_{iu}, \alpha_{iv}, p_{uv}^{[i]}$ $(i = 1, 2, \ldots, m)$.*

*Proof* Here $\|B\| = 1$ because the only non-zero elements of $B$ are ones in positions $(u, v)$ and $(v, u)$. By Lemma 6.3.1 and the observations preceding it, we know that $\mu_1(G + uv) = \mu_1(1) = \mu_1 + \sum_{r=1}^{\infty} c_r$ provided that $\mu_1 - \mu_2 > 2(1 + \sqrt{\|Bx\|})$. Here $\|Bx\| = \sqrt{\alpha_{1u}^2 + \alpha_{1v}^2}$, and so it remains to show that the $c_r$ are recursively defined functions of $\mu_i, \alpha_{iu}, \alpha_{iv}$ and $p_{uv}^{[i]}$ $(i = 1, \ldots, m)$. We describe a scalar as *known* if it is determined by these $4m$ invariants, and we take $u = 1, v = 2$ without loss of generality. We first show by induction on $t$ $(t \geq 0)$ that if $\mathbf{w}$ is a vector of the form

$$\mathbf{w} = E^{s_t} B E^{s_{t-1}} B \ldots E^{s_2} B E^{s_1} B \mathbf{x}, \qquad (6.3.10)$$

where each $s_i \in \mathbb{N}$, then the first two components of $\mathbf{w}$ are known. This is clear if $t = 0$ because $B\mathbf{x} = (x_2, x_1, 0, \ldots, 0)^T$ and $x_j = \|P_1 \mathbf{e}_j\|$ $(j = 1, 2)$. Let $E^s = (e_{ij}^{(s)})$, so that if $\mathbf{v} = (v_1, v_2, \ldots, v_n)^T$ then the first two components of $E^s B \mathbf{v}$ are $e_{11}^{(s)} v_2 + e_{12}^{(s)} v_1$ and $e_{21}^{(s)} v_2 + e_{22}^{(s)} v_1$. Now $e_{jk}^{(s)} = \sum_{i=2}^{m} (\mu_1 - \mu_i)^{-s} p_{jk}^{[i]}$ and so if the first two components of $\mathbf{v}$ are known then so are the first two components of $E^s B \mathbf{v}$. In particular, the first two components of $\mathbf{w}$ are known.

We can now show by induction on $r$ $(r \geq 1)$ that $c_r$ is known and $\mathbf{x}_r$ is a known linear combination of vectors of the form (6.3.10). This is clear for $r = 1$ because $c_1 = \mathbf{x}^T B \mathbf{x} = 2x_1 x_2$, while $\mathbf{x}_1 = EB\mathbf{x}$. Now suppose that $r > 1$ and consider $c_r = \mathbf{x}^T B \mathbf{x}_{r-1}$, where we may suppose that $\mathbf{x}_{r-1}$ is a known linear combination of vectors of the form (6.3.10). Then the first two components $x_{r-1,1}, x_{r-1,2}$ of $\mathbf{x}_{r-1}$ are known, and $c_r$ is known because $c_r = x_2 x_{r-1,1} + x_1 x_{r-1,2}$. Next consider the expression in (6.3.7) for $\mathbf{x}_r$ $(r \geq 2)$. In the right-hand side, each of $c_1, \ldots, c_{r-1}$ is known and each of $\mathbf{x}_1, \ldots, \mathbf{x}_{r-1}$ is a known linear combination of vectors of the form (6.3.10). If we premultiply a vector of this form by either $EB$ or $E$ then we obtain a vector of the same form: hence $\mathbf{x}_r$ is a known linear combination of vectors of the form (6.3.10). Thus the vectors $\mathbf{x}_r$, and hence the perturbation coefficients $c_r$, are determined recursively in terms of $\mu_i, \alpha_{iu}, \alpha_{iv}, p_{uv}^{[i]}$ $(i = 1, 2, \ldots, m)$. $\square$

In the situation of Theorem 6.3.2, we have by way of illustration $c_1 = 2x_1 x_2$ and $c_2 = \mathbf{x}^T B \mathbf{x}_1 = \mathbf{x}^T BEB\mathbf{x} = e_{11} x_2^2 + e_{12} x_2 x_1 + e_{21} x_1 x_2 + e_{22} x_1^2$. Thus in terms of $u$ and $v$,

$$c_1 = 2\alpha_{1u}\alpha_{1v}, \quad c_2 = \sum_{i=2}^{m}(\mu_1 - \mu_i)^{-1}(\alpha_{1v}^2 \alpha_{iu}^2 + 2\alpha_{1v}\alpha_{1u}p_{uv}^{[i]} + \alpha_{1u}^2 \alpha_{iv}^2). \quad (6.3.11)$$

For an explicit example, we take a graph whose symmetry makes it straightforward to calculate the invariants involved.

**6.3.3 Example** [Row8] Let $G$ be the graph whose complement $\overline{G}$ is the skeleton of an icosahedron. Here $m = 4$ and $\mu_1 = 6, \mu_2 = \sqrt{5} - 1, \mu_3 = 0, \mu_4 = -\sqrt{5} - 1$. In particular, $\mu_1 - \mu_2 > 4$ and so our condition for convergence (Theorem 6.3.2) is satisfied. In order to find the matrices $P_i$ $(i = 1, 2, 3, 4)$ it is convenient to consider the adjacency matrix $\overline{A}$ of $\overline{G}$, which (since $G$ is regular) is given by

$$\overline{A} = J - I - A = \overline{\mu}_1 P_1 + \overline{\mu}_2 P_2 + \overline{\mu}_3 P_3 + \overline{\mu}_4 P_4$$

where $\overline{\mu}_1 = 5, \overline{\mu}_2 = -\sqrt{5}, \overline{\mu}_3 = -1$ and $\overline{\mu}_4 = \sqrt{5}$. Since $\overline{G}$ is distance-transitive we can describe the $(u, v)$-entries of $I, \overline{A}, \overline{A}^2, \overline{A}^3, P_1, P_2, P_3, P_4$ as follows:

| $d_{\overline{G}}(u,v)$ | $I$ | $\overline{A}$ | $\overline{A}^2$ | $\overline{A}^3$ | $P_1$ | $P_2$ | $P_3$ | $P_4$ |
|---|---|---|---|---|---|---|---|---|
| 0 | 1 | 0 | 5 | 10 | $\frac{1}{12}$ | $\frac{1}{4}$ | $\frac{5}{12}$ | $\frac{1}{4}$ |
| 1 | 0 | 1 | 2 | 13 | $\frac{1}{12}$ | $\frac{\sqrt{5}}{20}$ | $\frac{-1}{12}$ | $\frac{-\sqrt{5}}{20}$ |
| 2 | 0 | 0 | 2 | 8 | $\frac{1}{12}$ | $\frac{-\sqrt{5}}{20}$ | $\frac{-1}{12}$ | $\frac{\sqrt{5}}{20}$ |
| 3 | 0 | 0 | 0 | 10 | $\frac{1}{12}$ | $\frac{-1}{4}$ | $\frac{5}{12}$ | $\frac{-1}{4}$ |

Here the entries of $\overline{A}^2$ and $\overline{A}^3$ are obtained directly from the icosahedron, while those of $P_2, P_3, P_4$ are calculated from the following cubic expression (see Section 1.1):

$$P_i = \prod_{j \neq i}(\overline{\mu}_i - \overline{\mu}_j)^{-1}(\overline{A} - \overline{\mu}_j I).$$

(We know that $P_1 = \frac{1}{12}J = I - P_2 - P_3 - P_4$.) Now to within isomorphism, there is only one graph $G + uv$, which arises when $d_{\overline{G}}(u,v) = 1$. If we use the second row of the foregoing table to substitute the appropriate values in equation (6.3.10), we find that $c_1 = \frac{1}{6}$ and $c_2 = \frac{49}{2376}$. Thus our first approximation to $\mu_1(G + uv)$ is 6.1667 and our second approximation is 6.1873, compared with an actual value of 6.1894 (to four decimal places). We remark that the condition $\mu_1 - \mu_2 > 2(1 + \sqrt[4]{\alpha_{1u}^2 + \alpha_{1v}^2})$ fails for the icosahedral graph $\overline{G}$, although it is possible that the series $\sum_r c_r$ nevertheless converges to $\mu_1(\overline{G} + uv) - \mu_1(\overline{G})$ when $d_{\overline{G}}(u,v) > 1$. □

The arguments used to prove Theorem 6.3.2 may be extended to deal with the attachment of a pendant edge to a connected graph $G$. One replaces $A$ with $A_0$, the adjacency matrix of the graph $G_0$ which consists of $G$ and an isolated vertex labelled 0. Although $G_0$ is not connected, $\mu_1$ remains a simple eigenvalue of $A_0$, and the vector x becomes $(0, \alpha_{11}, \alpha_{12}, \ldots, \alpha_{1n})^T$. If, without loss of generality, the pendant edge is attached at vertex 1 then the entries of $E$ which we need are

$$e_{01}^{(1)} = e_{10}^{(1)} = 0, \quad e_{00}^{(1)} = \mu_1^{-1} \text{ and } e_{11}^{(1)} = \sum_{i=2}^{m}(\mu_1 - \mu_i)^{-1}\alpha_{i1}^2.$$

If now we describe a scalar as known if it is determined by the $2m$ invariants $\mu_i, \alpha_{i1}$ $(i = 1, \ldots, m)$ then we can show that for odd $r$, we have $c_r = 0, \mathbf{x}_r = (a_r, 0, \ldots, 0)^T$ where $a_r$ is known; and for even $r$, $c_r$ is known while $\mathbf{x}_r = f_r(E)(0, 1, 0, \ldots, 0)^T$ where $f_r$ is a polynomial with known coefficients. Since $\|B\mathbf{x}\| = \alpha_{11}$, Lemma 6.3.1 yields the following analogue of Theorem 6.3.2.

**6.3.4 Theorem** *Let $G_u$ be the graph obtained from the connected graph $G$ by adding a pendant edge at vertex $u$. If $\mu_1 - \mu_2 > 2(1 + \sqrt{\alpha_{1u}})$ then $\mu_1(G_u) = \mu_1 + \sum_{r=1}^{\infty} c_r$ where the $c_r$ are recursively defined functions of $\mu_i, \alpha_{iu}$ $(i = 1, 2, \ldots, m)$.*

The perturbation coefficients which arise in Theorem 6.3.4 were obtained in [BeRo1] by examining power series solutions $\mu_1 + \sum_{r=1}^{\infty} c_r \zeta^r$ of the equation $\det(xI - A_0 - \zeta B) = 0$. Another condition for convergence which is established in that paper is $\mu_2(\mu_1 - \mu_2) > 4$. For odd $r$ we have $c_r = 0$, while for even $r$, the coefficient $c_r$ takes the form $\sum_{h=r/2}^{r-1} \frac{d_{hr}}{\mu_1^h}$, where each $d_{hr}$ is a function of

$$\alpha_{1u}, \ldots, \alpha_{mu}, \quad \mu_1 - \mu_2, \quad \mu_1 - \mu_3, \quad \ldots, \quad \mu_1 - \mu_m.$$

It remains to be seen whether the convergent series $\mu_1 + \sum\limits_{k=1}^{\infty} \sum\limits_{h=k}^{2k-1} \dfrac{d_{h,2k}}{\mu_1^h}$ may be rearranged to express $\mu_1(G_u)$ as a convergent power series of the form $\mu_1 + \sum\limits_{h=1}^{\infty} \dfrac{d_h'}{\mu_1^h}$, where each $d_h'$ is a function of $\alpha_{1u}, \ldots, \alpha_{mu}, \mu_1 - \mu_2, \ldots, \mu_1 - \mu_m$. The second approximation to $\mu_1(G_u)$ afforded by Theorem 6.3.4 is $\mu_1 + c_2 + c_4$, which turns out to be $\mu_1 + \dfrac{\alpha_{1u}^2}{\mu_1} + \dfrac{\alpha_{1u}^2 \tau_u}{\mu_1^2} - \dfrac{\alpha_{1u}^4}{\mu_1^3}$, where $\tau_u = \sum_{i=2}^{m}(\mu_1 - \mu_i)^{-1}\alpha_{iu}^2$. For the icosahedral graph of Example 6.2.5, a graph for which $\mu_2(\mu_1 - \mu_2) > 4$, this approximation is 5.01726, compared with $\mu_1(G_u) \approx 5.01728$.

## 6.4 An algebraic theory of perturbations

### 6.4.1 Introduction

The techniques of the previous section enable the indices of certain perturbed graphs to be computed to any degree of accuracy provided that (i) the spectrum and appropriate angles of the original graph are known, and (ii) $\mu_1 - \mu_2$ is large enough. If less information is given, typically the spectrum and relevant components $x_j (= \alpha_{1j})$ of the principal eigenvector **x**, then the methods of Section 6.3 can still be used to obtain bounds on the index of the perturbed graph. For example if $G'$ is obtained from $G$ by adding the edge $uv$ then (for large enough $\mu_1 - \mu_2$) the solution of equation (6.3.9) yields

$$\mu_1(G') < \mu_1 + \frac{1}{2e} - \frac{1}{2}\sqrt{\left(\frac{1}{2} - e\right)^2 - e^2 b} \qquad (6.4.1)$$

where $e = (\mu_1 - \mu_2)^{-1}$ and $b = \sqrt{x_u^2 + x_v^2}$. This however is a very crude bound and we can do better by invoking the algebraic theory of matrix perturbations exploited by Maas in [Maa2]. In this paper he shows that for $G' = G + uv$ as above, we have the following result.

**6.4.1 Theorem** *If $u, v$ are non-adjacent vertices of the connected graph $G$, then $\mu_1(G + uv) < \mu_1 + 1 + \delta - \gamma$ where $\delta > 0$ and*

$$\gamma = \frac{\delta(1 + \delta)(2 + \delta)}{(x_u + x_v)^2 + \delta(2 + \delta + 2x_u x_v)} = \mu_1 - \mu_2. \qquad (6.4.2)$$

(To see that Theorem 6.4.1 improves the inequality (6.4.1) for large enough $\mu_1 - \mu_2$, let $\mu_1 - \mu_2 \to \infty$.)

In Subsection 6.4.2 we outline the theory required to obtain Theorem

6.4.1 and similar results. Various applications and examples are given in Subsections 6.4.3 and 6.4.4. The techniques may be used to obtain an upper bound on any eigenvalue of a perturbed matrix $A + B$, not just the largest. We note here that analogous lower bounds for the index of a perturbed graph are obtained easily from the fact that $\mu_1(G') = \sup\{\mathbf{y}^T(A + B)\mathbf{y} : \|\mathbf{y}\| = 1\}$: on taking $\mathbf{y}$ to be the principal eigenvector of $G$, we obtain

$$\mu_1(G') \geq \mu_1 + \mathbf{x}^T B\mathbf{x}. \tag{6.4.3}$$

This lower bound is just $\mu_1 + c_1$ in the notation of Section 6.3. In the case that $G' = G + uv$, the inequality (6.4.3) becomes $\mu_1(G') \geq \mu_1 + 2x_u x_v$ (cf. Section 3.1). In Section 6.4 we shall compare the effects of two different perturbations on the index of a graph by showing that the lower limit for the index of one perturbed graph exceeds the upper limit for the index of the other.

### *6.4.2 Intermediate eigenvalue problems of the second type*

Here we outline the results needed from [WeSt] in terms of an $n$-dimensional Euclidean space $\mathscr{V}$ in which $(\mathbf{u}, \mathbf{v})$ denotes the inner product of vectors $\mathbf{u}$ and $\mathbf{v}$. The eigenvalues of a symmetric linear transformation $T$ of $\mathscr{V}$ are denoted by $\lambda_1(T), \ldots, \lambda_n(T)$ where now $\lambda_1(T) \leq \lambda_2(T) \leq \cdots \leq \lambda_n(T)$. Let $\tilde{A}$ be a symmetric linear transformation of $\mathscr{V}$ and let $\tilde{B}$ be a positive linear transformation of $\mathscr{V}$: the general problem is to find lower bounds for the eigenvalues $\lambda_i(\tilde{A} + \tilde{B})$, given $\tilde{B}$ and appropriate invariants of $\tilde{A}$.

We note first that another inner product may be defined on $\mathscr{V}$ by $[\mathbf{u}, \mathbf{v}] = (\tilde{B}\mathbf{u}, \mathbf{v})$. Now choose any basis $\{\mathbf{v}_1, \ldots, \mathbf{v}_n\}$ for $\mathscr{V}$ and, using this second inner product, let $Q_r$ be the orthogonal projection onto the subspace of $\mathscr{V}$ spanned by $\mathbf{v}_1, \ldots, \mathbf{v}_r$ $(r = 1, \ldots, n)$. Thus $Q_n = I$ and if we define $Q_0$ to be the zero transformation of $\mathscr{V}$ then $[Q_{r-1}\mathbf{v}, \mathbf{v}] \leq [Q_r\mathbf{v}, \mathbf{v}]$ for all $\mathbf{v} \in \mathscr{V}$ and all $r \in \{1, \ldots, n\}$. Hence $((\tilde{A} + \tilde{B}Q_{r-1})\mathbf{v}, \mathbf{v}) \leq ((\tilde{A} + \tilde{B}Q_r)\mathbf{v}, \mathbf{v})$ for all $\mathbf{v} \in \mathscr{V}$ and all $r \in \{1, \ldots, n\}$. It is straightforward to check that each $\tilde{B}P_r$ is a symmetric transformation of the original inner product space $\mathscr{V}$. Now for any symmetric transformation $T$ of $\mathscr{V}$, $\lambda_i(T)$ is the minimum of $\max\{(T\mathbf{v}, \mathbf{v}) : |\mathbf{v}| = 1, \mathbf{v} \in \mathscr{U}\}$ taken over all $i$-dimensional subspaces $\mathscr{U}$ of $\mathscr{V}$ (cf. equation (3.1.3)), and it follows that $\lambda_i(\tilde{A}) \leq \lambda_i(\tilde{A} + \tilde{B}Q_1) \leq \lambda_i(\tilde{A} + \tilde{B}Q_2) \leq \cdots \leq \lambda_i(\tilde{A} + \tilde{B}Q_{n-1}) \leq \lambda_i(\tilde{A} + \tilde{B})$ $(i = 1, \ldots, n)$.

The problem of determining the eigenvalues of $\tilde{A} + \tilde{B}Q_r$ for some integer $r \in \{1, \ldots, n - 1\}$ is called an intermediate eigenvalue problem of the second type. In practice we choose the basis $\{\mathbf{v}_1, \ldots, \mathbf{v}_n\}$ as follows, so that $\tilde{A} + \tilde{B}Q_r$ can be represented by a matrix of simple form. Let $\tilde{A}\mathbf{u}_i = \tilde{\lambda}_i\mathbf{u}_i$ $(i = 1, \ldots, n)$ where $\tilde{\lambda}_i = \lambda_i(\tilde{A})$ $(i = 1, \ldots, n)$ and $\mathbf{u}_1, \ldots, \mathbf{u}_n$ are orthonormal. If we now choose $\mathbf{v}_i = \tilde{B}^{-1}\mathbf{u}_i$ $(i = 1, \ldots, n)$ then $Q_r\mathbf{u}_j = \sum_{i=1}^{r} \gamma_{ij}\mathbf{v}_i$, where $(\gamma_{ij})^{-1}$ is the $r \times r$ Gram matrix $T_r$ with $(i, j)$-entry $[\mathbf{v}_i, \mathbf{v}_j]$ $(i, j = 1, \ldots, r)$; moreover the matrix of $\tilde{A} + \tilde{B}Q_r$ with respect to the basis $\{\mathbf{u}_1, \ldots, \mathbf{u}_r\}$ is

$$\begin{pmatrix} \tilde{\lambda}_1 & & & O \\ & \tilde{\lambda}_2 & & \\ & & \ddots & \\ O & & & \tilde{\lambda}_n \end{pmatrix} + \begin{pmatrix} T_r^{-1} & O \\ O & O \end{pmatrix}. \tag{6.4.4}$$

In our applications, we take $\mathscr{V} = \mathbb{R}^n, (\mathbf{u}, \mathbf{v}) = \mathbf{u}^T\mathbf{v}$, and we identify a linear transformation of $\mathbb{R}^n$ with its matrix with respect to the standard basis of $\mathbb{R}^n$. If, as usual, $A$ and $A + B$ are the adjacency matrices of a graph $G$ and its perturbation $G'$ then we take

$$\tilde{A} = -A - (\lambda_n(B) + \delta)I \quad \text{and} \quad \tilde{B} = (\lambda_n(B) + \delta)I - B, \text{ where } \delta > 0.$$

Thus $\tilde{B}$ is positive, $\tilde{A} + \tilde{B} = -A - B$ and $\mu_1(G') = -\lambda_1(\tilde{A} + \tilde{B}) \le -\lambda_1(\tilde{A} + \tilde{B}P_r)$. For a given value of $r$, the parameter $\delta$ is chosen to optimize this upper bound for $\mu_1(G')$. Results for various graph perturbations are given in the literature ([Maa2], [Row6], [Row8]), and in Subsection 6.4.3 we outline the calculations from [Row8] which deal with one of the more elaborate modifications. We shall then omit the calculations required for the generally simpler types of modification made to specific examples in Subsection 6.4.4.

### 6.4.3 A perturbation which preserves degrees

Let $G$ be a connected graph with vertices $t, u, v, w$ such that $t$ is adjacent to $w$ but not to $u$, and $v$ is adjacent to $u$ but not to $w$ (see Fig. 6.1). Here we discuss the perturbation $G'$ obtained from $G$ by replacing the edges $tw, uv$ with edges $tu, vw$, thereby preserving the degrees of all vertices of $G$. Thus $\mathbf{x}^T B\mathbf{x} = 2(x_t - x_v)(x_u - x_w)$ and it follows from the inequality (6.4.3) that if $(x_t - x_v)(x_u - x_w) > 0$ then $\mu_1(G') > \mu_1(G)$; and if $(x_t - x_v)(x_u - x_w) = 0$ then $\mu_1(G') \ge \mu_1(G)$. We use the results of Subsection 6.4.2 to see what can be said when $(x_t - x_v)(x_u - x_w) < 0$. Our complete result is the following.

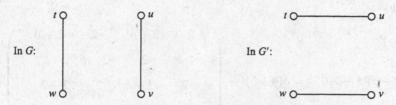

Fig. 6.1. A perturbation $G \to G'$ which preserves degrees

**6.4.2 Theorem** *Let $G'$ be the graph obtained from a connected graph $G$ by exchanging edges $tw, uv$ for non-edges $tu, vw$. If $(x_t - x_v)(x_u - x_w) > 0$ then $\mu_1(G') > \mu_1$. If $(x_t - x_v)(x_u - x_w) = 0$ then $\mu_1(G') \geq \mu_1$. If $(x_t - x_v)(x_u - x_w) < 0$ and $\mu_1 - \mu_2 > \dfrac{(x_t - x_v)^2 + (x_w - x_u)^2}{(x_t - x_v)(x_w - x_u)}$ then $\mu_1(G') < \mu_1$.*

*Proof* We take $t = 1, u = 2, v = 3, w = 4$ without loss of generality, so that

$$
B = \begin{pmatrix}
0 & 1 & 0 & -1 \\
1 & 0 & -1 & 0 & \vdots \\
0 & -1 & 0 & 1 \\
-1 & 0 & 1 & 0 \\
& & \cdots & & O
\end{pmatrix}.
$$

Now $U^T B U = \mathrm{diag}(2, -2, 0, 0, \ldots, 0)$, where

$$
U = \begin{pmatrix}
\frac{1}{2} & \frac{1}{2} & \frac{1}{2} & \frac{1}{2} \\
\frac{1}{2} & -\frac{1}{2} & \frac{1}{2} & -\frac{1}{2} & \vdots \\
-\frac{1}{2} & -\frac{1}{2} & \frac{1}{2} & \frac{1}{2} \\
-\frac{1}{2} & \frac{1}{2} & \frac{1}{2} & -\frac{1}{2} \\
& & \cdots & & I
\end{pmatrix}.
$$

Hence $\lambda_n(B) = 2, \tilde{B} = (\delta + 2)I - B$ and

$$
\tilde{B}^{-1} = (2 + \delta)^{-1} I + U \begin{pmatrix} \delta^{-1} - (2 + \delta)^{-1} \\ & O \end{pmatrix} U^T.
$$

Thus

$$\delta(2+\delta)(4+\delta)\tilde{B}^{-1} = \delta(4+\delta)I + \begin{pmatrix} 2 & 2+\delta & -2 & -2-\delta & \\ 2+\delta & 2 & -2-\delta & -2 & \vdots \\ -2 & -2-\delta & 2 & 2+\delta & \\ -2-\delta & -2 & 2+\delta & 2 & \\ & & \cdots & & 0 \end{pmatrix}.$$

We now apply the results of Subsection 6.4.2 with $r = 1$ and $\mathbf{u}_1 = \mathbf{x}$. Since $\tilde{\lambda}_i = \lambda_i(-A - (2+\delta)I)$ $(i = 1, 2, \ldots, n)$, we have $\tilde{\lambda}_1 = -\mu_1 - 2 - \delta$ and $\tilde{\lambda}_2 = -\mu_2 - 2 - \delta$. The matrix $\tilde{A} + \tilde{B}Q_1$ is similar to $\mathrm{diag}(\tilde{\lambda}_1 + \gamma, \tilde{\lambda}_2, \tilde{\lambda}_3, \ldots, \tilde{\lambda}_n)$, where $\gamma^{-1} = \gamma_{11}^{-1} = [\mathbf{v}_1, \mathbf{v}_1] = [\tilde{B}^{-1}\mathbf{u}_1, \tilde{B}^{-1}\mathbf{u}_1] = \mathbf{x}^T\tilde{B}^{-1}\mathbf{x}$. Having calculated $\tilde{B}^{-1}$ above, we know that

$$\gamma = \frac{\delta(2+\delta)(4+\delta)}{\delta(4+\delta) + 2(x_1 + x_2 - x_3 - x_4)^2 + 2\delta(x_1 - x_3)(x_2 - x_4)}.$$

Now

$$\lambda_1(\tilde{A} + \tilde{B}Q_1) = \begin{cases} \tilde{\lambda}_1 + \gamma & \text{if } \gamma \le \tilde{\lambda}_2 - \tilde{\lambda}_1, \\ \tilde{\lambda}_2 & \text{if } \gamma > \tilde{\lambda}_2 - \tilde{\lambda}_1, \end{cases}$$

and so

$$\mu_1(G') \le \begin{cases} \mu_1 + 2 + \delta - \gamma & \text{if } \gamma \le \mu_1 - \mu_2, \\ \mu_2 + 2 + \delta & \text{if } \gamma > \mu_1 - \mu_2. \end{cases}$$

As a function of $\delta$ $(\delta > 0)$, $\gamma$ has range $(0, \infty)$ and so we may choose $\delta > 0$ such that $\gamma = \mu_1 - \mu_2$. This determines our upper bound for $\mu_1(G')$, namely $\mu_2 + 2 + \delta$, and it remains to obtain conditions under which this upper bound is less than $\mu_1$. Let us write $\alpha$ for $(x_1 + x_2 - x_3 - x_4)^2$ and $\beta$ for $(x_1 - x_3)(x_4 - x_2)$: thus $\alpha \ge 0$ and $\beta > 0$. Now $\mu_2 + 2 + \delta < \mu_1$ if and only if $\gamma - (2 + \delta) > 0$; and $\gamma - (2 + \delta) = \dfrac{-2(\alpha - \delta\beta)(2 + \delta)}{\delta(4 + \delta) + 2(\alpha - \delta\beta)} = \dfrac{-2(\alpha - \delta\beta)\gamma}{\delta(4 + \delta)}$. If $\gamma > 2 + \alpha\beta^{-1}$ then we have

$$-2(\alpha - \delta\beta) = \gamma^{-1}\delta(4 + \delta)\{\gamma - (2 + \delta)\} > \gamma^{-1}\delta(4 + \delta)(\alpha\beta^{-1} - \delta)$$
$$= \gamma^{-1}\delta(4 + \delta)\beta^{-1}(\alpha - \delta\beta),$$

whence $(\alpha - \delta\beta)\{2 + \gamma^{-1}\delta(4 + \delta)\beta^{-1}\} < 0$. Hence if $\gamma > 2 + \alpha\beta^{-1}$, then $\alpha - \delta\beta < 0$ and so $\gamma - (2 + \delta) > 0$, $\mu_1(G') < \mu_1$. Since $2 + \alpha\beta^{-1} = \dfrac{(x_1 - x_3)^2 + (x_4 - x_2)^2}{(x_1 - x_3)(x_4 - x_2)}$, our theorem is proved. $\qquad\qquad\square$

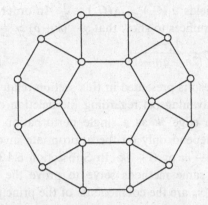

Fig. 6.2. The graph $G_6$ of Example 6.4.4.

### 6.4.4 Further applications and examples

The techniques of the previous subsection may be used to prove Theorem 6.4.1, which concerns the addition of an edge, and the following result (stated in [Row6]) which deals with the deletion of an edge.

**6.4.3 Theorem** *Let $G$ be a connected graph with principal eigenvector $(x_1,\ldots,x_n)^T$. If the graph $G'$ is obtained from $G$ by deleting the edge $hi$ then $\mu_1(G') < \mu_1 + 1 + \delta - \gamma$ where*

$$\gamma = \frac{\delta(1+\delta)(2+\delta)}{(x_h-x_i)^2 + \delta(2+\delta-2x_hx_i)} = \mu_1 - \mu_2.$$

Of course, in the situation of Theorem 6.4.3, we already know that $\mu_1(G') < \mu_1$ and so the result provides new information only when $\gamma > 1 + \delta$, equivalently when $2x_hx_i > (x_h - x_i)^2$. This condition is certainly satisfied in the following example, taken from [Row6], because the vertices $h$, $i$ in question lie in the same orbit of $\text{Aut}(G)$.

**6.4.4 Example** Let $G_n$ $(n \geq 3)$ be the $3n$-vertex graph consisting of a $2n$-cycle $C_{2n}$ with vertices $1, 2, \ldots, 2n$ (in order), an $n$-cycle $C_n$ with vertices $2n+1, \ldots, 3n$ (in order) and edges from $2n+i$ to $2i-1$ and $2i$ $(i = 1,\ldots,n)$. The graph $G_6$ is shown in Fig. 6.2.

The orbits of $\text{Aut}(G_n)$ are $\{1, 2, \ldots, 2n\}$ and $\{2n + 1, \ldots, 3n\}$, and we have $x_1 = \cdots = x_{2n} = 1/2\sqrt{n}$, $x_{2n+1} = \cdots = x_{3n} = 1/\sqrt{2n}$. If we remove an edge of $C_n$ we obtain a graph $G'_n$ such that $\mu_1(G'_n) < \mu_1(G_n)+1+\delta-\gamma$ where $\delta, \gamma$ are given by Theorem 6.4.3. If we remove an edge of $C_{2n}$ then there are two possibilities for the resulting graph $G''_n$: in either case, the inequality (6.4.3) yields $\mu_1(G_n'') > \mu_1(G_n) - \frac{1}{2n}$. In order to conclude that $\mu_1(G_n'') > \mu_1(G_n')$ it suffices to show that $\gamma - (1 + \delta) > \frac{1}{2n}$; but this is clear since $\gamma - (1 + \delta) = \dfrac{1 + \delta}{(2n + \delta n - 1)}$. $\qquad\square$

The remaining results presented in this section relate to the relocation of an edge. The advantage in regarding the deletion of an edge $hi$ and the addition of an edge $jk$ as a single perturbation of $G$ is that the bounds obtained depend only on the appropriate invariants of $G$, and not on those of $G - hi$ or $G + jk$. In Subsection 6.4.3, *two* edges were relocated and the same methods serve to prove the next two results, where again $x_1, \ldots, x_n$ are the components of the principal eigenvector of the connected graph $G$.

**6.4.5 Theorem** [Row6] *Let $G$ be a connected graph with distinct vertices $h, i, j, k$ such that $h$ is adjacent to $i$, $j$ is not adjacent to $k$. Let $G'$ be the graph obtained from $G$ by replacing edge $hi$ with $jk$. If $x_j x_k \geq x_h x_i$ then $\mu_1(G') > \mu_1$. If $x_j x_k < x_h x_i$ and $\mu_1 - \mu_2 > \dfrac{x_h^2 + x_i^2 + x_j^2 + x_k^2}{2(x_h x_i - x_j x_k)}$ then $\mu_1(G') < \mu_1$.*

One detail which remains to be checked here is the case $x_j x_k = x_i x_h$, where inequality (6.4.3) yields only $\mu_1(G') \geq \mu_1(G)$. If however $\mu_1(G') = \mu_1(G)$ in this situation, we have $A'\mathbf{x} = \mu_1\mathbf{x} = A\mathbf{x}$ where $A'$ is the adjacency matrix of $G'$; but a comparison of the $h$-th components of $A'\mathbf{x}$ and $A\mathbf{x}$ shows that this is impossible.

We note in passing that if $x_j x_k < x_h x_i$ then the index may increase or decrease when $hi$ is replaced by $jk$, as illustrated by the following example.

**6.4.6 Example** Let $G$ be the labelled graph shown in Fig. 6.3. We have $\mu_1(G) = 3.9781$, $x_2 x_6 < x_1 x_3$ and $x_2 x_7 < x_1 x_3$. If edge 13 is replaced by 26 the index increases to 4.0465, while if edge 13 is replaced by 27 the index decreases to 3.8791. $\qquad\square$

The last result deals with the situation in which the 'old' edge and the 'new' edge have a vertex in common.

**6.4.7 Theorem** [Row6] *Let $G$ be a connected graph with distinct vertices $h, i, k$ such that $h$ is adjacent to $i$ but not to $k$. Let $G'$ be the graph obtained from $G$ by replacing edge $hi$ with $hk$. If $x_k \geq x_i$ then $\mu_1(G') > \mu_1$. If $x_k < x_i$ and $\mu_1 - \mu_2 > \dfrac{2x_h^2 + (x_i - x_k)^2}{2x_h(x_i - x_k)}$ then $\mu_1(G') < \mu_1$.*

We illustrate Theorems 6.4.5 and 6.4.7 with an example from [Row6].

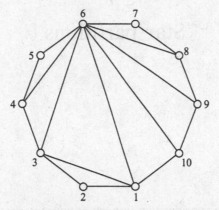

Fig. 6.3. The graph $G$ of Example 6.4.6.

**6.4.8 Example** Let $G$ be the wheel $W_{n+1}$ $(n \geq 4)$ with vertex 0 adjacent to each vertex of the cycle $123...n1$. The eigenvalues of $G$ are $1 \pm \sqrt{n+1}$ and $2\cos\frac{2\pi i}{n}$ $(i = 1,...,n-1)$ (see Theorem 2.8 of [CvDS]). Thus $\mu_1 = 1 + \sqrt{n+1}$ and $\mu_1 - \mu_2 > \mu_1 - 2$. The principal eigenvector of $G$ is $(\alpha, \beta, ..., \beta)^T$ where $n\beta = \mu_1\alpha$ and $\alpha^2 + n\beta^2 = 1$. Let $G'$ be the graph obtained from $G$ by replacing the spoke 02 with the chord 13. To apply Theorem 6.4.5 with $h = 0, i = 2, j = 1, k = 3$ we require $\mu_1 - \mu_2 > \dfrac{\alpha^2 + 3\beta^2}{2(\alpha\beta - \beta^2)}$. It is sufficient to show that $\mu_1 - 2 > \dfrac{n^2 + 3\mu_1^2}{2\mu_1(n - \mu_1)}$; but this holds for large enough $n$ (certainly for $n \geq 15$) because $\dfrac{n^2 + 3\mu_1^2}{2\mu_1(n - \mu_1)(\mu_1 - 2)} \to \frac{1}{2}$ as $n \to \infty$. Then $\mu_1(G') < \mu_1(G)$ by Theorem 6.4.5. Now let $G''$ be the graph obtained from $G$ by replacing the spoke 02 with the chord 24. To apply Theorem 6.4.7 with $h = 2, i = 0, k = 4$ we require $\mu_1 - \mu_2 > \dfrac{2\beta^2 + (\alpha - \beta)^2}{2\beta(\alpha - \beta)}$, a condition which holds for $n \geq 15$ as a consequence of the corresponding inequality for $G'$. Thus $\mu_1(G'') < \mu_1(G)$.

In this example we have $x_1 = x_2 = \cdots = x_n$. Accordingly the arguments show that if the spoke of a (large enough) wheel is replaced by any chord then the index decreases. On the other hand (applying the first parts of Theorems 6.4.5 and 6.4.7) if any other edge is replaced by a chord then the index increases. $\qquad\square$

# 7
# Star partitions

In this chapter we consider two important concepts: *star bases* and their combinatorial counterpart *star partitions*. These concepts were introduced recently in [CvRS1] as a means of extending spectral methods in graph theory, and they provide a strong link between graphs and linear algebra. This connection is promising in that it not only reflects the geometry of eigenspaces but also extends to combinatorial aspects such as matching theory. Star bases were originally introduced as a means of investigating the complexity of the graph isomorphism problem (Chapter 8), but it turned out that the direct relation between graph structure and the underlying star partitions could be exploited to advantage. In particular, there are connections with dominating sets and implications for cubic graphs, and these are two of the topics discussed here.

## 7.1 Introduction

A graph is determined by its eigenvalues and eigenspaces, but not in general by its eigenvalues and angles. In seeking further algebraic invariants we may look to bases of eigenspaces, but of course for eigenspaces of dimension greater than 1 there is not a natural choice of basis. We can however focus our attention on *star bases*, which as we shall see are related to the geometry of finite-dimensional Euclidean spaces. The key notion which underlines star bases, and which is of wider interest as well, is that of a *star partition*. In this section we introduce both star partitions and star bases, and prove that they exist for any graph (indeed, for any real symmetric matrix).

Recall that a set of vectors in an inner product space is *spherical* if all of its vectors have the same length. Following [Sei3] we have

150

**7.1.1 Definition** *Let $\mathcal{U}$ be a non-trivial subspace of the finite-dimensional inner product space $\mathcal{V}$. A eutactic star in $\mathcal{U}$ is the orthogonal projection onto $\mathcal{U}$ of an orthogonal spherical basis of $\mathcal{V}$.*

The word 'eutactic' means 'well-situated', and the vectors (or *arms*) of a eutactic star are well-situated in that, to within a common scalar multiple, they are the orthogonal projection of an orthonormal basis.

Now let $\mathcal{V} = \mathbb{R}^n$ with $(\mathbf{x}, \mathbf{y}) = \mathbf{x}^T \mathbf{y}$ (also written $\mathbf{x} \cdot \mathbf{y}$) as the inner product of two (column) vectors $\mathbf{x}, \mathbf{y}$; and let $\{\mathbf{e}_1, \dots, \mathbf{e}_n\}$ be the standard (orthonormal) basis of $\mathbb{R}^n$. If $A$ is an $n \times n$ symmetric matrix with real entries and spectral decomposition $A = \mu_1 P_1 + \cdots + \mu_m P_m$ (see Section 1.1), then for each $i = 1, \dots, m$, the vectors $P_i \mathbf{e}_1, \dots, P_i \mathbf{e}_n$ are the arms of a eutactic star $\mathcal{S}_i$ which spans the eigenspace $\mathcal{E}(\mu_i)$. Recall that the length of the arm $P_i \mathbf{e}_j$ is just the angle $\alpha_{ij}$ defined in Chapter 4.

In the case that $A$ is the adjacency matrix of a graph $G$ with vertex set $V(G) = \{1, 2, \dots, n\}$ we have ([CvRS1])

**7.1.2 Definition** *A partition $X_1 \dot\cup \cdots \dot\cup X_m$ of $V(G)$ is a star partition if the vectors $P_i \mathbf{e}_j$ ($j \in X_i$) form a basis for $\mathcal{E}(\mu_i)$, for each $i = 1, \dots, m$.*

A simple counting argument shows that each of the following provides an equivalent definition:

(i) *the set of vectors $P_i \mathbf{e}_j$ ($j \in X_i$) is linearly independent for each $i = 1, \dots, m$, or*

(ii) *the set of vectors $P_i \mathbf{e}_j$ ($j \in X_i$) spans $\mathcal{E}(\mu_i)$ for each $i = 1, \dots, m$.*

Note that the size of a *star cell* $X_i$ is the multiplicity of $\mu_i$ as an eigenvalue of $G$, here denoted by $k_i$. Any partition $Y_1 \dot\cup \cdots \dot\cup Y_m$ with $|Y_i| = k_i$ ($i = 1, \dots, m$) will be called a *feasible partition*. We assume as usual that $\mu_1 > \cdots > \mu_m$, and so we regard star partitions and feasible partitions as ordered partitions. Since $X_i$ is always associated with $\mu_i$ we say that $X_i$ is a $\mu_i$-cell and we sometimes write $X(\mu_i)$ instead of $X_i$.

The notion of a star partition extends to any real symmetric matrix $A$: in this case the partitioned set $\{1, 2, \dots, n\}$ is regarded as the index set of the columns of $A$. We now prove that star partitions always exist [CvRS1]. The proof reproduced here follows the one from [Row11]. Another proof is given in Subsection 7.8.1.

**7.1.3 Theorem** *For any real symmetric matrix $A$, there exists at least one star partition.*

*Proof* Let $\{\mathbf{x}_1, \dots, \mathbf{x}_n\}$ be a basis of $\mathbb{R}^n$ obtained by stringing together arbitrary fixed bases of $\mathscr{E}(\mu_1),\ldots,\mathscr{E}(\mu_m)$; say $\mathscr{E}(\mu_i)$ has basis $\{\mathbf{x}_h \ : \ h \in R_i\}$, where $R_1 \dot{\cup} \cdots \dot{\cup} R_m$ is a fixed partition of $\{1,\ldots,n\}$. Let

$$\mathbf{e}_j = \sum_{h=1}^{n} t_{hj}\mathbf{x}_h \ (j = 1,\ldots,n), \tag{7.1.1}$$

where $T = (t_{hj})$ is the transition matrix from the basis $\{\mathbf{x}_1,\ldots,\mathbf{x}_n\}$ to the basis $\{\mathbf{e}_1,\ldots,\mathbf{e}_n\}$. Then, on projecting orthogonally into $\mathscr{E}(\mu_i)$, we have

$$P_i\mathbf{e}_j = \sum_{h \in R_i} t_{hj}\mathbf{x}_h. \tag{7.1.2}$$

Let $C_1 \dot{\cup} \cdots \dot{\cup} C_m$ be any feasible partition of $\{1,\ldots,n\}$, and let $T_i$ be the $k_i \times k_i$ submatrix of $T$ whose rows are indexed by $R_i$ and whose columns are indexed by $C_i$. By considering the corresponding multiple Laplacian development of $T$, it follows that $\det T = \sum \{\pm \prod_{i=1}^{m} \det T_i\}$ where the sum is taken over all $\frac{n!}{k_1! \cdots k_m!}$ feasible partitions $C_1 \dot{\cup} \cdots \dot{\cup} C_m$. Since $T$ is invertible, some term $\prod_{i=1}^{m} \det T_i$ is non-zero, say that determined by the partition $X_1 \dot{\cup} \cdots \dot{\cup} X_m$. This partition is a star partition because in view of (7.1.1), the invertibility of $T_i$ guarantees that each $\mathbf{x}_h$ ($h \in R_i$) is a linear combination of the vectors $P_i\mathbf{e}_j$ ($j \in X_i$).                        □

**7.1.4 Remark** Let $\{\mathbf{x}_1,\ldots,\mathbf{x}_n\}$ be an orthonormal basis of $\mathbb{R}^n$ chosen so that $\{\mathbf{x}_h \ : \ h \in X_i\}$ is a basis of $\mathscr{E}(\mu_i)$ $(i = 1,\ldots,m)$. Then the orthogonal matrix $E$ whose columns are $\mathbf{x}_1,\ldots,\mathbf{x}_n$ is the inverse of the transition matrix $T = (t_{hj})$ defined by (7.1.1). Moreover, if $E_{ii}$ denotes the principal submatrix of $E$ determined by $X_i$, then $E_{ii}$ is the transition matrix from the basis $\{P_i\mathbf{e}_j \ : \ j \in X_i\}$ of $\mathscr{E}(\mu_i)$ to the basis $\{\mathbf{x}_j \ : \ j \in X_i\}$ $(i = 1,\ldots,m)$. □

The existence of star partitions enables us to define star bases.

**7.1.5 Definition** *If $X_1 \dot{\cup} \cdots \dot{\cup} X_m$ is a star partition for a graph (or real symmetric matrix) then $\{P_i\mathbf{e}_j : j \in X_i\}$ is called a star basis of $\mathscr{E}(\mu_i)$ and $\bigcup_{i=1}^{m}\{P_i\mathbf{e}_j \ : \ j \in X_i\}$ is called a star basis of $\mathbb{R}^n$.*

More details on star bases (including algorithmic aspects) are given in Chapter 8, where *canonical star bases* are introduced. The motivation for investigating star bases is the graph isomorphism problem, where we seek a complete set of spectral invariants, but in the remaining sections of the present chapter we shall focus our attention on star partitions.

## 7.2 Characterizations of star partitions

In this section we characterize star partitions of graphs in terms of (i) decompositions of $\mathbb{R}^n$, (ii) eigenvalues of induced subgraphs, (iii) entries of eigenvectors. An induced subgraph of the graph $G$ may be obtained from $G$ by the successive deletion of vertices. Note that if we delete a single vertex $v$ from $G$ then the multiplicity of $\mu_i$ as an eigenvalue of $G - v$ is at least $k_i - 1$. To see this, observe that $\dim(\mathscr{E}(\mu_i) \cap \mathscr{U}) \geq k_i - 1$ where $\mathscr{U} = \{(x_1, x_2, \ldots, x_n)^T \in \mathbb{R}^n : x_v = 0\}$. It follows that if $S$ is a subset of vertices such that $\mu_i$ is not an eigenvalue of $G - S$ then $|S| \geq k_i$. We shall see that always there exists a $k_i$-element set $X_i$ such that $\mu_i$ is not an eigenvalue of $G - X_i$, equivalently $\mu_i$ is not a root of the characteristic polynomial of $G - X_i$. These observations motivate the following definition (see also Sections 7.4 and 7.8).

**7.2.1 Definition** *A partition* $X_1 \dot{\cup} \cdots \dot{\cup} X_m$ *of* $V(G)$ *is a polynomial partition if for each* $i \in \{1, \ldots, m\}$, *$\mu_i$ is not an eigenvalue of* $G - X_i$.

We shall see that the polynomial partitions of $G$ are precisely the star partitions. First we establish a characterization of star partitions in terms of direct sum decompositions of $\mathbb{R}^n$.

**7.2.2 Theorem** [CvRS1] *For any real symmetric matrix $A$ of size $n \times n$,* $X_1 \dot{\cup} \cdots \dot{\cup} X_m$ *is a star partition if and only if, for each* $i = 1, \ldots, m$,

$$\mathbb{R}^n = \mathscr{E}(\mu_i) \oplus \mathscr{V}_i,$$

*where* $\mathscr{V}_i = \langle \mathbf{e}_j : j \notin X_i \rangle$.

*Proof* We first prove necessity. Since $\dim \mathscr{E}(\mu_i) = k_i$ and $\dim \mathscr{V}_i = n - k_i$, it suffices to show that $\mathscr{E}(\mu_i) \cap \mathscr{V}_i = \{\mathbf{0}\}$. Let $\mathbf{x} \in \mathscr{E}(\mu_i) \cap \mathscr{V}_i$. Then $\mathbf{x} = P_i \mathbf{x}$ and $\mathbf{x}^T \mathbf{e}_j = 0$ for all $j \in X_i$. Hence, $\mathbf{x}^T (P_i \mathbf{e}_j) = \mathbf{x}^T (P_i^T \mathbf{e}_j) = (P_i \mathbf{x})^T \mathbf{e}_j = 0$ for all $j \in X_i$. Thus $\mathbf{x} \in \langle P_i \mathbf{e}_j : j \in X_i \rangle^\perp = \mathscr{E}(\mu_i)^\perp$ and $\mathbf{x} = \mathbf{0}$.

Conversely, if $\mathbb{R}^n = \mathscr{E}(\mu_i) \oplus \mathscr{V}_i$, then $\mathbb{R}^n = \mathscr{E}(\mu_i)^\perp \oplus \mathscr{V}_i^\perp$ and so $\mathscr{E}(\mu_i) = P_i(\mathbb{R}^n) = P_i(\mathscr{V}_i^\perp) = \langle P_i \mathbf{e}_j : j \in X_i \rangle$. Hence $X_1 \dot{\cup} \cdots \dot{\cup} X_m$ is a star partition if $\mathbb{R}^n = \mathscr{E}(\mu_i) \oplus \mathscr{V}_i$ for each $i$. $\square$

We can now prove that star partitions are the same as polynomial partitions.

**7.2.3 Theorem** [CvRS1] *$X_1 \dot{\cup} \cdots \dot{\cup} X_m$ is a star partition of a graph $G$ if and only if, for each* $i = 1, \ldots, m$, *$\mu_i$ is not an eigenvalue of* $G - X_i$.

*Proof* Suppose first that $X_1 \dot{\cup} \cdots \dot{\cup} X_m$ is a star partition. Without loss of generality take $i = 1$. Let $H = G - X_1$, and let $A'$ be the adjacency matrix of $H$. Suppose that $A'\mathbf{x}' = \mu_1\mathbf{x}'$. If $\mathbf{y} = \begin{pmatrix} 0 \\ \mathbf{x}' \end{pmatrix}$, then

$$A\mathbf{y} = \begin{pmatrix} * & * \\ * & A' \end{pmatrix} \begin{pmatrix} 0 \\ \mathbf{x}' \end{pmatrix} = \begin{pmatrix} * \\ \mu_i\mathbf{x}' \end{pmatrix}.$$

Now let $\mathbf{x} \in \mathscr{V}_1 (= \langle e_j : j \notin X_1 \rangle)$. Then

$$\mathbf{x}^T A\mathbf{y} = \begin{pmatrix} 0 \\ \mathbf{z} \end{pmatrix}^T \begin{pmatrix} * \\ \mu_i\mathbf{x}' \end{pmatrix} = \mu_1\mathbf{z}^T\mathbf{x}' = \mu_1\mathbf{x}^T\mathbf{y}$$

and so $(A - \mu_1 I)\mathbf{y} \in \mathscr{V}_1^\perp$. On the other hand, if $\mathbf{x} \in \mathscr{E}(\mu_1)$, then $\mathbf{x}^T A\mathbf{y} = \mathbf{x}^T A^T\mathbf{y} = (A\mathbf{x})^T\mathbf{y} = (\mu_1\mathbf{x})^T\mathbf{y} = \mu_1\mathbf{x}^T\mathbf{y}$ and so $(A - \mu_1 I)\mathbf{y} \in \mathscr{E}(\mu_1)^\perp$. Hence $(A - \mu_1 I)\mathbf{y} \in \mathscr{V}_1^\perp \cap \mathscr{E}(\mu_1)^\perp = (\mathscr{E}(\mu_1) + \mathscr{V}_1)^\perp$, which is trivial by Theorem 7.2.2. Hence $\mathbf{y} \in \mathscr{E}(\mu_1)$. But $\mathbf{y} \in \mathscr{V}_1$ and since $\mathscr{E}(\mu_1) \cap \mathscr{V}_1 = \{\mathbf{0}\}$ by Theorem 7.2.2 we have $\mathbf{y} = \mathbf{0}$, and hence $\mathbf{x}' = \mathbf{0}$. Thus $\mu_1$ is not an eigenvalue of $G - X_1$.

To prove the converse, it suffices to prove that $\langle P_i e_j : j \in X_i \rangle = \mathscr{E}(\mu_i)$ for each $i \in \{1, \ldots, m\}$. Again, without loss of generality, we may assume that $i = 1$. Suppose, by way of contradiction, that $\langle P_1 e_j : j \in X_1 \rangle \subset \mathscr{E}(\mu_1)$. Then there is a non-zero vector $\mathbf{x} \in \mathscr{E}(\mu_1) \cap \langle P_1 e_j : j \in X_1 \rangle^\perp$. Thus $\mathbf{x}^T P_1 e_j = 0$ for all $j \in X_1$. Hence $(P_1\mathbf{x})^T e_j = (\mathbf{x}^T P_1)e_j = 0$ for all $j \in X_1$. Consequently $P_1\mathbf{x} \in \langle e_j : j \in X_1 \rangle^\perp = \langle e_s : s \notin X_1 \rangle = \mathscr{V}_1$. But $\mathbf{x} = P_1\mathbf{x}$ and so we have non-zero $\mathbf{x} \in \mathscr{E}(\mu_1) \cap \mathscr{V}_1$. Since $\mathbf{x} = \begin{pmatrix} 0 \\ \mathbf{x}' \end{pmatrix}$ with $\mathbf{x}' \neq \mathbf{0}$ it follows that $\mathbf{x}'$ is an eigenvector of $G - X_1$, a contradiction. $\square$

**7.2.4 Corollary** [CvRS1] *If $S \subseteq X_i$, then $\mu_i$ is an eigenvalue of $G - S$ with multiplicity $k_i - |S|$.*

*Proof* Removal of a vertex cannot reduce the multiplicity of $\mu_i$ by more than 1; but after the $k_i$ vertices of $X_i$ are removed, $\mu_i$ has a multiplicity 0. Hence the multiplicity of $\mu_i$ is reduced by 1 on removal of any vertex of $X_i$ (in any order). $\square$

A star partition provides a natural way (in general non-unique) of ascribing the eigenvalues of a graph to its vertices. Thus for each of the star partitions illustrated in Fig. 7.1 the vertices of $X_i$ are labelled $\mu_i$. By Corollary 7.2.4, the removal of any $r$ vertices labelled $\mu_i$ $(0 < r < k_i)$ results in a graph which has $\mu_i$ as an eigenvalue of multiplicity $k_i - r$.

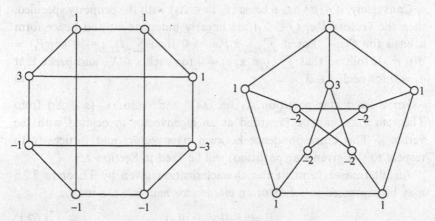

Fig. 7.1. Two star partitions.

Here, as in the following example, Theorem 7.2.3 provides a means of checking whether or not a given feasible partition is a star partition.

**7.2.5 Example** Consider a graph $G = P_n$ (a path on $n$ vertices) whose vertices are labelled in a natural way from 1 to $n$. For $n = 4$, one can easily check that any feasible partition (of $P_4$) is a star partition. On the other hand, if $n = 5$ then the vertex labelled by 2 (or 4) cannot constitute a star cell for the eigenvalue 0. (Recall that all eigenvalues of $P_n$ are distinct, and 0 appears as an eigenvalue if and only if $n$ is odd.) For a more general analysis see Proposition 7.7.3. □

We shall now establish another characterization of star partitions, this time in terms of eigenvectors. The vectors involved are computed explicitly in Section 7.8, where the result is viewed from a different perspective.

**7.2.6 Theorem** [Row11] $X_1 \dot\cup \cdots \dot\cup X_m$ *is a star partition if and only if, for each* $i = 1, \ldots, m$, $\mathscr{E}(\mu_i)$ *has a basis of eigenvectors* $\{\mathbf{x}_s : s \in X_i\}$ *such that* $\mathbf{x}_s^T \mathbf{e}_t = \delta_{st}$ *whenever* $s, t \in X_i$.

*Proof* Assume that $X_1 \dot\cup \cdots \dot\cup X_m$ is a star partition and let $\mathscr{V}_i = \langle \mathbf{e}_j : j \notin X_i \rangle$. If $Q_i$ represents the orthogonal projection of $\mathbb{R}^n$ onto $\mathscr{V}_i^\perp$ then (by Theorem 7.2.2) we have

$$\langle \mathbf{e}_j : j \in X_i \rangle = \mathscr{V}_i^\perp = Q_i(\mathbb{R}^n) = Q_i(\mathscr{E}(\mu_i) \oplus \mathscr{V}_i) = Q_i(\mathscr{E}(\mu_i)).$$

Hence for each $s \in X_i$ there exists a vector $\mathbf{x}_s \in \mathscr{E}(\mu_i)$ such that $Q_i(\mathbf{x}_s) = \mathbf{e}_s$.

Conversely, if $\mathscr{E}(\mu_i)$ has a basis $\{\mathbf{x}_s : s \in X_i\}$ with the property specified, then the vectors $P_i \mathbf{e}_j$ $(j \in X_i)$ are linearly independent and hence form a basis for $\mathscr{E}(\mu_i)$. For if $\sum_{j \in X_i} \alpha_j P_i \mathbf{e}_j = \mathbf{0}$ then $\sum_{j \in X_i} \alpha_j \mathbf{e}_j \in \ker(P_i) = \mathscr{E}(\mu_i)^{\perp}$; it follows that $\sum_{j \in X_i} \alpha_j \mathbf{x}_s^T \mathbf{e}_j = \mathbf{0}$ for each $s \in X_i$, and hence that $\alpha_j = 0$ for each $j \in X_i$. $\qquad\square$

For a fixed star partition $X_1 \dot\cup \cdots \dot\cup X_m$, the vector $\mathbf{x}_s$ $(s \in X_i)$ from Theorem 7.2.6 can be regarded as an eigenvector associated with the vertex $s$. This correspondence between eigenvectors and vertices (with respect to the given star partition) will be used in Section 7.5.

An alternative form of the characterization given by Theorem 7.2.6 may be expressed in terms of an *eigenvector matrix*, i.e. a matrix

$$E = (E_1 | E_2 | \cdots | E_m), \qquad (7.2.1)$$

where for each $i = 1, \ldots, m$, $E_i$ is an $n \times k_i$ matrix whose columns form a basis of $\mathscr{E}(\mu_i)$.

Note that an eigenvector matrix $E$ is just the transition matrix from the standard basis of $\mathbb{R}^n$ to a basis of eigenvectors, while the transition matrix $T$ used in the proof of Theorem 7.1.3 is the inverse of such a matrix.

**7.2.7 Corollary** $X_1 \dot\cup \cdots \dot\cup X_m$ *is a star partition if and only if an eigenvector matrix, after an appropriate labelling of rows (i.e. an appropriate ordering of vertices of G) has the for m*

$$
\begin{array}{cccc}
\mathscr{E}(\mu_1) & \mathscr{E}(\mu_2) & \cdots & \mathscr{E}(\mu_m)
\end{array}
$$

$$
E = \begin{pmatrix}
E_{11} & * & \cdots & * \\
* & E_{22} & \cdots & * \\
\vdots & \vdots & \ddots & \vdots \\
* & * & \cdots & E_{mm}
\end{pmatrix}
\begin{array}{c}
X_1 \\
X_2 \\
\vdots \\
X_m
\end{array}
\qquad (7.2.2)
$$

*where, for each $i = 1, \ldots m$, $E_{ii}$ is an invertible matrix and $*$ denotes a block matrix of an appropriate size.*

*Proof* If each block $E_{ii}$ is a unit matrix of appropriate size the conclusion follows directly from Theorem 7.2.6. Otherwise the result follows by making use of elementary column transformations. $\qquad\square$

The next theorem summarizes all the characterizations obtained so far.

**7.2.8 Theorem** *Each of the following assertions is a necessary and sufficient* *condition for the partition* $X_1 \dot{\cup} \cdots \dot{\cup} X_m$ *of* $V(G)$ *to be a star partition of* $G$:

*(1) for each $i$, $\{P_i e_j : j \in X_i\}$ is a basis of $\mathcal{E}(\mu_i)$;*

*(2) for each $i$, $\mathbb{R}^n = \mathcal{E}(\mu_i) \oplus \mathcal{V}_i$, where $\mathcal{V}_i = \langle e_j : j \notin X_i \rangle$;*

*(3) for each $i$, $\mu_i$ is not an eigenvalue of $G - X_i$;*

*(4) for each $i$, $\mathcal{E}(\mu_i)$ has a basis of eigenvectors $\{x_s : s \in X_i\}$ such that $x_s^T e_t = \delta_{st}$ whenever $s, t \in X_i$;*

*(5) an eigenvector matrix, after an appropriate choice of eigenvectors, and an appropriate ordering of vertices of $G$, has the form (7.2.2), where $E_{ii}$ is an invertible matrix of order $k_i$ $(i = 1, \ldots, m)$.*

Finally, we remark that the proofs of Theorems 7.2.2, 7.2.3 and 7.2.6 establish the following result.

**7.2.9 Theorem** *For an eigenvalue $\mu_i$ of $G$ and a subset $X$ of $V(G)$ the following are equivalent:*

*(1) $\{P_i e_j : j \in X\}$ is a basis of $\mathcal{E}(\mu_i)$;*

*(2) $\mathbb{R}^n = \mathcal{E}(\mu_i) \oplus \mathcal{V}$ where $\mathcal{V} = \langle e_j : j \notin X \rangle$;*

*(3) $|X| = \dim \mathcal{E}(\mu_i)$ and $\mu_i$ is not an eigenvalue of $G - X$;*

*(4) $\mathcal{E}(\mu_i)$ has a basis of eigenvectors $x_s$ $(s \in X)$ such that $x_s^T e_t = \delta_{st}$ whenever $s, t \in X$.*

## 7.3 Structural considerations

In this section we explore the relation between star partitions and graph structure, and the relation between a star cell $X_i$ and the spectrum of the subgraph induced by $X_i$.

Our first observation (see [CvRS1]) provides a crucial restriction on star cells in terms of forbidden subsets.

**7.3.1 Theorem** *Let $X_1 \dot{\cup} \cdots \dot{\cup} X_m$ be a star partition of a graph $G$ and let $s \in X_i$. Then either $s$ is adjacent to some vertex $t \notin X_i$, or $s$ is isolated and moreover $\mu_i = 0$.*

*Proof* Since $\mu_i P_i e_s = A P_i e_s = P_i A e_s = P_i(\sum_{t \sim s} e_t) = \sum_{t \sim s} P_i e_s$, we have: if $s$ is not isolated, the vectors $\{P_i e_s\} \cup \{P_i e_t : t \sim s\}$ are linearly dependent and so not every $t$ adjacent to $s$ can be in $X_i$. On the other hand, if $s$ is isolated then $\mu_i = 0$ since $P_i e_s \neq 0$. $\qquad \square$

**7.3.2 Remark** It follows from the proof of Theorem 7.3.1 that if $\mu_i \neq 0$ then there exists a vertex $t \notin X_i$ such that $P_i e_t \neq 0$. $\qquad \square$

We now know that with the exception of isolated vertices, a vertex and its neighbourhood cannot be contained in a single star cell; in other words, if $s \in X_i$, then $\Delta^*(s)$ is forbidden as a subset of $X_i$ – here, in contrast to $\Delta(v)$, which denotes the *open neighbourhood* of $v$, $\Delta^*(v)$ denotes the *closed neighbourhood* of $v$; thus $\Delta^*(v) = \Delta(v) \cup \{v\}$.

For some graphs, this constraint alone is sufficient to determine the form of star partitions; for example, if $G = K_{m,n}$ $(m, n \geq 3)$, then a star partition necessarily has the form $\{u\} \, \dot\cup \, (V(G) \backslash \{u, v\}) \, \dot\cup \, \{v\}$, where $u \sim v$. Notice that the restriction here on $m$ and $n$ may be dropped by making use of more powerful tools such as Theorem 7.2.3.

**7.3.3 Remark** The argument of Theorem 7.3.1 extends to any polynomial $f$ in $A$, since $f(\mu_i)P_i\mathbf{e}_s = P_i f(A)\mathbf{e}_s$. In particular, if $f(A) = A^2$ and $s$ is non-isolated then we find that the vertices in $W_2(s)$ (those reachable from $s$ by walks of length 2) form a forbidden set unless $W_2(s) = \{s\}$ and $\mu_i^2 = \deg(s)$. In the exceptional case, if $G$ is connected then $G$ is a star $K_{1,m}$ with centre $s$ and $\mu_i = \pm\sqrt{m}$.                              □

**7.3.4 Remark** Another consequence of Theorem 7.3.1 concerns cubic graphs: here the subgraph induced by any star cell has maximal degree at most 2, hence has known structure, namely a union of paths and cycles. See Section 7.7 for a description of all the subgraphs which can arise in this way from a star partition of the Petersen graph.          □

In what follows we make use of the following notation. For $s \in X_i$, let $\Gamma(s) = \Delta(s) \cap X_i$ and $\overline{\Gamma}(s) = \Delta(s) \cap \overline{X}_i$ (where $\overline{X}_i = V(G) \backslash X_i$). The next theorem excludes another configuration when $\mu_i \notin \{-1, 0\}$.

**7.3.5 Theorem** [CvRS1] *Let $s, t$ be distinct vertices in $X_i$. If $\overline{\Gamma}(s) = \overline{\Gamma}(t)$ then either*

*(a)* $s \sim t$, $\mu_i = -1$ *and* $\Delta^*(s) = \Delta^*(t)$, *or*
*(b)* $s \not\sim t$, $\mu_i = 0$ *and* $\Delta(s) = \Delta(t)$.

*Proof* We have $\mu_i P_i \mathbf{e}_s = \sum_{j \in \Gamma(s)} P_i \mathbf{e}_j + \sum_{j \in \overline{\Gamma}(s)} P_i \mathbf{e}_j$ together with a similar expression for $\mu_i P_i \mathbf{e}_t$. Since $\overline{\Gamma}(s) = \overline{\Gamma}(t)$ it follows that $\mu_i P_i \mathbf{e}_s - \sum_{j \in \Gamma(s)} P_i \mathbf{e}_j = \mu_i P_i \mathbf{e}_t - \sum_{j \in \Gamma(t)} P_i \mathbf{e}_j$.

When $s \sim t$ this becomes

$$(\mu_i + 1)P_i\mathbf{e}_s - (\mu_i + 1)P_i\mathbf{e}_t - \sum_{j \in \Gamma(s)\backslash\{t\}} P_i\mathbf{e}_j + \sum_{j \in \Gamma(t)\backslash\{s\}} P_i\mathbf{e}_j = \mathbf{0}.$$

When $s \not\sim t$ we have instead

$$\mu_i P_i \mathbf{e}_s - \mu_i P_i \mathbf{e}_t - \sum_{j \in \Gamma(s)} P_i \mathbf{e}_j + \sum_{j \in \Gamma(t)} P_i \mathbf{e}_j = \mathbf{0}.$$

The respective conclusions now follow from linear independence of the vectors $P_i \mathbf{e}_j$ $(j \in X_i)$. $\square$

**7.3.6 Corollary** *If $\mu_i \notin \{-1, 0\}$ then the sets $\overline{\Gamma}(s)$ $(s \in X_i)$ are distinct subsets of $\overline{X}_i$.*

**7.3.7 Remark** Making use of the same arguments as in the proof of Theorem 7.3.5 we obtain:

  (i) if $s \sim t$ then the symmetric difference $\Delta(s) + \Delta(t)$ is forbidden (as a subset of $X_i$) unless $\mu_i = -1$; if $\mu_i = -1$, $\Delta^*(s) + \Delta^*(t)$ is forbidden unless it is empty;

  (ii) if $s \not\sim t$, then $\Delta^*(s) + \Delta^*(t)$ is forbidden unless $\mu_i = 0$; if $\mu_i = 0$, $\Delta(s) + \Delta(t)$ is forbidden unless it is empty. $\square$

**7.3.8 Corollary** [Row15] *For a given positive integer $m$, there are only finitely many graphs with an eigenvalue $\mu \notin \{-1, 0\}$ for which the eigenspace $\mathcal{E}(\mu)$ has codimension $m$.*

*Proof* If $|\overline{X}_i| = m$, then clearly $|V(G)| < m + 2^m$ by Theorem 7.3.5. $\square$

The next theorem (see [CvRS1], [Row11]) gives a global structural restriction on a star cell (in contrast to local restrictions, which refer to particular vertices of a star cell).

**7.3.9 Theorem** *Let $X_1 \dot{\cup} \cdots \dot{\cup} X_m$ be a star partition of a graph $G$, and let $k_i'$ (possibly zero) be the multiplicity of $\mu_i$ as an eigenvalue of $G - \overline{X}_i$. Then the number of vertices in $\overline{X}_i$ which are adjacent to some vertex of $X_i$ is at least $k_i - k_i'$.*

*Proof* Let $\overline{\Gamma}(X_i) = \bigcup_{j \in X_i} \overline{\Gamma}(j)$, and let $A_i$ be the adjacency matrix of $G - \overline{X}_i$. From the relation

$$\mu_i P_i \mathbf{e}_s = AP_i \mathbf{e}_s = P_i A \mathbf{e}_s = \sum_{t \sim s} P_i \mathbf{e}_t = \sum_{t \in \Gamma(s)} P_i \mathbf{e}_t + \sum_{t \in \overline{\Gamma}(s)} P_i \mathbf{e}_t,$$

we obtain

$$\mu_i P_i \mathbf{e}_s - \sum_{t \in \Gamma(s)} P_i \mathbf{e}_t = \sum_{t \in \overline{\Gamma}(s)} P_i \mathbf{e}_t \qquad (s \in X_i).$$

On the other hand $\mu_i I - A_i$ is the matrix with respect to the basis $\{P_i \mathbf{e}_s : s \in X_i\}$ of the linear transformation of $\mathscr{E}(\mu_i)$ defined by

$$P_i \mathbf{e}_s \mapsto \mu_i P_i \mathbf{e}_s - \sum_{t \in \Gamma(s)} P_i \mathbf{e}_t.$$

Since the rank of $\mu_i I - A_i$ is $k_i - k_i'$, the vectors $\sum_{t \in \bar{\Gamma}(s)} P_i \mathbf{e}_t$ span a $(k_i - k_i')$-dimensional subspace of the space $\langle P_i \mathbf{e}_j : j \in \bar{\Gamma}(X_i) \rangle$. This space has dimension at most $|\bar{\Gamma}(X_i)|$, and so we have $|\bar{\Gamma}(X_i)| \geq k_i - k_i'$, as required.                                                                      □

In particular, when $k_i' = 0$ we have

**7.3.10 Corollary** *If $\mu_i$ is not an eigenvalue of $G - \overline{X}_i$, then there are at least $k_i$ vertices of $\overline{X}_i$ which are adjacent to some vertex of $X_i$.*

In this situation we can deduce the following result concerning columns from the corresponding projection matrix.

**7.3.11 Theorem** *For $j \in X_i$, let $\mathbf{c}_j$ be the column of length $n - k_i$ consisting of the entries of $P_i \mathbf{e}_j$ indexed by $\overline{X}_i$. If $\mu_i$ is not an eigenvalue of $G - \overline{X}_i$ then the $k_i$ columns $\mathbf{c}_j$ $(j \in X_i)$ are linearly independent.*

*Proof* We take $i = 1$ and $X_1 = \{1, \ldots, k_1\}$ without loss of generality. Suppose that $\sum_{j \in X_1} \alpha_j \mathbf{c}_j = \mathbf{0}$ where not all of the $\alpha_j$ $(j \in X_1)$ are zero. Then $\sum_{j \in X_1} \alpha_j P_1 \mathbf{e}_j$ has the form $\begin{pmatrix} \mathbf{x} \\ \mathbf{0} \end{pmatrix}$, where $\mathbf{x} \neq \mathbf{0}$ because the vectors $P_1 \mathbf{e}_j$ $(j \in X_1)$ are linearly independent. Since $A \begin{pmatrix} \mathbf{x} \\ \mathbf{0} \end{pmatrix} = \mu_1 \begin{pmatrix} \mathbf{x} \\ \mathbf{0} \end{pmatrix}$, it follows that $\mathbf{x}$ is an eigenvector of $G - \overline{X}_1$ corresponding to $\mu_1$.                                                     □

When $k_i' \neq 0$ we have the following result as a partial converse of Theorem 7.3.11.

**7.3.12 Theorem** [Row11] *For $j \in X_i$, let $\mathbf{c}_j$ be the column of length $n - k_i$ consisting of the entries of $P_i \mathbf{e}_j$ indexed by $\overline{X}_i$. If $\mu_i$ is an eigenvalue of $G - \overline{X}_i$ then either the $k_i$ columns $\mathbf{c}_j$ $(j \in X_i)$ are linearly dependent or $\mu_i$ lies strictly between the smallest and largest eigenvalues of $G - X_i$.*

*Proof* We take $i = 1$ and $X_1 = \{1, \ldots, k_1\}$ without loss of generality, and we write $A = \begin{pmatrix} A_1 & B_1^T \\ B_1 & C_1 \end{pmatrix}$ where $A_1$, $C_1$ are the adjacency matrices of $G - \overline{X}_1$, $G - X_1$, respectively. Let $A_1 \mathbf{x} = \mu_1 \mathbf{x}$ where $\mathbf{x} \neq \mathbf{0}$. By

Theorem 7.2.2, we have $\mathbb{R}^n = \mathscr{E}(\mu_1) \oplus \mathscr{V}_1$ and so there exists a unique vector $\begin{pmatrix} 0 \\ y \end{pmatrix} \in \mathscr{V}_1$ such that $\begin{pmatrix} x \\ y \end{pmatrix} \in \mathscr{E}(\mu_1)$. The equation $\begin{pmatrix} \mu_1 x \\ \mu_1 y \end{pmatrix} =$ $\begin{pmatrix} A_1 & B_1^T \\ B_1 & C_1 \end{pmatrix} \begin{pmatrix} x \\ y \end{pmatrix}$ yields $B_1^T y = 0$ and $\mu_1 y = B_1 x + C_1 y$. Consequently we have $y^T (\mu_1 I - C_1) y = 0$.

If $y = 0$ then $\begin{pmatrix} x \\ 0 \end{pmatrix} \in \langle P_1 e_j \ : \ j \in X_1 \rangle$ and (since $x \neq 0$) it follows that the columns $c_j$ are linearly dependent. If $y \neq 0$ then the eigenvalues of the symmetric matrix $\mu_1 I - C_1$ are neither all negative nor all positive, and so $\mu_1$ lies in the spectral range of $C_1$. By Theorem 7.2.3, $\mu_1$ is not an eigenvalue of $G - X_1$, and the result follows. $\qquad\qquad \square$

## 7.4 Reconstruction and extension

One of the most natural questions regarding any combinatorial object concerns the relation between structure and substructure. Here we bring together several results which address questions of this sort. For example, we ask when a given subset of vertices can be extended to a star cell corresponding to a given eigenvalue. But first we prove a remarkable result which shows that a certain part of a graph can be reconstructed from the remaining part and an associated eigenvalue. The reconstructed part can be a significant portion of the graph (see Example 8.6.5). The theorem, first proved in [CvRS1], reveals the role of an individual eigenvalue in the structure of a graph. The proof here is taken from [Row13].

**7.4.1 Theorem** (The Reconstruction Theorem). *Let $X_i$ be a star cell of the graph $G$ corresponding to the eigenvalue $\mu_i$ of $G$. If $G - \overline{X}_i$, $G - X_i$ have adjacency matrices $A_i$, $C_i$ respectively then $G$ has an adjacency matrix of the form*

$$A = \begin{pmatrix} A_i & B_i^T \\ B_i & C_i \end{pmatrix},$$

*where $\mu_i I - A_i = B_i^T (\mu_i I - C_i)^{-1} B_i$.*

**Proof** Clearly we may take $A = \begin{pmatrix} A_i & B_i^T \\ B_i & C_i \end{pmatrix}$. We have

$$\mu_i I - A = \begin{pmatrix} \mu_i I - A_i & -B_i^T \\ -B_i & \mu_i I - C_i \end{pmatrix},$$

where $\mu_i I - C_i$ is invertible by Theorem 7.2.3. In particular, the matrix $(-B_i \mid \mu_i I - C_i)$ has rank $n - k_i$; but $\mu_i I - A$ has rank $n - k_i$ and so the rows of $(-B_i \mid \mu_i I - C)$ form a basis for the row space of $\mu_i I - A$. Hence there exists a $k_i \times (n - k_i)$ matrix $L$ such that $(\mu_i I - A_i \mid -B_i^T) = L(-B_i \mid \mu_i I - C_i)$. Now $\mu_i I - A_i = -LB_i, -B_i^T = L(\mu_i I - C_i)$ and the result follows.                                                                    □

**7.4.2 Corollary** *Let* $G, H$ *be graphs with star cells* $X, Y$ *corresponding to a common eigenvalue, and let* $E(X), E(Y)$ *be the corresponding edge sets. If there is a bijection* $\phi : V(G) \mapsto V(H)$ *such that*

 *(i)* $\phi(X) = Y$,
 *(ii)* $\phi$ *reduces to an isomorphism* $G - X$ *to* $H - Y$,
 *(iii)* $\phi$ *reduces to an isomorphism* $G - E(X)$ *to* $H - E(Y)$,

*then* $\phi$ *is an isomorphism from* $G$ *to* $H$.

Theorem 7.4.1 asserts that $G$ is reconstructible from the eigenvalue $\mu_i$, the subgraph $G - X_i$ and the set of edges between $X_i$ and $\overline{X}_i$. Taken together with Corollary 7.3.8, this result suggests the possibility of characterizing a graph (or family of graphs) by properties of $\overline{X}_i$ which have implications for the set of edges between $X_i$ and $\overline{X}_i$. One such property, discussed in Section 7.6, is the minimality of $\overline{X}_i$ as a dominating set. Another is the regularity of the subgraph induced by $\overline{X}_i$ in a graph $G$ which is itself regular. In this situation we have the following result:

**7.4.3 Theorem** [Row15] *Let* $G$ *be a* $k$*-regular graph* $(k > 0)$ *with a star cell* $X_i$ *corresponding to the eigenvalue* $\mu_i$. *If* $\overline{X}_i$ *induces a regular subgraph of degree* $r$ *then one of the following holds:*

 *(a)* $\mu_i = k$, $r = k - 1$ *and each component of* $G$ *is a complete graph on* $k + 1$ *vertices,*
 *(b)* $\mu_i \neq k$ *and* $X_i$ *induces a regular subgraph of degree* $\mu_i + k - r$.

*In the latter case,* $\mu_i$ *is an integer and* $r - k \leq \mu_i \leq r - 1$.

*Proof* Suppose first that $\mu_i = k$. Then $|X_i|$ is the number of components of $G$ (Theorem 2.1.4). On the other hand, $\mu_i$ is not an eigenvalue of $G - X_i$ and so $X_i$ contains a vertex from each component of $G$. It follows that $|X_i \cap C| = 1$ for each component $C$; moreover, by regularity of $G - X_i$ each vertex of $\overline{X}_i \cap C$ is adjacent to the vertex in $X_i \cap C$. Since $G$ is $k$-regular, conclusion (a) follows.

Now suppose that $\mu_i \neq k$. For each vertex $u$ of $G$ we have $\mu_i P_i \mathbf{e}_u = \sum_{v \sim u} P_i \mathbf{e}_v$, which may be written

$$\mu_i P_i \mathbf{e}_u = \sum \{P_i \mathbf{e}_v : v \in \Delta(u) \cap X_i\} + \sum \{P_i \mathbf{e}_v : v \in \Delta(u) \cap \overline{X_i}\}.$$

Summing over all $u \in X_i$ we obtain

$$\mu_i \sum_{u \in X_i} P_i \mathbf{e}_u = \sum_{u \in X_i} d_u P_i \mathbf{e}_u + \sum_{v \in \overline{X_i}} (k-r)P_i \mathbf{e}_v, \qquad (7.4.2)$$

where $d_u$ denotes the degree of $u$ as a vertex of $G - \overline{X_i}$.

If $\mu_i \neq k$ we have $P_i \mathbf{j} = \mathbf{0}$, equivalently

$$\sum_{v \in \overline{X_i}} P_i \mathbf{e}_v = -\sum_{u \in X_i} P_i \mathbf{e}_u,$$

and so equation (7.4.1) may be written

$$\sum_{u \in X_i} (\mu_i + k - r - d_u)P_i \mathbf{e}_u = \mathbf{0}.$$

Since the vectors $P_i \mathbf{e}_u$ ($u \in X_i$) are linearly independent, we have $d_u = \mu_i + k - r$ for each $u \in X_i$, and so (b) holds. In view of Theorem 7.3.1 the integer $\mu_i + k - r$ lies between 0 and $k-1$, and so $\mu_i$ is an integer between $r - k$ and $r - 1$. $\qquad \square$

In the situation of Theorem 7.4.3, if $k$ is given there are only finitely many possibilities for $r$ and $\mu_i$, since $r \leq k - 1$. The first non-trivial case is $k = 3$, and then the subgraphs induced by $X_i$ and $\overline{X_i}$ have known structure because they are regular of degree at most 2. The cubic graphs which arise are investigated in [Row15], where it is shown that for $(r, \mu_i) \notin \{(2, -1), (2, 0)\}$ the only connected examples are the Petersen graph, with $(r, \mu_i) \in \{(1, -2), (2, 1)\}$ (see Fig. 7.1), the complete graph $K_4$ with $(r, \mu_i) \in \{(2, 3), (0, -1)\}$, and the complete bipartite graph $K_{3,3}$, with $(r, \mu_i) = (1, 0)$. It turns out that when $(r, \mu_i) \in \{(2, -1), (2, 0)\}$ there are infinitely many connected cubic graphs satisfying the hypotheses of Theorem 7.4.3. Such graphs are constructed by finding infinite families of adjacency matrices which satisfy the conclusions of the Reconstruction Theorem: there are infinitely many solutions of the equation

$$\mu_i I - A_i = B_i^T (\mu_i I - C_i)^{-1} B_i$$

and we need a converse of Theorem 7.4.1 to show that each solution corresponds to a graph with an appropriate star partition. For this we first need the following observation.

**7.4.4 Proposition** [Row15] *If the graph $G$ has an adjacency matrix $A = \begin{pmatrix} A_i & B_i^T \\ B_i & C_i \end{pmatrix}$, where $A_i$ has size $k_i \times k_i$ and $\mu_i I - A_i = B_i^T(\mu_i I - C_i)^{-1} B_i$, then $\mu_i$ is an eigenvalue of $G$ with multiplicity $k_i$.*

*Proof* The nullspace of $\mu_i I - A$ consists of the vectors $\begin{pmatrix} \mathbf{x} \\ (\mu_i I - C_i)^{-1} B_i \mathbf{x} \end{pmatrix}$, where $\mathbf{x} \in \mathbb{R}^{k_i}$. $\square$

Now a graph $G$ which satisfies the hypotheses of Proposition 7.4.4 has a set $S$ of vertices with the properties (i) $|S|$ is the multiplicity of the eigenvalue $\mu_i$, (ii) $\mu_i$ is not an eigenvalue of $G - S$. The question which remains is whether $G$ has a star partition $X_1 \dot\cup \cdots \dot\cup X_m$ with $X_i = S$. This may be regarded as an extension problem, and our next result shows that $\{S\}$ can always be extended to a star partition.

**7.4.5 Theorem** [Row15] *If $S \subseteq V(G), |S| = k_i$ and $\mu_i$ is not an eigenvalue of $G - S$, then $G$ has a star partition $X_1 \dot\cup X_2 \dot\cup \cdots \dot\cup X_m$ with $X_i = S$.*

*Proof* As in (7.1.1), let $\mathbf{e}_j = \sum_{h=1}^n t_{hj} \mathbf{x}_j \ (j = 1, \ldots, n)$, where $\mathscr{E}(\mu_i) = \langle \mathbf{x}_h : h \in R_i \rangle \ (i = 1, \ldots, m)$. From the proof of Theorem 7.1.3 we see that it suffices to prove that the matrices $(t_{hj})_{h \in R_i, j \in S}, (t_{hj})_{h \notin R_i, j \notin S}$ are invertible. As in the proof of Theorem 7.2.3 we have $\mathscr{E}(\mu_i) = \langle P_i \mathbf{e}_j : j \in S \rangle$, and so the vectors $P_i \mathbf{e}_j \ (j \in S)$ are linearly independent. Since $P_i \mathbf{e}_j = \sum_{h \in R_i} t_{hj} \mathbf{x}_h \ (j \in S)$, the matrix $(t_{hj})_{h \in R_i, j \in S}$ is invertible. The complementary matrix $(t_{hj})_{h \notin R_i, j \notin S}$ represents a linear map $\langle \mathbf{e}_j : j \notin S \rangle \rightarrow \langle \mathbf{x}_h : h \notin R_i \rangle$ given by $\mathbf{x} \mapsto (I - P_i)\mathbf{x}$; and this map is invertible because its kernel is the trivial space $\mathscr{E}(\mu_i) \cap \langle \mathbf{e}_j : j \notin S \rangle$ (cf. Theorem 7.2.2). $\square$

We can now give two examples of infinite families of connected cubic graphs which (as may be verified using Theorem 7.4.5) satisfy the conditions of Theorem 7.4.3. These are illustrated in Figs. 7.2 and 7.3, where the vertices of the star cell $X_i$ are shown in black. In the first case, $\mu_i = -1$, and in the second case, $\mu_i = 0$.

A slight variation of the proof of Theorem 7.4.5 yields the following theorem, which is an analogue for star bases of the fundamental result that, in a finite-dimensional vector space, any set of linearly independent vectors can be extended to a basis.

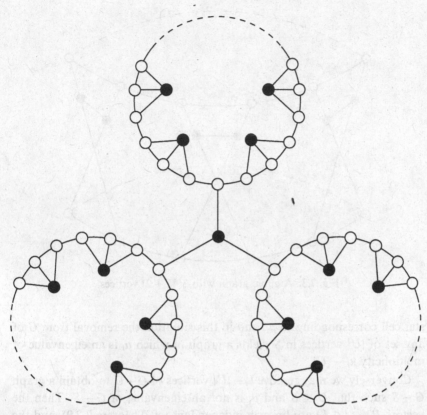

Fig. 7.2. A cubic graph with $4(3k + 1)$ vertices.

**7.4.6 Theorem** *Let $U$ be a subset of $V(G)$ such that the vectors $P_i \mathbf{e}_j$ $(j \in U)$ are linearly independent. Then $G$ has a star partition $X_1 \dot\cup X_2 \dot\cup \cdots \dot\cup X_m$ such that $U \subseteq X_i$.*

*Proof* Since $\mathscr{E}(\mu_i)$ is spanned by the vectors $P_i \mathbf{e}_1, P_i \mathbf{e}_2, \ldots, P_i \mathbf{e}_n$, we can extend the set of vectors $P_i \mathbf{e}_j$ $(j \in U)$ to a basis $\{P_i \mathbf{e}_j : j \in S\}$ of $\mathscr{E}(\mu_i)$. The rest of the argument follows the proof of Theorem 7.4.5. □

**7.4.7 Corollary** *The vectors $P_i \mathbf{e}_j$ $(j \in U)$ are linearly independent if and only if $G - U$ has $\mu_i$ as an eigenvalue of multiplicity $k_i - |U|$.*

*Proof* If the vectors $P_i \mathbf{e}_j$ $(j \in U)$ are linearly independent then we may extend them to a basis $\{P_i \mathbf{e}_j : j \in S\}$ of $\mathscr{E}(\mu_i)$. By Theorem 7.4.5, $S$ is a

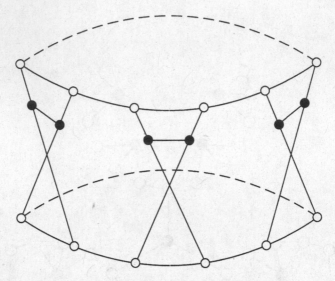

Fig. 7.3. A cubic graph with $3(4k+2)$ vertices.

star cell corresponding to $\mu_i$, and in this situation the removal from $G$ of any set of $|U|$ vertices in $S$ yields a graph in which $\mu_i$ is an eigenvalue of multiplicity $k_i - |U|$.

Conversely we may remove $k_i - |U|$ vertices of $G - U$ to obtain a graph $G - S$ such that $U \subseteq S$ and $\mu_i$ is not an eigenvalue of $G - S$. Then the vectors $P_i e_j$ ($j \in S$) are linearly independent (cf. Theorem 7.2.9), and the result follows.                                                                $\square$

Another consequence of Theorem 7.4.6 is that, for a fixed eigenvalue $\mu_i$, the union of all star cells corresponding to $\mu_i$ consists of the vertices $j$ for which $P_i e_j \neq 0$. The next result describes the intersection of such star cells.

**7.4.8 Proposition** *The vertex $v$ lies in every star cell corresponding to $\mu_i$ if and only if $\mu_i = 0$ and $v$ is isolated.*

*Proof* By Theorem 7.4.6, if $P_j e_v \neq 0$ for $j \neq i$ then there exists a star partition $X_1 \dot\cup X_2 \dot\cup \cdots \dot\cup X_m$ with $v \notin X_i$. It follows that $v$ lies in every star cell corresponding to $\mu_i$ if and only if $P_j e_v = 0$ for each $j \neq i$. In this situation, the relations

$$P_1 + P_2 + \cdots + P_m = I, \quad \mu_1 P_1 + \mu_2 P_2 + \cdots + \mu_m P_m = A$$

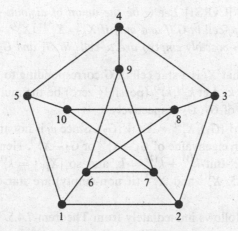

Fig. 7.4. The graph of Example 7.4.9.

show that $P_i \mathbf{e}_v = \mathbf{e}_v$ and $\mu_i P_i \mathbf{e}_v = A \mathbf{e}_v$. Hence $\mu_i \mathbf{e}_v = \sum_{u \sim v} \mathbf{e}_u$, from which it follows that $\mu_i = 0$ and $v$ is isolated.

Conversely if $\mu_i = 0$ and $v$ is isolated then, by Theorem 7.3.1, $v$ cannot lie in a star cell corresponding to a non-zero eigenvalue. It follows that $v$ lies in every star cell corresponding to $\mu_i$.  □

We conclude this section which an example which shows that we cannot expect to extend Theorem 7.4.5 to a result in which more than one star cell is prescribed.

**7.4.9 Example** Let $G$ be the Petersen graph as depicted in Fig. 7.4. Let $S_1 = \{1\}$ and $S_2 = \{2, 6, 7, 8, 10\}$. Then $S_1$ satisfies the conditions of Theorem 7.4.5 in respect of $\mu_1 = 3$, and $S_2$ satisfies them in respect of $\mu_2 = 1$. But $G$ has no star partition $X_1 \dot\cup X_2 \dot\cup X_3$ with $X_1 = S_1$ and $X_2 = S_2$ because $X_3 \neq \Delta^*(4)$ by Theorem 7.3.1.  □

## 7.5 Operations on graphs

In this section we discuss the behaviour of star partitions under various graph operations. We shall see the advantage of having a variety of characterizations of star partitions at our disposal. The first operation to consider is the union of two graphs, and here there are no surprises:

**7.5.1 Proposition** [CvRS1] *Let $G$ be the union of disjoint graphs $G_1$ and $G_2$. Then $X_i$ is a $\mu_i$-cell in $G$ if and only if $X_i = X_i^{(1)} \cup X_i^{(2)}$, where $X_i^{(1)}$ and $X_i^{(2)}$ (one of them possibly empty) are $\mu_i$-cells in $G_1$ and $G_2$ respectivley.*

*Proof* Suppose that $X_i$ is a star cell in $G$ corresponding to the eigenvalue $\mu_i$ of multiplicity $k_i$. Let $k_i^{(1)}, k_i^{(2)}$ (possibly zero) be the multiplicities of $\mu_i$ as an eigenvalue of $G_1, G_2$ respectively.

Let $X_i^{(1)} = X_i \cap V(G_1), X_i^{(2)} = X_i \cap V(G_2)$. Since $\mu_i$ is not an eigenvalue of $G - X_i$, it is not an eigenvalue of $G_1 - X_i^{(1)}$ or $G_2 - X_i^{(2)}$. Hence $|X_i^{(1)}| \geq k_i^{(1)}$ and $|X_i^{(2)}| \geq k_i^{(2)}$. But $k_i^{(1)} + k_i^{(2)} = k_i$ and so $|X_i^{(1)}| = k_i^{(1)}, |X_i^{(2)}| = k_i^{(2)}$. By Theorem 7.4.5, $X_i^{(1)}$ and $X_i^{(2)}$ (if non-empty) are star cells in $G_1, G_2$ respectively.

The converse follows immediately from Theorem 7.4.5. □

For remarks concerning the complement of a graph we shall need the following lemma.

**7.5.2 Lemma** *If the eigenspaces of the (labelled) graphs $G$ and $G'$ coincide, i.e. if $\mathcal{E}_A(\mu_i) = \mathcal{E}_{A'}(\mu_i')$ $(i = 1, 2, \ldots, m)$, where $\mu_i$ is not necessarily equal to $\mu_i'$, then the star partitions of $G$ are precisely the star partitions of $G'$.*

*Proof* For $G$ and $G'$ we obtain the same eutactic stars by projecting $e_1, e_2, \ldots, e_n$ orthogonally onto the eigenspaces. □

**7.5.3 Theorem** *Let $G$ be a connected regular graph whose complement is also connected. Then the star partitions of $\overline{G}$ are precisely the star partitions of $G$.*

*Proof* By Theorem 2.1.3 the index of each of $G, \overline{G}$ is a simple eigenvalue. Accordingly, if $G$ has distinct eigenvalues $\mu_1, \mu_2, \ldots, \mu_m$ in decreasing order then in the notation of Section 1.1 we have $\mathcal{E}_A(\mu_1) = \langle \mathbf{j} \rangle = \mathcal{E}_{J-A-I}(n-1-\mu_1)$ and $\mathcal{E}_A(\mu_i) = \mathcal{E}_{J-A-I}(-\mu_i-1)$ $(i = 2, \ldots, m)$. The result now follows from Lemma 7.5.2. □

Note that the conclusion of Theorem 7.5.3 may not hold when $G$ and $\overline{G}$ are not both connected. For example if $n \geq 3$ then $K_n$ has more than one star partition, while $\overline{K}_n$ has only one.

Next we consider how the star partitions of a regular graph behave under Seidel switching.

**7.5.4 Theorem** *If $G, G'$ are connected, switching-equivalent regular graphs of the same degree then the star partitions of $G$ are precisely the star partitions of $G'$.*

*Proof* Suppose that $G$ and $G'$ are regular of degree $r$, with Seidel matrices $S$ and $S'$ respectively. As noted in Section 1.1, there exists a diagonal matrix $D$, with diagonal entries $\pm 1$, such that $S' = D^{-1}SD$. Since $G$ and $G'$ are connected, they are cospectral; indeed for any eigenvalue $\mu_i$ of $G$ other than $r$ we have

$$D\mathscr{E}_{A'}(\mu_i) = D\mathscr{E}_{S'}(-2\mu_i - 1) = \mathscr{E}_S(-2\mu_i - 1) = \mathscr{E}_A(\mu_i).$$

Moreover, if $P_i$ represents the orthogonal projection onto $\mathscr{E}_A(\mu_i)$ then the orthogonal projection onto $\mathscr{E}_{A'}(\mu_i)$ is represented by $D^{-1}P_iD$. The result now follows from the observation that the columns of $P_i$ (namely the vectors $P_ie_1, P_ie_2, \ldots, P_ie_n$) satisfy exactly the same linear dependence relations as the columns of $D^{-1}P_iD$. □

For the remainder of this section we shall be concerned with the general graph operation of NEPS (non-complete extended $p$-sum) defined in Section 2.3. Since the vertex set of the resulting graph is the Cartesian product of the vertex sets of the graphs on which the operation is performed, it is natural to ask whether this property extends to star cells. The fact that the eigenvectors of a NEPS are Kronecker products of eigenvectors of the original graphs leads us to expect an affirmative answer. In fact, the best possible result in this respect is the next proposition (a stronger version of Theorem 4.2 from [CvSi2]).

**7.5.5 Proposition** *Let $G$ be a graph obtained from $G_1, \ldots, G_k$ as a NEPS with basis $\mathscr{B}$, and let*

$$X_1^{(1)} \,\dot{\cup}\, \cdots \,\dot{\cup}\, X_{m_1}^{(1)}, \quad \ldots \quad , X_1^{(k)} \,\dot{\cup}\, \cdots \,\dot{\cup}\, X_{m_k}^{(k)}$$

*be star partitions of the graphs $G_1, \ldots, G_k$ respectively. Then*

$$X_{i_1}^{(1)} \times \cdots \times X_{i_k}^{(k)} \quad (1 \le i_1 \le m_1, \ldots, 1 \le i_k \le m_k)$$

*is a star cell of $G$ corresponding to the eigenvalue $\mu_{i_1,\ldots,i_k}$ (obtained from the eigenvalues $\mu_{i_1}, \ldots, \mu_{i_k}$) provided that this eigenvalue does not coincide with some other, i.e. provided that $\mu_{i_1,\ldots,i_k} \ne \mu_{j_1,\ldots,j_k}$ whenever $(i_1,\ldots,i_k) \ne (j_1,\ldots,j_k)$.*

*Proof* Suppose that $(s_1,\ldots,s_k)$ is a vertex of $G$, where $s_j \in X_{i_j}^{(j)}$ for each $j \in \{1,\ldots,k\}$. We ascribe an eigenvector $\mathbf{x}_{s_j}^{(j)}$ to each vertex $s_j$ of $G_j$

as in Theorem 7.2.6. Then the vectors $\mathbf{x}_{s_1}^{(1)} \otimes \cdots \otimes \mathbf{x}_{s_k}^{(k)}$ are eigenvectors of $G$ whose inner products with $\mathbf{e}_1, \ldots, \mathbf{e}_n$ satisfy condition (4) of Theorem 7.2.9 in respect of the set $X_i^{(1)} \times \cdots \times X_{i_k}^{(k)}$. Moreover they form a basis of $\mathscr{E}(\mu_{i_1,\ldots,i_k})$ because $\mu_{i_1,\ldots,i_k}$ does not coincide with any other eigenvalue of $G$. The result now follows from Theorems 7.2.9 and 7.4.6. $\square$

Recall from Section 4.3 that we say a NEPS is coincidence-free if no two eigenvalues coincide when obtained as in Theorem 2.3.4. The result from [CvSi2] can now be restated as

**7.5.6 Corollary** *If $G$ is a graph obtained from graphs $G_1, \ldots, G_k$ as a coincidence-free NEPS, then the Cartesian product of any star partitions of graphs $G_1, \ldots, G_k$ (with cells as described in Proposition 7.5.5) is a star partition of $G$.*

Notice that a star partition in NEPS does not depend on the basis of NEPS, as long as the NEPS remains coincidence-free.

We shall now show that the conclusions of Proposition 7.5.5 are indeed the best possible. In other words, we shall show that generally it is not true that a star partition of a NEPS can be obtained by amalgamating Cartesian products corresponding to coincident eigenvalues.

**7.5.7 Example** We have $K_2 \times K_2 = K_2 \cup K_2$. The only star partition of $K_2$ consists of two trivial cells, corresponding to eigenvalues 1 or $-1$, and we have coincidence of eigenvalues in $K_2 \times K_2$ since $1 = 1 \times 1 = (-1) \times (-1)$. By amalgamating Cartesian products we obtain each copy of $K_2$ in the resulting graph; but neither is a star cell. $\square$

This phenomenon is, at first glance, unexpected in view of the behaviour of angles and main angles under the coincidence of eigenvalues of a NEPS (see Section 4.3). The explanation is as follows. When two eigenvalues coincide, the corresponding eigenspaces are joined in the sense of direct sum. The union of star bases of these eigenspaces is indeed a basis of the resulting eigenspace, but not necessarily a star basis. A star basis might well correspond to vertices having no relation to the union of cells corresponding to original eigenspaces.

Notice that a graph may have several representations as a NEPS. For example, we have $C_4 = C_4 + K_1 = K_2 + K_2$. In view of this, the property of being coincidence-free makes sense only with respect to a fixed representation. For example, $C_4$ is coincidence-free with respect to the first representation, but not with respect to the second one. Therefore

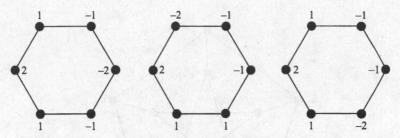

Fig. 7.5. Some star partitions of $C_6$.

it is reasonable now to pose the following question: is it true that any star partition in a coincidence-free NEPS is induced by the Cartesian product of some star partitions of the graphs? A negative answer to this question is given by the following example.

**7.5.8 Example** We have $K_3 \times K_2 = C_6$ and a few star partitions of $C_6$ are displayed in Fig. 7.5. Vertices in a star cell are labelled by the corresponding eigenvalue.

Only the first one is induced by (essentially unique) star partitions of $K_3$ and $K_2$, while the others are not. The fact that the given partitions are indeed star partitions can easily be verified by Theorem 7.2.3.    □

## 7.6 Application to graph dominance

In this section we shall develop some ideas which originated with the notion of a star partition, and which give rise to some new and improved results on dominating sets. In particular we find upper bounds for the domination number and location-domination number of a graph in terms of multiplicities of eigenvalues. Although these bounds are (in general) crude, they are of interest in that the problem of finding the domination number of a graph is known to be NP-complete (see, for example, [GaJo]). In addition, we shall characterize the Petersen graph among connected regular graphs by a property considered in Section 7.3, namely the minimality of $\overline{X_i}$ as a dominating set, where $X_i$ is a star cell corresponding to an eigenvalue $\mu_i$ other than $-1$ or $0$. This section is written mainly in accordance with [Row13].

Recall that a *dominating set* in the graph $G$ is a subset $D$ of $V(G)$ such that each vertex of $\overline{D}$ is adjacent to a vertex of $D$. The dominating set $D$ is *minimal* if for each $v \in D$, $D \setminus \{v\}$ is not a dominating set: in this

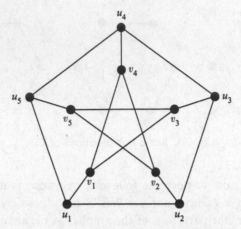

Fig. 7.6. The Petersen graph.

situation either $v$ is isolated in $G - \overline{D}$ or there exists a vertex $u$ in $\overline{D}$ such that $\Delta(u) \cap D = \{v\}$ (see [Ore], Theorem 13.1.2). The dominating set $D$ is a *location-dominating set* if $\Delta(u_1) \cap D \neq \Delta(u_2) \cap D$ whenever $u_1, u_2$ are distinct vertices in $\overline{D}$. The *domination number* (respectively, *location-domination number*) of $G$ is the least cardinality of a dominating set (location-dominating set), here denoted by $\beta(G)$ (by $\hat{\beta}(G)$). The concept of a location-dominating set was apparently introduced by Slater [Sla], who explains its relevance to safeguards analysis and the design of surveillance systems.

**7.6.1 Example** We have $\beta = 3$, $\hat{\beta} = 4$ for the Petersen graph, which is illustrated in Fig. 7.6: here $\{u_2, u_5, v_1\}$ is a *minimum dominating set* (i.e. a dominating set of least cardinality), whereas $\{u_3, u_4, v_1, v_2\}$ is a *minimum location-dominating set*. Notice also that $\{v_1, v_2, v_3, v_4, v_5\}$ is just a minimal dominating set.

The following proposition shows how dominating sets arise naturally from star partitions.

**7.6.2 Proposition** [Row13] *Let* $X_1 \dot{\cup} \cdots \dot{\cup} X_m$ *be a star partition of the graph* $G$, *and suppose that* $G$ *has no isolated vertices. Then*

*(i) for each* $i \in \{1, \ldots, m\}$, $\overline{X}_i$ *is a dominating set for* $G$;
*(ii) if* $\mu_i \notin \{-1, 0\}$ *then* $\overline{X}_i$ *is a location-dominating set for* $G$.

*Proof* The proposition is a restatement of Theorem 7.3.1 and Corollary 7.3.6. □

We give two examples to show that the cases $\mu_i = -1$, $\mu_i = 0$ are essential exceptions in Proposition 7.6.2(ii). First, if $G$ is the complete graph $K_n$ $(n > 1)$ then $G$ has eigenvalues $\mu_1 = n - 1$, $\mu_2 = -1$ with multiplicities $k_1 = 1$, $k_2 = n-1$; any partition with $|X_1| = 1$, $|X_2| = n-1$ is a star partition, and $\Delta(u) \cap \overline{X}_2 = X_1$ for all $u \in X_2$. Secondly, if $G$ is the complete bipartite graph $K_{m,n}$ $(m > 2, n > 2)$ then $G$ has eigenvalues $\mu_1 = \sqrt{mn}$, $\mu_2 = 0$, $\mu_3 = -\sqrt{mn}$ with multiplicities $k_1 = 1$, $k_2 = m + n - 2$, $k_3 = 1$. By Theorem 7.3.1, a star partition $X_1 \dot\cup X_2 \dot\cup X_3$ has $X_1 = \{v\}$, $X_3 = \{w\}$ where $v, w$ are adjacent vertices; and for $u \in X_2$ the only possibilities for $\Delta(u) \cap \overline{X}_2$ are $X_1$ and $X_3$.

**7.6.3 Corollary** *Let $G$ be a graph with $n$ vertices, none of them isolated. Let $k$ be the largest multiplicity of an eigenvalue of $G$, and let $k'$ be the largest multiplicity of an eigenvalue other than $-1$ or $0$. Then*

*(i) $\beta(G) \leq n - k$, and*
*(ii) $\hat\beta(G) \leq n - k'$.*

Corollary 7.6.3(i) improves a result of van Nuffelen [Nuf] who showed that $\beta(G) \leq r(G)$ where $r(G)$ denotes the rank of $A$, that is, $\beta(G) \leq n - k_0$ where $k_0$ is the multiplicity of zero as an eigenvalue. Nevertheless Corollary 7.6.3(i) can be of use only when $k > \frac{1}{2}n$ because always $\beta(G) \leq \frac{1}{2}n$ for a graph $G$ without isolated vertices: this is a consequence of the observation that then the complement of a minimal dominating set is itself a dominating set (see [Ore], Theorem 13.1.4). Note that both of the upper bounds in Corollary 7.6.3 are attained when $G = K_n$.

If $\beta(G) = n - k_i$ then $\overline{X}_i$ is a minimum dominating set; but rather than pursue the case of equality in Corollary 7.6.3(i) we investigate the more general situation in which the dominating set $\overline{X}_i$ is merely minimal. The motivation is the example of the Petersen graph (with $\mu_i = 1$).

In the rest of this section we suppose that

- $G$ has no isolated vertices,
- the dominating set $\overline{X}_i$ is minimal, and
- $\mu_i \notin \{-1, 0\}$.

For $v \in V(G)$, we let $\Gamma(v) = \Delta(v) \cap X_i$, $\overline{\Gamma}(v) = \Delta(v) \cap \overline{X}_i$.

**7.6.4 Lemma** *If $v$ is a non-isolated vertex of $G - X_i$ then there exists a unique $u \in X_i$ such that $\overline{\Gamma}(u) = \{v\}$.*

*Proof* The existence of a vertex $u \in X_i$ such that $\overline{\Gamma}(u) = \{v\}$ is a property of minimal dominating sets already noted. Uniqueness follows from Proposition 7.6.2(ii). □

**7.6.5 Lemma** *If $G - X_i$ has no isolated vertices then the set of all edges between $X_i$ and $\overline{X}_i$ is a perfect matching for $G$.*

*Proof* Let $\overline{X}_i = \{v_1, \ldots, v_d\}$, where $d = n - k_i$. By Lemma 7.6.4, for each $j \in \{1, \ldots, d\}$ there exists a unique $u_j \in X_i$ such that $\overline{\Gamma}(u_j) = \{v_j\}$. Suppose by way of contradiction that there exists a vertex $u \in X_i - \{u_1, u_2, \ldots, u_d\}$. Without loss of generality, $\overline{\Gamma}(u) = \{v_1, v_2, \ldots, v_r\}$, where $r > 1$. Since $\mu_i P_i \mathbf{e}_u = \sum_{k \sim u} P_i \mathbf{e}_k$ we have

$$\mu_i P_i \mathbf{e}_u = \sum_{k \in \Gamma(u)} P_i \mathbf{e}_k + \sum_{j=1}^{r} P_i \mathbf{e}_{v_j},$$

and similarly

$$\mu_i P_i \mathbf{e}_{u_j} = P_i \mathbf{e}_{v_j} + \sum_{k \in \Gamma(u_j)} P_i \mathbf{e}_k \quad (j = 1, \ldots, r),$$

whence

$$\mu_i \sum_{j=1}^{r} P_i \mathbf{e}_{u_j} - \mu_i P_i \mathbf{e}_u = \sum_{j=1}^{r} \sum_{k \in \Gamma(u_j)} P_i \mathbf{e}_k - \sum_{k \in \Gamma(u)} P_i \mathbf{e}_k.$$

Since the vectors $P_i \mathbf{e}_k$ ($k \in X_i$) are linearly independent we may equate coefficients of $P_i \mathbf{e}_u$ to deduce that $-\mu_i$ is the number of vertices $u_1, \ldots, u_r$ adjacent to $u$. Let $d_j$ be the number of these vertices adjacent to $u_j$. On equating coefficients of $P_i \mathbf{e}_{u_j}$ we find that $\mu_i$ is $d_j - 1$ or $d_j$ according as $u$ is or is not adjacent to $u_j$; in particular, $\mu_i \geq -1$. But $\mu_i$ is a non-positive integer and so $\mu_i \in \{-1, 0\}$, contrary to assumption. It follows that $X_i = \{u_1, \ldots, u_d\}$ and the edges $u_j v_j$ ($j \in \{1, \ldots, d\}$) are the only edges between $X_i$ and $\overline{X}_i$. $\qquad\square$

**7.6.6 Lemma** *If $G - X_i$ has a non-isolated vertex $v$ then*

*(i)* $\mu_i = 1$,
*(ii)* *$G$ has no 4-cycle $vv'u'uv$ with $\overline{\Gamma}(u') = \{v'\}$ and $\overline{\Gamma}(u) = \{v\}$.*

*Proof* Let $\overline{\Gamma}(v) = \{v_1, v_2, \ldots, v_r\}$. By Lemma 7.6.4, there exists a unique $u \in X_i$ such that $\overline{\Gamma}(u) = \{v\}$; moreover, for each $j \in \{1, \ldots, r\}$, $v_j$ is non-isolated in $G - X_i$ and so there exists a unique $u_j \in X_i$ such that $\overline{\Gamma}(u_j) = \{v_j\}$. Now we have

$$\mu_i P_i \mathbf{e}_{u_j} = P_i \mathbf{e}_{v_j} + \sum_{k \in \Gamma(u_j)} P_i \mathbf{e}_k,$$

$$\mu_i P_i e_u = P_i e_v + \sum_{k \in \Gamma(u)} P_i e_k,$$

$$\mu_i P_i e_v = \sum_{k \in \Gamma(v)} P_i e_k + \sum_{j=1}^{r} P_i e_{v_j}.$$

We use the first two equations to substitute for $P_i e_v, P_i e_{v_j}$ in the third, and obtain

$$\mu_i^2 P_i e_u - \mu_i \sum_{k \in \Gamma(u)} P_i e_k = \sum_{k \in \Gamma(v)} P_i e_k + \sum_{j=1}^{r} (\mu_i P_i e_{u_j} - \sum_{k \in \Gamma(u_j)} P_i e_k).$$

On equating coefficients of $P_i e_u$ we find that $\mu_i^2 = 1 - m_v$ where $m_v$ is the number of vertices $u_1, u_2, \ldots, u_r$ adjacent to $u$. Since $\mu_i \notin \{-1, 0\}$, we must have $\mu_i = 1$ and $m_v = 0$. This proves both parts of our lemma because any 4-cycle $vv'u'uv$ with $\overline{\Gamma}(u') = \{v'\}, \overline{\Gamma}(u) = \{v\}$ necessarily has $v' = v_j, u' = u_j$ for some $j \in \{1, \ldots, r\}$. $\qquad \square$

**7.6.7 Theorem** *Let $G$ be a regular graph with star partition $V(G) = X_1 \dot\cup \cdots \dot\cup X_m$, and suppose that $G$ has an eigenvalue $\mu_i \notin \{-1, 0\}$ for which the corresponding dominating set $\overline{X}_i$ is minimal.*

(i) *If each vertex of $\overline{X}_i$ is isolated in $G - X_i$ then $\mu_i = 1$ and each component of $G$ is isomorphic to $K_2$.*

(ii) *If no vertex of $\overline{X}_i$ is isolated in $G - X_i$ then $\mu_i = 1$ and each component of $G$ is isomorphic to the Petersen graph.*

*Proof* Let $G$ be regular of degree $\rho$. (Note that $\rho \neq 0$ because $G$ has $\mu_i$ as a non-zero eigenvalue; hence $G$ has no isolated vertices, and $\overline{X}_i$ is a dominating set by Proposition 7.6.2(i).) We show first that if $\mu_i = \rho$ then $X_i$ contains precisely one vertex from each component of $G$. Since $G - X_i$ does not have $\rho$ as an eigenvalue (by Theorem 7.2.3), no component of $G$ is a component of $G - X_i$. Hence $X_i$ contains at least one vertex from each component of $G$; but the number of vertices in $X_i$ is the multiplicity of $\rho$ and this is just the number of components of $G$ (see Theorem 2.1.4). Our assertion follows, and we now deal separately with the two cases of the theorem.

(i) Suppose by way of contradiction that $\mu_i \neq \rho$ when $\overline{X}_i$ is an independent set in $G$. The eigenspace $\mathscr{E}(\rho)$ is orthogonal to $\mathscr{E}(\mu_i)$ and contains the all-1 vector $\mathbf{j}$: hence $P_i \mathbf{j} = \mathbf{0}$. Now $\mu_i P_i e_t = \sum_{j \in \Gamma(t)} P_i e_j$ for each $t \in \overline{X}_i$, and $\mu_i P_i e_s = \sum_{k \in \Gamma(s)} P_i e_u + \sum_{t \in \overline{\Gamma}(s)} P_i e_t$ for each $s \in$

$X_i$. It follows that $\mu_i^2 P_i e_s = \mu_i \sum_{k \in \Gamma(s)} P_i e_u + \sum_{t \in \overline{\Gamma}(s)} \sum_{j \in \Gamma(t)} P_i e_j$, whence $\mu_i^2 = |\overline{\Gamma}(s)|$. On the other hand, since $P_i \mathbf{j} = \mathbf{0}$, we have $-\mu_i \sum_{s \in X_i} P_i e_s = \mu_i \sum_{t \in \overline{X}_i} P_i e_t = \sum_{t \in \overline{X}_i} \sum_{j \in \Gamma(t)} P_i e_j$. On equating coefficients of $P_i e_s$, we find that $-\mu_i = |\overline{\Gamma}(s)|$. Hence $\mu_i^2 + \mu_i = 0$, contrary to assumption. Accordingly $\mu_i = \rho$ and $X_i$ contains precisely one vertex from each component of $G$; but each vertex in $\overline{X}_i$ is adjacent to precisely $\rho$ vertices of $X_i$. Therefore, $\mu_i = \rho = 1$ and the result follows.

(ii) Since no vertex in $\overline{X}_i$ is isolated in $G - X_i$, the edges between $X_i$ and $\overline{X}_i$ constitute a perfect matching for $G$ by Lemma 7.6.5. One consequence is that $\mu_i \neq \rho$ for otherwise our initial remarks show that each vertex in $X_i$ is isolated in $G - \overline{X}_i$; it follows that $\rho = 1$ and hence that each vertex of $\overline{X}_i$ is isolated in $G - X_i$, contrary to assumption. Hence $\mu_i \neq \rho$ and we have $P_i \mathbf{j} = \mathbf{0}$ as before.

In view of the perfect matching let $h = |X_i| = |\overline{X}_i| = \frac{1}{2}n$, let $\overline{X}_i = \{v_1, \ldots, v_h\}$ and for $j \in \{1, \ldots, h\}$, let $u_j$ be the unique vertex in $X_i$ such that $\overline{\Gamma}(u_j) = \{v_j\}$. We have $\mu_i P_i e_{u_j} = \sum_{k \in \Gamma(u_j)} P_i e_u + P_i e_{v_j}$ $(j = 1, \ldots, h)$, and since $P_i \mathbf{j} = \mathbf{0}$ we obtain

$$\sum_{j=1}^{h} (\mu_i + 1) P_i e_{u_j} = \sum_{j=1}^{h} \sum_{k \in \Gamma(u_j)} P_i e_k.$$

On equating coefficients of $P_i e_{u_j}$ we find that $\mu_i + 1$ is the degree of $u_j$ in $G - \overline{X}_i$. By Lemma 7.6.6(i), $\mu_i = 1$ and so $G - \overline{X}_i$ is regular of degree 2. By Lemma 7.6.5, $\rho = 3$ and $G - X_i$ is regular of degree 2; moreover the matrix $B_i$ of Theorem 7.4.1 is just the identity matrix when the vertices of $G$ are labelled in the natural way. In the notation of Theorem 7.4.1, $I - A_i = (I - C_i)^{-1}$ and so the involutory map $\lambda \mapsto \lambda/(\lambda - 1)$ is a one-to-one correspondence between the spectrum of $A_i$ and the spectrum of $C_i$. Each component of $G - \overline{X}_i$, and each component of $G - X_i$, is a cycle, and so the eigenvalues of both $A_i$ and $C_i$ are of the form $2\cos\theta$ where for an $m$-cycle, $\theta = 2\pi j/m$, $j \in \{1, \ldots, m\}$ (see [CvDS], p. 53). Since $|\lambda/(\lambda - 1)| \leq 2$ for each eigenvalue $\lambda$, either $\cos\theta = 1$ or $\cos\theta \leq \frac{1}{3}$; but $\cos\frac{2\pi}{m} \leq \frac{1}{3}$ if and only if $m \leq 5$. Moreover $m \notin \{3, 4\}$ for when $\lambda \in \{2\cos\frac{2\pi}{3}, 2\cos\frac{4\pi}{4}\}$, $\lambda/(\lambda - 1)$ is not an eigenvalue of a cycle of length $\leq 5$. Thus each component of $G - \overline{X}_i$, and of $G - X_i$, is a 5-cycle.

Now let $u_1 u_2 u_3 u_4 u_5 u_1$ be a 5-cycle in $G - \overline{X}_i$. Since $\mu_i = 1$ we have $P_i e_{u_j} = P_i e_{v_j} + \sum_{k \in \Gamma(u_j)} P_i e_k$. Since $\sum_{j=1}^{5} \sum_{k \in \Gamma(u_j)} P_i e_k = 2\sum_{j=1}^{5} P_i e_{u_j}$ it follows that $\sum_{j=1}^{5} P_i \mathbf{e}_{u_j} + \sum_{j=1}^{5} P_i \mathbf{e}_{v_j} = \mathbf{0}$. Since $P_i \mathbf{e}_{v_j} = P_i \mathbf{e}_{u_j} + \sum_{k \in \overline{\Gamma}(v_j)} P_i \mathbf{e}_k$, we have

$$\sum_{j=1}^{5} \sum_{k \in \overline{\Gamma}(v_j)} P_i \mathbf{e}_k = 2 \sum_{j=1}^{5} P_i \mathbf{e}_{v_j}. \tag{7.6.1}$$

Now the linear transformation of $\mathcal{E}(\mu_i)$ given by

$$P_i \mathbf{e}_{u_j} \mapsto P_i \mathbf{e}_{v_j} \quad (= P_i \mathbf{e}_{u_j} - \sum_{k \in \Gamma(u_j)} P_i \mathbf{e}_k)$$

has matrix $I - A_i$ with respect to the basis $\{P_i \mathbf{e}_u : u \in X_i\}$. In view of Theorem 7.4.1 (with $B_i = I$), this linear transformation is invertible and so the image vectors $P_i \mathbf{e}_v$ ($v \in \overline{X}_i$) are linearly independent. Accordingly we may equate coefficients of $P_i \mathbf{e}_{v_j}$ in (7.6.1) to deduce that $v_1, v_2, v_3, v_4, v_5$ induce a regular subgraph of degree 2, necessarily a 5-cycle. By Lemma 7.6.6(ii), $v_1$ is not adjacent to $v_2$ or $v_5$ and so $\overline{\Gamma}(v_1) = \{v_3, v_4\}$. Similar arguments for the remaining vertices $v_2, v_3, v_4, v_5$ show that $u_1, u_2, u_3, u_4, u_5, v_1, v_2, v_3, v_4, v_5$ induce the Petersen graph, which is necessarily a component of $G$. It follows that each component of $G$ is a Petersen graph. $\quad\square$

**7.6.8 Corollary** *Let $G$ be a connected regular graph for which the dominating set $\overline{X}_i$ is minimal, with $\mu_i \notin \{-1, 0\}$. If $G - X_i$ has no isolated vertices then $G$ is the Petersen graph.*

Although one motive for studying eigenspaces was the search for additional algebraic invariants which might help to characterize graphs, what we have in Corollary 7.6.8 is the construction of a graph directly from a single eigenspace.

## 7.7 Some enumerative considerations

So far we have proved that any graph has at least one star partition. In this section we find the number of star partitions of some particular graphs, and we list all non-isomorphic star partitions of the Petersen graph.

Let $SP(G)$ denote the number of star partitions of a graph $G$. For example, it is easy to see that $SP(K_n) = n$, whereas $SP(K_{m,n}) = 2mn$. Two star partitions of a graph $G$ are *isomorphic* if there is an automorphism of $G$ taking one partition to another. Let $NSP(G)$ denote the number of non-isomorphic star partitions of $G$. Of course, $NSP(G) \leq SP(G)$. For example, $NSP(K_n) = 1$, whereas $NSP(K_{m,n}) = 2$ for $m \neq n$ and $NSP(K_{n,n}) = 1$.

**7.7.1 Proposition** [CvRS2] *For a graph G on n vertices with eigenvalue multiplicities* $k_1, \ldots, k_m$, *we have*

$$1 \leq SP(G) \leq \frac{n!}{k_1! \ldots k_m!}. \tag{7.7.1}$$

*Proof* The upper bound is just the number of (ordered) feasible partitions of $G$. $\qquad \square$

The next proposition describes the graphs for which the lower bound in (7.7.1) is attained.

**7.7.2 Proposition** [CvRS2] *We have* $SP(G) = 1$ *if and only if* $G = nK_1$ ($n \geq 1$).

*Proof* If $G = nK_1$ then clearly $SP(G) = 1$. If $G \neq nK_1$ then $G$ has at least two distinct eigenvalues, and the same holds for some non-trivial component of $G$, say $H$. From Theorem 7.4.5 (applied to the index of $H$) we know that $SP(H) > 1$. Given a star partition of each component we obtain a star partition of $G$ by combining star cells corresponding to the same eigenvalue (see Proposition 7.5.1). Thus $SP(G) > 1$. $\qquad \square$

The upper bound in (7.7.1) is attained when $G$ is a complete graph. If all eigenvalues of $G$ are simple then we have $SP(G) \leq n!$, and the next proposition provides an infinite family of graphs for which this bound is attained.

**7.7.3 Proposition** [CvRS2] *Let* $P_n$ *be a path on n vertices. We have* $SP(P_n) = n!$ *if and only if* $n + 1$ *is a prime number.*

*Proof* Let $G = P_n$. Then the eigenvalues $\mu_i$ of $G$ are simple and $\mu_i = 2\cos\frac{\pi i}{n+1}$ ($1 \leq i \leq n$). By Theorem 7.2.3, $SP(G) = n!$ if and only if no $\mu_i$ is an eigenvalue of $G - k$, for any $k \in \{1, \ldots, n\}$. Since $G - k$ is a graph with each component a path, this is true if and only if

$$2\cos\frac{\pi i}{n+1} \neq 2\cos\frac{\pi j}{m+1}$$

for each $m, i$ and $j$ ($1 \leq m < n; 1 \leq i \leq n; 1 \leq j \leq m$). The latter is possible if and only if $n + 1$ is a prime number. $\qquad \square$

**7.7.4 Remark** If $n + 1$ ($n \neq 1$) is prime then $NSP(P_n) = n!/2$ since $P_n$ has only one non-trivial automorphism when $n > 1$. $\qquad \square$

For graphs having only simple eigenvalues we have the following necessary and sufficient condition for every feasible partition to be a star partition.

**7.7.5 Proposition** *For a graph $G$ with only simple eigenvalues, any feasible partition is a star partition if and only if all entries of an eigenvector matrix are non-zero.*

*Proof* Note that all eigenvector matrices of $G$ have the same pattern of zeros. Now the result follows immediately from Corollary 7.2.7. $\square$

More generally, for a graph having only simple eigenvalues, the problem of enumerating the star partitions becomes equivalent to the problem of enumerating the perfect matchings in the König digraph $K(E)$ with an associated eigenvector matrix $E$ (see Definition 1.1.1). For if $E$ has size $n \times n$ then by Corollary 7.2.7 we need to find $n$ non-zero entries exactly one of which appears in each row and each column. This is precisely a perfect matching in $K(E)$ and so $SP(G)$ becomes equal to the number of perfect matchings in $K(E)$. Enumeration of perfect matchings in a bipartite graph can be performed in several ways (see, for example, [LoPl]).

In the rest of this section we describe the construction in [Cve21] of all star partitions of the Petersen graph. It is well known that this graph appears in many situations in graph theory as an example or counterexample. The same applies in the context of star partitions, as we have already seen; accordingly it merits further investigation here.

The Petersen graph $P$ is a transitive strongly regular graph with eigenvalues $\mu_1 = 3, \mu_2 = 1, \mu_3 = -2$ of multiplicities 1, 5, 4, respectively. We shall see that it has just 10 non-isomorphic star partitions, illustrated by the first 10 copies of $P$ shown in Fig. 7.7.

Let $X_1 \dot\cup X_2 \dot\cup X_3$ be a star partition of $P$, so that $|X_1| = 1, |X_2| = 5$ and $|X_3| = 4$. Since $P$ is transitive we may take the single vertex in $X_1$ to be the central vertex in our illustrations. In order to describe the possibilities for $X_2$ and $X_3$, it is convenient to denote the graphs $C_5, P_5, P_4 \cup K_1, 2K_2 \cup K_1$ by $a, b, c, d$ and the graphs $P_4, P_3 \cup K_1, K_2 \cup 2K_1, 4K_1$ by $1, 2, 3, 4$ respectively. By Theorem 7.2.3, each of the first 15 copies of $P$ in Fig. 7.7 illustrates a star partition for which the label (e.g. $a1, b1, \ldots$) refers to the subgraphs induced by $X_2$ and $X_3$. The subgraph induced by $X_2$ is shown with white vertices and broken edges, and the subgraph induced by $X_3$ is shown with black vertices and solid edges.

The first ten star partitions are non-isomorphic, while the partitions 11-15 are isomorphic to predecessors with the same labels.

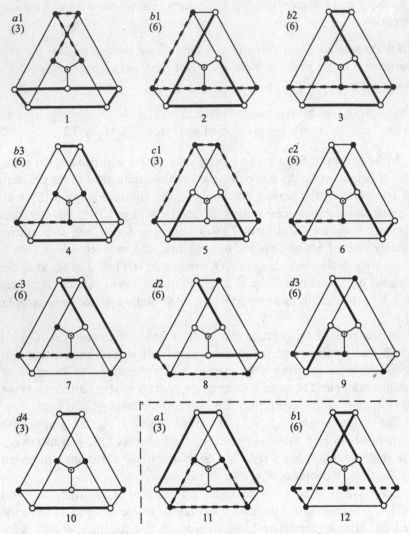

Fig. 7.7. Star partitions of the Petersen graph.

In order to show that we have all star partitions of $P$ up to isomorphism, consider the copies of $P$ numbered 16-24 in Fig. 7.7. In each case the induced subgraph with black vertices and solid edges is a graph which has an eigenvalue $\lambda$ equal to $-2$ or $1$ as indicated. It follows from Theorem 7.2.3 that the white vertices do not form a star cell corresponding to $\lambda$. Among the 24 feasible partitions illustrated in Fig. 7.7, none can

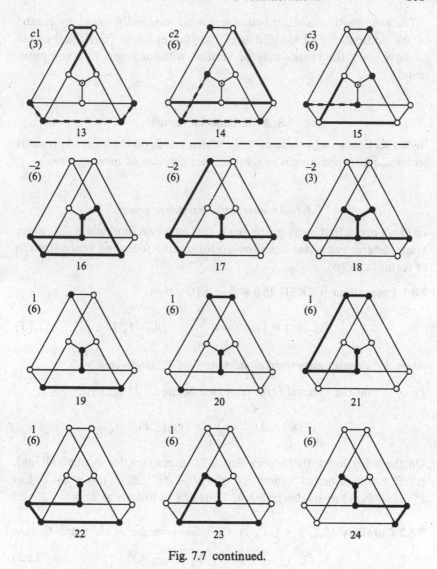

Fig. 7.7 continued.

be transformed to another by rotations and reflections. If we take these into account then the number of feasible partitions we obtain (shown in parentheses in each case) is 126. In other words we have generated all $\binom{9}{4}$ feasible partitions and only those 75 arising from partitions 1-15 are star partitions. It follows that $P$ has 750 star partitions in all, divided into the ten isomorphism classes represented by partitions 1-10.

The star partitions described above were originally found by means of the computer package GRAPH (see for example [CvKS]), but, as we have seen, the results may be verified without resort to a computer search.

## 7.8 Miscellaneous results

In this section we add some material which builds on the ideas of previous sections, and which might suggest further avenues of investigation.

### 7.8.1 An alternative existence proof

Here we prove that every graph has a star partition (and hence that every graph determines a star basis) by exploiting the following generalization of formula (4.3.5).

**7.8.1 Proposition** [CvRS1] *If $\emptyset \neq S \subset V(G)$, then*

$$P_{G-S}(x) = P_G(x) \det(\sum_{i=1}^{m}(x - \mu_i)^{-1}P_i^S), \qquad (7.8.1)$$

*where $P_i^S$ denotes the principal submatrix of $P_i$ determined by $S$.*

*Proof* From the spectral form of $A$ (see Section 1.1) we have

$$(xI - A)^{-1} = \sum_{i=1}^{m}(x - \mu_i)^{-1}P_i.$$

On the other hand, by Jacobi's Ratio Theorem (see, for example, [Gan], p. 21) the principal submatrix of $(xI - A)^{-1}$ determined by $S$ has $P_{G-S}(x)/P_G(x)$ as its determinant. Thus (7.8.1) follows at once.  □

**7.8.2 Corollary** *If $X_1 \dot\cup \cdots \dot\cup X_m$ is a feasible partition of the graph $G$, then*

$$P_{G-X_i}(\mu_i) = \det(P_i^{X_i}) \prod_{s \neq i}(\mu_i - \mu_s)^{k_s}. \qquad (7.8.2)$$

*Proof* In (7.8.1) we use determinant multilinearity with respect to rows (or columns), and take the limit of $P_{G-X_i}(x)$ as $x \to \mu_i$.  □

**7.8.3 Remark** Since $P_i^2 = P_i = P_i^T$ for each $i$, $\det P_i^{X_i}$ can be interpreted as the volume of the parallelepiped generated by vectors $P_i e_s$ ($s \in X_i$). Thus $\det P_i^{X_i} \neq 0$ if and only if the vectors $P_i e_s$ $s \in X_i$ are linearly independent. Since $\prod_{s \neq i}(\mu_i - \mu_s)^{k_s} \neq 0$, we see that the feasible partition $X_1 \dot\cup \cdots \dot\cup X_m$

is a star partition if and only if $P_{G-X_i}(\mu_i) \neq 0$ for all $i$ (another proof of Theorem 7.2.3). $\qquad\qquad\qquad\qquad\qquad\qquad\qquad\qquad\qquad\qquad\qquad$ □

**7.8.4 Corollary** *If $X_1 \dot{\cup} \cdots \dot{\cup} X_m$ is a feasible partition of the graph $G$, then the volume $V(X_1, \ldots, X_m)$ of the parallelepiped generated by the vectors $P_i e_s$ ($i = 1, \ldots, m; s \in X_i$) is given by*

$$V(X_1, \ldots, X_m) = \sqrt{C \prod_{i=1}^{m} P_{G-X_i}(\mu_i)}, \qquad (7.8.3)$$

*where $C$ depends only on the spectrum of $G$ and not on the particular partition.*

*Proof* Since $P_i^2 = P_i = P_i^T$ for each $i$, the Gram matrix of the vectors $P_i e_s$ ($s \in X_i$) ($i = 1, \ldots, m$) has the form

$$M = P_1^{X_1} \dotplus \cdots \dotplus P_m^{X_m},$$

where $\dotplus$ denotes the direct sum of matrices. Now, since $V(X_1, \ldots, X_m) = \sqrt{\det M}$, we easily obtain (7.8.3) by making use of (7.8.2). $\qquad\qquad$ □

Two questions which may be posed here are:

(1) To what extent is a graph $G$ determined by its spectrum and the submatrices $P_1^{X_1}, P_2^{X_2}, \ldots, P_m^{X_m}$?

(2) If $X_1 \dot{\cup} \cdots \dot{\cup} X_m, Y_1 \dot{\cup} \cdots \dot{\cup} Y_m$ are feasible partitions of $G, H$ respectively, and if $P_{G-X_i}(x) = P_{H-Y_i}(x)$ ($i = 1, 2, \ldots, m$), does it follow that $P_G(x) = P_H(x)$?

The second problem may well be intractable in general because in the case that $G$ and $H$ have only simple eigenvalues, it becomes the non-trivial case of the polynomial reconstruction problem [GuCv]. This is the problem of reconstructing the characteristic polynomial of a graph from the collection of characteristic polynomials of its vertex-deleted subgraphs.

### 7.8.2 More on polynomial partitions

Polynomial partitions (of the vertex set of a graph) were introduced in Definition 7.2.1, and we saw in Section 7.2 that they coincide with star partitions. The motivation for introducing polynomial partitions can be traced back to an attempt to compute the eigenvectors of a graph starting from the eigenvalue equations

$$\sum_{j=1}^{n} a_{ij}x_j = \lambda x_i \ (i = 1, 2, \ldots, n).$$

This approach leads naturally to the concepts of *partial $\lambda$-eigenvectors* and *exit-values* (as considered by Neumaier in [Neu]).

Let $\lambda \in \mathbb{R}$ and let $r \in \{1, \ldots, n\}$. Following Neumaier we say that $\mathbf{y} = (y_1, \ldots, y_n)^T \in \mathbb{R}^n$ is a partial $\lambda$-eigenvector of an $n \times n$ matrix $A = (a_{ij})$ with respect to $r$, if $y_r = 1$ and $\mathbf{y}$ satisfies all eigenvalue equations except the $r$-th, i.e. if

$$\sum_{j=1}^{n} a_{ij}y_j = \lambda y_i \ (i \neq r). \tag{7.8.4}$$

The corresponding exit-value is defined by

$$\phi_r(\lambda) = \lambda - \sum_{j=1}^{n} a_{rj}y_j. \tag{7.8.5}$$

Note that if $\phi_r(\tilde{\lambda}) = 0$ then $\tilde{\lambda}$ is an eigenvalue of $A$.

Now assume that $A$ is the adjacency matrix of a graph $G$, and let $r$ be some fixed vertex of $G$. If (7.8.4) and (7.8.5) are interpreted in $G$, we have

$$\sum_{j \sim i} y_j = \lambda y_i \ (i \neq r), \tag{7.8.6}$$

and for $i = r$,

$$\phi_r(\lambda) = \lambda - \sum_{j \sim r} y_j. \tag{7.8.7}$$

We now proceed to find the partial $\lambda$-eigenvectors of a graph $G$ (i.e. of its adjacency matrix) corresponding to the fixed vertex $r$. Without loss of generality, let $r = n$. Starting from the eigenvalue system

$$(\lambda - a_{11})x_1(\lambda) - a_{12}x_2(\lambda) - \cdots - a_{1n}x_n(\lambda) = 0,$$
$$-a_{21}x_1(\lambda) + (\lambda - a_{22})x_2(\lambda) - \cdots - a_{2n}x_n(\lambda) = 0,$$

$$\vdots$$

$$-a_{n1}x_1(\lambda) - a_{n2}x_2(\lambda) - \cdots + (\lambda - a_{nn})x_n(\lambda) = 0$$

it immediately follows (by ignoring the last equation and applying Cramer's rule to the system in $x_1(\lambda), \ldots, x_{n-1}(\lambda)$) that

$$P_{G-n}(\lambda)x_i(\lambda) = x_n(\lambda)\theta_{ni}(\lambda I - A) \quad (i \in \{1, \ldots, n-1\}),$$

where $\theta_{st}(X)$ denotes the $(s, t)$-cofactor of a matrix $X$. Now from [Row3] (cf. (5.1.5)) we obtain

$$\theta_{ni}(\lambda I - A) = \begin{cases} \frac{1}{2}[P_{G-in}(\lambda) + P_{G-i-n}(\lambda) - P_G(\lambda)] & \text{if } i \sim n, \\[2mm] \frac{1}{2}[P_G(\lambda) - P_{G-i-n}(\lambda) - P_{G+in}(\lambda)] & \text{if } i \not\sim n. \end{cases}$$

When $P_{G-n}(\lambda) \neq 0$, we set $x_n(\lambda) = 2P_{G-n}(\lambda)$ to obtain (for any $r$, not necessarily $r = n$)

$$x_i(\lambda) = x_i(G;r,\lambda) = \left\{ \begin{array}{ll} P_{G-ir}(\lambda) + P_{G-i-r}(\lambda) - P_G(\lambda) & \text{if } i \sim r, \\ P_G(\lambda) - P_{G-i-r}(\lambda) - P_{G+ir}(\lambda) & \text{if } i \not\sim r, \\ 2P_{G-r}(\lambda) & \text{if } i = r. \end{array} \right\} \quad (7.8.8)$$

For the partial $\lambda$-eigenvector $\mathbf{y} = (y_1, \ldots, y_n)^T$ we find that

$$y_i = y_i(\lambda) = \frac{x_i(\lambda)}{x_r(\lambda)} \quad (i = 1, \ldots, n).$$

We can now prove the following.

**7.8.5 Proposition** [Neu] *Suppose that the real number $\lambda$ is not an eigenvalue of $G - r$. Then $G$ has a unique partial $\lambda$-eigenvector with respect to $r$, and the corresponding exit-value is given by*

$$\phi_r(\lambda) = \frac{P_G(\lambda)}{P_{G-r}(\lambda)}. \tag{7.8.9}$$

*Proof* Let $\mathbf{e}_r$ be the $r$-th column of the identity matrix $I$. Then $\mathbf{y}$ is a partial $\lambda$-eigenvector and $\phi$ the corresponding exit-value if and only if $\mathbf{y}$ is a solution of the homogeneous equation

$$(\lambda I - A - \phi \mathbf{e}_r \mathbf{e}_r^T)\mathbf{y} = 0 \tag{7.8.10}$$

with side condition

$$\mathbf{e}_r^T \mathbf{y} = 1. \tag{7.8.11}$$

We distinguish two cases:

(i) $\lambda I - A$ is non-singular. Then (7.8.10) and (7.8.11) imply that $\mathbf{y} = \phi(\lambda I - A)^{-1}\mathbf{e}_r(\mathbf{e}_r^T \mathbf{y}) = \phi(\lambda I - A)^{-1}\mathbf{e}_r$, whence $\phi^{-1} = \mathbf{e}_r^T(\lambda I - A)^{-1}\mathbf{e}_r = P_{G-r}(\lambda)/P_G(\lambda)$, by Cramer's rule. Hence $\phi$ and $\mathbf{y}$ are unique; moreover the foregoing expressions satisfy (7.8.10) and (7.8.11), and so (7.8.9) holds.

(ii) $\lambda I - A$ is singular. Then $\lambda$ is an eigenvalue of $G$. Let $\mathbf{y}$ be a corresponding eigenvector. If $y_r = 0$, then the remaining entries of $\mathbf{y}$ determine a $\lambda$-eigenvector of $G - r$, a contradiction. Hence we may normalize $\mathbf{y}$ so that $y_r = 1$; then $\mathbf{y}$ is a partial eigenvector with exit-value $\phi = 0$, and (7.8.9) holds. Finally, if $\mathbf{y'}$ is another partial $\lambda$-eigenvector, then $\mathbf{y'} - \mathbf{y}$ is a $\lambda$-eigenvector of $G - r$, contrary to assumption.            $\square$

We shall now use the above formulas to compute the eigenvectors of $G$. More precisely, we shall compute, for a given vertex $u$ of $G$, an eigenvector corresponding to $u$ associated with some fixed star partition (cf. Theorem 7.2.6). We take $U$ to be the star cell containing $u$, and $\mu_s$ the associated eigenvalue. With these assumptions, if $(x_1, \ldots, x_n)^T$ denotes the corresponding eigenvector $\mathbf{x}^u$, we have the following possibilities.

**Case 1:** $\mu_s$ is a simple eigenvalue of $G$ $(k_s = 1)$. Then $P_{G-u}(\mu_s) \neq 0$. Now (7.8.8) yields

$$x_i = x_i(G; u, \mu_s) \quad (i \in V(G)). \tag{7.8.12}$$

**Case 2:** $\mu_s$ is a multiple eigenvalue of $G$ $(k_s > 1)$. Then $P_{G-U}(\mu_s) \neq 0$. Now (7.8.8), applied to the subgraph $H_u$ of $G$ induced by the vertex set $\{u\} \cup (V(G) \setminus U)$ $(u \in U)$, yields

$$x_i = \begin{cases} 0 & i \notin V(H_u), \\ x_i(H_u; u, \mu_s) & i \in V(H_u). \end{cases} \tag{7.8.13}$$

Notice that for a fixed eigenvalue $\mu_s$, the $k_s$ vectors $\mathbf{x}^{(u)}$ $(u \in U)$ specified above are linearly independent. To see that they are eigenvectors of $G$, observe that each $x^{(u)}$ represents a solution of the $n - k_s$ equations of the eigenvalue system (with $x = \mu_s$), or, more precisely, a solution of those equations which are indexed by $V(G) \setminus U$. Since the rank of $\mu_s I - A$ is $n - k_s$ and $P_{G-U}(\mu_s) \neq 0$, these $n - k_s$ equations are equivalent to the whole system. Thus the above set of vectors forms a basis for the corresponding eigenspace $\mathscr{E}(\mu_s)$.

We summarize our conclusions as follows:

**7.8.6 Theorem** *Let* $X_1 \dot\cup X_2 \dot\cup \ldots \dot\cup X_m$ *be a star partition of* $G$. *For each* $s \in \{1, 2, \ldots, m\}$, *the vectors* $\mathbf{x}^{(u)}(u \in X_s)$, *with components given by (7.8.12) or (7.8.13), form a basis for* $\mathscr{E}(\mu_s)$.

From (7.8.13) we see that the components of $\mathbf{x}^{(u)}$ $(u \in U)$ do not depend on the structure of the subgraph induced by vertices from $U$. This phenomenon is already explained by the Reconstruction Theorem (Theorem 7.4.1).

Although polynomial partitions and star partitions are one and the same, it is worth mentioning at this point that the idea of polynomial partitions stemmed from an attempt to establish a one-to-one correspondence betwen the vertices of graph and a basis of eigenvectors. In the notation above, formulas (7.8.12) and (7.8.13) enable us to associate a $\mu_s$-eigenvector with each of the $k_s$ vertices in the $U$ provided that $P_{G-U}(\mu_s) \neq 0$. Thus in order to construct a basis of eigenvectors in this way we need a polynomial partition of $G$.

We now deduce some consequences of the above considerations. First recall that two (distinct) eigenvalues of a graph are *conjugate* if they are conjugate as algebraic numbers (equivalently, if they have the same minimal polynomial over the rationals). Now we can also say that two star cells (within a fixed star partition), say $X_i$ and $X_j$, are *conjugate* if and only if the corresponding eigenvalues $\mu_i$ and $\mu_j$ are conjugate. We have

**7.8.7 Proposition** *Let $\mu_i$ and $\mu_j$ be conjugate eigenvalues of a graph $G$. Then $U$ is a $\mu_i$-cell if and only if it is a $\mu_j$-cell.*

*Proof* Suppose that $U$ is a $\mu_i$-cell. By Theorem 7.2.9, $\mu_i$ is not an eigenvalue of $G - U$. Since $\mu_i$ and $\mu_j$ are conjugate, we also have that $\mu_j$ is not an eigenvalue of $G - U$. By Theorem 7.4.5, $U$ is a $\mu_j$-cell. The converse is true by symmetry. $\qquad\qquad\square$

From this proposition it follows that we can obtain a new star partition by interchanging all vertices between two conjugate star cells. A similar result, with an identical proof, holds for bipartite graphs in respect of opposite eigenvalues.

**7.8.8 Proposition** [CvRS1] *Let $\mu_i$ and $\mu_j$ be opposite eigenvalues of a bipartite graph $G$ (i.e. $\mu_i = -\mu_j$). Then $U$ is a $\mu_i$-cell if and only if it is a $\mu_j$-cell.*

**7.8.9 Example** Let $G$ be the corona $K_n \circ K_1$; in other words, $G$ consists of a complete graph on $n$ vertices and a pendant edge at each vertex. From [CvGS] we know that $G$ has conjugate eigenvalues $\sigma = \frac{1}{2}(\sqrt{5} - 1)$ and $\tau = -\frac{1}{2}(\sqrt{5} + 1)$, each of multiplicity $n - 1$. Their minimal polynomial is $x^2 + x - 1$. Now, for example, any collection of $n - 1$ vertices of degree 1 (or $n$) is a star cell for both eigenvalues. Notice also that neither $\sigma$ nor $\tau$ is a main eigenvalue. (To see this, construct $n - 1$ linearly independent eigenvectors, each having non-zero components corresponding to the vertices of an induced subgraph isomorphic to $P_4$ and zero components elsewhere.) $\qquad\qquad\square$

**7.8.10 Proposition** *Let $\mu_i$ and $\mu_j$ be conjugate eigenvalues of a graph G. Then $\mu_i$ is a main eigenvalue of G if and only if $\mu_j$ is a main eigenvalue of G.*

*Proof* In view of Proposition 7.8.7, we may assume that $U$ is a star cell for both eigenvalues. Then we can associate with each vertex $u$ of $U$ a $\mu_i$-eigenvector and a $\mu_j$-eigenvector as specified by Proposition 7.8.5. Then these eigenvectors involve the same polynomial functions for each component, evaluated at different points. Let $\mathbf{x}^{(u)}(\lambda)$ be the corresponding vector function of $\lambda$ and consider $p_u(\lambda)$, where $p_u(\lambda) = (\mathbf{x}^{(u)}(\lambda))^T\mathbf{j}$ and $\mathbf{j}$ is the all-1 vector. Clearly, $p_u(\lambda)$ is a polynomial in $\lambda$. Since $\mu_i$ and $\mu_j$ are conjugate, this implies that $p_u(\mu_i) = 0$ if and only if $p_u(\mu_j) = 0$. Thus the vector $\mathbf{j}$ is orthogonal to each eigenspace (since it is orthogonal to a basis). It follows that $\mathbf{j}$ is orthogonal to $\mathscr{E}(\mu_i)$ if and only if it is orthogonal to $\mathscr{E}(\mu_j)$. $\qquad\square$

Finally we deduce (cf. [CvRS1]) a recursive formula for computing the characteristic polynomial of any graph. It is worth mentioning that all the graphs involved in this formula are local modifications of the graph in question, in contrast to the formulas of Schwenk (see (5.3.5) or [Sch2]). For some other formulas involving local modifications see [CvDGT], or (5.1.4) and (5.3.6).

**7.8.11 Proposition** *Let G be a graph with a vertex u whose degree differs from 2, i.e. $\deg(u) \neq 2$. Then*

$$P_G(x) = \frac{1}{\deg(u)-2}\left(\sum_{i\sim u} P_{G-iu}(x) + \sum_{i\sim u} P_{G-i-u}(x) - 2xP_{G-u}(x)\right). \quad (7.8.14)$$

*Proof* Let $\lambda$ be a real number not in the spectrum of $G-u$. From (7.8.7), (7.8.8) and (7.8.9) we have

$$P_G(\lambda) = \lambda P_{G-u}(\lambda) - \frac{1}{2}\sum_{i\sim u} x_i(\lambda).$$

Next, by (7.8.8), we obtain

$$(\deg(u) - 2)P_G(\lambda) = \sum_{i\sim u} P_{G-iu}(\lambda) + \sum_{i\sim u} P_{G-i-u}(\lambda) - 2\lambda P_{G-u}(\lambda). \quad (7.8.15)$$

We may now replace $\lambda$ with $x$ to obtain the result. $\qquad\square$

**7.8.12 Remark** The formula (7.8.14) can also be derived by combining the aforementioned formulas of Schwenk. Note also that if $\deg(u) = 2$ in

(7.8.15), then the equation reduces to a simple consequence of Heilbronner's formulas (see, for example, [CvDS], p. 59). □

### 7.8.3 Line star partitions

Here we introduce the concept of *line star partitions*, which were first defined in [Cve21], and which will be useful in further considerations (see Section 8.6).

Let $G$ be a graph and $L(G)$ its line graph. A star partition of $L(G)$ determines a partition of the edge set of $G$ in a natural way. We shall identify edges of $G$ with vertices of $L(G)$ and use the same notation for a star partition of $L(G)$ as for the corresponding edge partition of $G$.

A partition of the edge set of a graph $G$, which is determined by a star partition of $L(G)$, is called a *line star partition* of $G$. Cells of a line star partition are called *line star cells* of $G$.

With Theorems 1.3.16 and 2.6.6 in mind, we formulate the following theorem.

**7.8.13 Theorem** *Let $G$ be a connected graph with an even cycle or two odd cycles. Let $F$ be a spanning subgraph of $G$ whose edges complement a line star cell of $G$ corresponding to the eigenvalue $-2$. If $G$ is bipartite then $F$ is a tree. If $G$ is not bipartite then each component of $F$ is a unicyclic graph with a cycle of odd length.*

*Proof* If $X$ is the given line star cell then $|X| = k$, where $k$ is the least number of edges of $G$ whose removal results in a graph $L(G) - X$ which does not have $-2$ as an eigenvalue. By Theorem 1.3.16, $L(G) - X$ has this property if and only if each component of $F$ has no even cycle and at most one odd cycle. Let $|E(G)| = m$, $|V(G)| = n$. If $G$ is bipartite then $k = m - n + 1$ and $F$ is a spanning tree. If $G$ is non-bipartite then $k = m - n$ and each component of $F$ is a unicyclic graph in which the cycle has odd length. □

**7.8.14 Example** The graph $L(K_{n,n})$ has eigenvalues $2n - 2$, $n - 2$ and $-2$ with multiplicities 1, $2n - 2$ and $(n - 1)^2$ respectively (see, for example, [CvDS], p. 174). The union of line star cells belonging to eigenvalues $2n - 2$ and $n - 2$ forms a spanning tree of $K_{n,n}$. □

**7.8.15 Example** The graph $L(K_n)$ has eigenvalues $2n - 4$, $n - 4$ and $-2$

with multiplicities 1, $n - 1$ and $n(n - 3)/2$, respectively. Let $F$ be the spanning subgraph of $K_n$ formed by the union of line star cells belonging to the eigenvalues $2n - 4$ and $n - 4$. Then the components of $F$ are unicyclic graphs with an odd cycle.    □

# 8

# Canonical star bases

We associate with an $n$-vertex graph a uniquely determined star basis of $\mathbb{R}^n$ which is canonical in the sense that two cospectral graphs are isomorphic if and only if they determine the same canonical star basis. Such a basis was introduced in [CvRS1] as a means of investigating the complexity of the graph isomorphism problem. Here we first present a polynomial algorithm [CvRS2] for constructing a star partition of $G$, and hence a corresponding star basis of $\mathbb{R}^n$. Thereafter we describe a procedure, based on [CvRS2] and [Cve21], for constructing the canonical star basis, with emphasis on three special cases: graphs with distinct eigenvalues, graphs with bounded eigenvalue multiplicities, and strongly regular graphs. The approach provides an alternative proof of a result of Babai et al. [BaGM] that isomorphism testing for graphs with bounded eigenvalue multiplicities can be performed in polynomial time.

Since the material presented in this chapter is the subject of current research, changes and improvements to the procedure for constructing a canonical basis may well emerge in due course. The chapter is included nevertheless because it represents the original motivation for some of the work discussed earlier in the book.

## 8.1 Introduction

There are only finitely many star bases associated with a given graph $G$; for there are only finitely many star partitions of $V(G)$, and if $|V(G)| = n$ then each star partition determines $n!$ star bases of $\mathbb{R}^n$ (one for each labelling of the vertices). Accordingly we could take as a canonical basis the one which is largest with respect to some lexicographical ordering. This however would be highly inefficient, and our objective here is to construct a canonical star basis in significantly fewer steps. We begin

191

with an expression for the spectral decomposition of a real symmetric matrix in terms of bases of its eigenspaces.

**8.1.1 Theorem** *Let $A$ be a real symmetric matrix whose distinct eigenvalues are $\mu_1, ..., \mu_m$. For each $i \in \{1, ..., m\}$ let $S_i$ be a matrix whose columns form a basis for $\mathscr{E}_A(\mu_i)$. Then*

$$A = \mu_1 S_1 (S_1^T S_1)^{-1} S_1^T + \cdots + \mu_m S_m (S_m^T S_m)^{-1} S_m^T. \qquad (8.1.1)$$

*Proof* Let $Q_i = S_i (S_i^T S_i)^{-1} S_i^T$, where $S_i$ has size $n \times k_i$. Since $Q_i^2 = Q_i = Q_i^T$, $Q_i$ represents the orthogonal projection of $\mathbb{R}^n$ onto $Q_i(\mathbb{R}^n)$, which is a subspace of $\mathscr{E}_A(\mu_i)$. Since $Q_i$ has rank $k_i$, we have $Q_i(\mathbb{R}^n) = \mathscr{E}_A(\mu_i)$ and so (8.1.1) is the spectral decomposition of $A$. $\qquad\qquad\square$

Now let $G$ be a graph with $n$ vertices and distinct eigenvalues $\mu_1, \mu_2, \ldots, \mu_m$ in decreasing order. We shall associate with $G$ an ordering of vertices, unique to within an automorphism of $G$, which affords a unique matrix $S(G)$ whose columns form a star basis of $\mathbb{R}^n$. If $A$ is the adjacency matrix of $G$ then $S(G) = (S_1 | S_2 | \cdots | S_m)$ where the columns of $S_i$ form a star basis of $\mathscr{E}_A(\mu_i)$. In view of Theorem 8.1.1, $G$ is determined by $\mu_1, \mu_2, \ldots, \mu_m$ and $S_1, S_2, \ldots, S_m$. Accordingly these $2m$ invariants are called the *canonical code* of $G$, while the columns of $S(G)$ constitute the *canonical star basis* of $\mathbb{R}^n$ determined by $G$. We may now rephrase our remarks as follows.

**8.1.2 Theorem** *Two graphs are isomorphic if and only if they share the same eigenvalues and the same canonical basis; equivalently, if and only if they share the same spectral moments and the same canonical basis.*

To construct canonical star bases, we introduce in Section 8.2 a total ordering of graphs called the *canonical graph ordering* (CGO) and a relation on vertices called *canonical vertex ordering* (CVO). For a graph $G$, CVO on $V(G)$ is an orbit of total orders under the automorphism group of $G$. Our procedure for constructing $S(G)$ should be regarded as a paradigm for similar approaches; for example, although in Definition 8.2.5 we first order graphs lexicographically by their spectral moments, we might first order them lexicographically by their eigenvalues instead. The usefulness of ordering graphs by spectral moments has however been noted in [CvPe2], [CvRa2], [CvDGT], [CvPe3].

Both CGO and CVO are defined recursively in terms of graphs with a smaller number of vertices, and the process requires several optimizations. For some classes of graphs these include instances of well-known combinatorial optimization problems such as finding a perfect matching, a minimal cut or a maximal clique in a graph. In such situations we obtain information on the complexity of our procedure, and hence on the complexity of the graph isomorphism problem for the class of graphs in question. In some cases the optimization problems belong to the class P, while others are known to be in the class NP. From the point of view of practical computation it is usually not necessary to know whether the graph isomorphism problem is NP-complete or belongs to P: experience has shown that any 'reasonable' algorithm for graph isomorphism testing performs well 'on average'. However, the question has great theoretical significance: leaving aside implications for the theory of complexity of algorithms, one can say that understanding the difficulties arising in the graph isomorphism problem enables one to understand those that arise in graph theory problems in general. We note in passing that other procedures for testing graph isomorphism, based on the spectral decomposition of matrices, are described in [Živ] and [Lin]. Other spectrally based algortithms can be found in [Che], [JoLe], [Kuh], [SeFK] and [WiKe].

We remark that it is often the case that graph invariants which provide structural information (and are therefore useful for the graph isomorphism problem) are obtained by solving some kind of optimization problem. For example, eigenvalues are obtained in this way (as extrema of the Rayleigh quotient) and the same holds for angles of a graph (Section 4.2). See [ChPh], [Ver] for other examples.

The rest of the chapter is organized as follows. Section 8.2 contains definitions and results needed for the construction of canonical star bases. In particular we define a canonical ordering of weighted graphs and a canonical ordering of vertices in a weighted graph. We also introduce the notions of an *orthodox* star partition and an *orthodox* star basis. In Section 8.3 we describe two algorithms from [CvRS2] which show that, given an $n$-vertex graph $G$, it is possible to construct one star partition of $V(G)$ in time bounded by a polynomial function of $n$. In Section 8.4 we consider graphs without repeated eigenvalues: in this case the task of finding a canonical basis is reduced to the problem of finding a perfect matching of maximal weight in a bipartite graph. In Section 8.5 the problem of finding an orthodox star basis is reduced to the problem of finding a maximal clique in a weighted graph. If eigenvalue multiplicities are bounded, a maximal clique can be found in polynomial time and in this way we confirm a result by Babai *et al.* [BaGM] that isomorphism testing for graphs with bounded multiplicities can be performed in polynomial time. Strongly regular graphs are discussed in Section 8.6; in some cases the problem of finding an orthodox star basis is reduced to the problem of dividing a vertex-deleted subgraph into two parts with given cardinalities such that the number of edges between the parts is minimal (the minimal cut problem). Then the canonical star basis is related to a decomposition of the graph which is determined by two strongly regular induced subgraphs.

## 8.2 Canonical star bases and weighted graphs

In this section we give recursive definitions of canonical graph ordering (CGO) and canonical vertex ordering (CVO). If these have been defined for graphs with $k$ or fewer vertices then we can define a canonical ordering (CWGO) of weighted graphs with $k$ vertices, and a canonical vertex ordering (CWGVO) for a weighted graph with $k$ vertices. We are then in a position to define the canonical star basis of an $n$-vertex graph in terms of weighted graphs with fewer than $n$ vertices. (In a weighted graph, each vertex may have a loop of non-zero weight attached.)

Let $G$ be a graph whose vertices are labelled $1, 2, \ldots, n$, and whose distinct eigenvalues are $\mu_1, \mu_2, \ldots, \mu_m$ in decreasing order. Recall that if $X_1 \dot\cup X_2 \dot\cup \cdots \dot\cup X_m$ is a star partition for $G$ then a corresponding star basis of $\mathbb{R}^n$ consists of the vectors $P_i e_j$ ($j \in X_i; i = 1, 2, \ldots, m$). We regard bases as *ordered* bases, and so it will be necessary to order the vectors concerned. We shall always order them first by $i$, and secondly by specifying an order of the vertices within each star cell. The corresponding eigenvector matrix has the form $S = (S_1 | S_2 | \cdots | S_m)$ where the columns of $S_i$ form a star basis for $\mathscr{E}_A(\mu_i)$ ($i = 1, 2, \ldots, m$). It is important to distinguish between the ordering and relabelling of vertices. If the vertices of $X_i$ are reordered then $S_i$ is postmultiplied by a permutation matrix and (as can be seen from equation (8.1.1)) $A$ remains unchanged. On the other hand, if the vertices $1, 2, \ldots, n$ are relabelled $\pi(1), \pi(2), \ldots, \pi(n)$ then $S$ is premultiplied by $Q$, where $Q$ is the $n \times n$ permutation matrix ($\delta_{i\pi(j)}$). In this situation, the spectral decomposition (8.1.1) becomes

$$A' = \sum_{i=1}^{m} \mu_i (QS_i)((QS_i)^T (QS_i))^{-1} (QS_i)^T.$$

Thus $A' = QAQ^T = \sum_{i=1}^{m} \mu_i P_i'$, where $P_i' = QP_iQ^T$ and $\mathscr{E}_{A'}(\mu_i) = Q\mathscr{E}_A(\mu_i)$ ($i = 1, 2, \ldots, m$).

Let $W_i$ be the Gram matrix of the vectors $P_i e_j$ ($j \in X_i$). Thus

$W_i = S_i^T S_i$, and $W_i$ represents a weighted graph: here and subsequently we shall identify a real symmetric matrix with the corresponding weighted graph. If $m = 1$ then $G = nK_1$ and we take this as the smallest graph in our total ordering (CGO) of $n$-vertex graphs. If $m > 1$ then each $W_i$ has fewer vertices than $G$ and so if CGO has been defined for graphs with fewer than $n$ vertices then the graphs $W_1, \ldots, W_m$ can be ordered by CWGO. Since CWGO is not in general a total ordering (Remark 8.2.7) we first define *orthodox* star partitions as follows.

**8.2.1 Definition** *The star partition $X_1 \dot\cup \cdots \dot\cup X_m$ for $G$ is said to be orthodox if the sequence $(W_1, \ldots, W_m)$ is maximal in the lexicographical ordering induced by CWGO.*

Again, if $m > 1$ the vertices of each $W_i$ may be ordered by CWGVO. Such an ordering induces an ordering of $X_1 \dot\cup \cdots \dot\cup X_m$, called an *admissible* ordering of $V(G)$.

**8.2.2 Definition** *Let $X_1 \dot\cup \cdots \dot\cup X_m$ be an orthodox star partition for $G$, with an admissible ordering of vertices. Then the corresponding star basis of $\mathbb{R}^n$ is called an orthodox star basis.*

Now suppose that the eigenvector matrix $S$ represents an orthodox star basis corresponding to the orthodox star partition $X_1 \dot\cup X_2 \dot\cup \cdots \dot\cup X_m$. If we relabel vertices $1, 2, \ldots, n$ as $\pi(1), \pi(2), \ldots, \pi(n)$ and if $Q = (\delta_{i\pi(j)})$, then $QS$ is an eigenvector matrix which represents an orthodox star basis corresponding to the orthodox star partition $\pi(X_1) \dot\cup \pi(X_2) \dot\cup \cdots \dot\cup \pi(X_m)$. Since the rows of $S$ are linearly independent there is a unique permutation $\pi$ of $\{1, 2, \ldots, n\}$ such that the rows of $QS$ are in decreasing lexicographical order. The star basis of $\mathbb{R}^n$ obtained in this way is called the *quasi-canonical* star basis determined by $S$.

**8.2.3 Definition** *For a given graph $G$, we order quasi-canonical star bases lexicographically: the largest such basis is called the canonical star basis determined by $G$.*

Note that the lexicographical ordering of star bases can be performed efficiently only if the vectors in a star basis are ordered in a prescribed way (by CWGVO for smaller graphs in our case). Otherwise we would have to consider all possible orderings of the basis vectors in star bases, and the situation would be very similar to that in which adjacency matrices are ordered by a binary number as described in Section 1.4.

Different orthodox star partitions, or different admissible orderings of vertices in the same orthodox star partition, can give rise to the same quasi-canonical basis. In particular, the canonical star basis may arise as the $n$-tuple of columns in both $QS$ and $Q^*S^*$ where $Q^* = (\delta_{i\pi^*(j)})$. In this situation $\pi$ and $\pi^*$ differ by an automorphism of $G$. To see this, note that (8.1.1) holds in respect of $S^* = (S_1^*|S_2^*|\cdots|S_m^*)$ as well as $S$, and so $Q^{-1}Q^*$ commutes with $A$; in other words, $\pi^{-1}\pi^*$ is an automorphism of $G$.

**8.2.4 Definition** *Let $G$ be a graph with vertices $1, 2, \ldots, n$, and let $\pi$ be a permutation of $1, 2, \ldots, n$ associated with the canonical star basis of $G$. A canonical vertex ordering (CVO) for $G$ is a total order $\pi\sigma(1) \prec \pi\sigma(2) \prec \ldots \prec \pi\sigma(n)$, where $\sigma$ is any automorphism of $G$.*

Note that if $G$ has automorphism group $\Gamma$ and if the vertices are relabelled $\pi(1), \ldots, \pi(n)$ then the automorphism group becomes $\pi\Gamma\pi^{-1}$, and the canonical vertex orderings are equivalent under the action of this group.

**8.2.5 Definition** *Ordering graphs lexicographically first by spectral moments and then by canonical bases defines the canonical graph ordering (CGO).*

It remains to define a canonical ordering of weighted graphs on a given number of vertices (CWGO) and a canonical vertex ordering in a weighted graph (CWGVO).

For a weighted graph $W$ let $\mathcal{D}$ be the family of diagonal entries and let $\mathcal{O}$ be the family of off-diagonal entries of the corresponding weight matrix, both ordered in non-increasing order. We order weighted graphs lexicographically by their families $\mathcal{D}$ and then refine the ordering by $\mathcal{O}$. If both $\mathcal{D}$ and $\mathcal{O}$ are the same for two graphs we consider modified weight-generated graphs, here defined slightly differently compared with [CvRS1].

**8.2.6 Definition** *Let $W$ be a real symmetric matrix in which $\alpha_1, \ldots, \alpha_s$ are the distinct off-diagonal entries. Then $W$ is expressible as $D + \sum_{i=1}^{s} \alpha_i A_i$ where $D$ is diagonal and each $A_i$ is the adjacency matrix of a graph $H_i$. We refer to $H_i$ as an $\alpha_i$-graph, and to $H_1, \ldots, H_s$ as the weight-generated graphs associated with $W$. Let $G_i$ be the graph with adjacency matrix $A_1 + \cdots + A_i$ $(1 = 1, 2, \ldots, s)$. Then the graphs $G_1, \ldots, G_s$ are called the modified weight-generated graphs associated with $W$.*

Weighted graphs with the same families $\mathcal{D}$ and $\mathcal{O}$, in which members of $\mathcal{D}$ are mutually equal, are ordered lexicographically using the sequence $G_1, \ldots, G_s$ of modified weight-generated graphs ordered by CGO.

If members of $\mathcal{D}$ are not mutually equal we reorder rows and columns of $W$ so that the members of $\mathcal{D}$ form a non-increasing sequence. We split $W$ into blocks $W_{ij}$ in such a way that on the main (block) diagonal we have square blocks $W_{11}, \ldots, W_{pp}$, with constant diagonal entries. We then order such weighted graphs $W$ lexicographically by the sequence $W_{11}, \ldots, W_{pp}$ and refine the ordering by weighted bipartite graphs $\begin{pmatrix} O & W_{ij} \\ W_{ji} & O \end{pmatrix}$ taken in some order.

In this way we have defined an order which is called a *canonical ordering of weighted graphs* (CWGO).

**8.2.7 Remark** Note that CWGO is not a total order. Indeed, one can easily construct two weighted graphs such that neither is before the other in CWGO. In view of Definition 8.2.2, this implies that we might have orthodox bases with non-isomorphic weighted graph sequences $W_1, \ldots, W_m$ in the same graph. $\square$

Let us turn finally to the ordering of vertices in a weighted graph $W$. We order the vertices of $W$ first by the diagonal entries of $W$. For each group of mutually equal diagonal entries we consider the corresponding principal submatrix $M$ of $W$. Let $\alpha_1, \ldots, \alpha_k$ ($\alpha_1 > \cdots > \alpha_k$) be the off-diagonal entries of $M$, and let $H_i$ be the $\alpha_i$-graph of $M$ ($i = 1, 2, \ldots, k$). For any set $X$ of vertices we write $M^X$ for the corresponding principal submatrix of $M$, and $H_i^X$ for the $\alpha_i$-graph of $M^X$. Moreover, for any ordering $\pi$ of vertices, $\pi_X$ denotes the ordering of $X$ induced by $\pi$. Now we define a nested sequence $\Pi_1 \supseteq \Pi_2 \supseteq \cdots \supseteq \Pi_k$, where $\Pi_1$ is the set of canonical vertex orderings of $H_1$. Suppose that $i > 1$ and $\Pi_{i-1}$ has been defined. If $\Pi_{i-1}$ is a singleton then we set $\Pi_i = \Pi_{i-1}$. Otherwise, let $\Pi_i'$ consist of those $\pi \in \Pi_{i-1}$ such that $\pi_X$ is a CVO of $H_i^X$ for each orbit $X$ of the automorphism group of $H_{i-1}$. If $\Pi_i'$ is non-empty then we set $\Pi_i = \Pi_i'$, while if $\Pi_i' = \emptyset$ then we set $\Pi_i = \Pi_{i-1}$. Finally the set $\Pi_k$ of total vertex orderings (which may or may not be a singleton) is defined to be CWGVO for the weighted graph $M$. By stringing together such orderings, we can define CWGVO for the weighted graph $W$.

**8.2.8 Remark** Comparing CWGO for two given weighted graphs or finding CWGVO for a given weighted graph can be reduced in polynomial time to the corresponding problems for non-weighted graphs (comparing CGO and finding CVO). $\square$

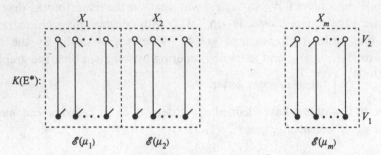

Fig. 8.1. A König digraph.

## 8.3 Algorithms for finding a star partition

Before we discuss algorithms to find a canonical star basis and an associated star partition we direct our attention to the more modest task of finding just one (arbitrary) star partition.

In this section we show that the problem of finding a star partition (and hence a star basis) is a polynomial one. We discuss two algorithms which are investigated in [CvRS2]: the first is based on the maximum matching problem for graphs, and the second invokes an algorithm for matroid intersection. The complexity of the first algorithm is estimated as $O(n^4)$, and that of the second as $O(n^5)$, where $n$ is the number of vertices of an instance graph.

Some basic requirements for the first algorithm are summarized in the next few lines. The keystone is the following form of an eigenvector matrix (see (7.2.1)) whose existence is guaranteed by Corollary 7.2.7.

$$E^* = \begin{array}{c} \mathscr{E}(\mu_1)\ \mathscr{E}(\mu_2)\ldots\mathscr{E}(\mu_m) \\ \left( \begin{array}{cccc} I_{k_1} & * & & * \\ * & I_{k_2} & & * \\ \vdots & \vdots & \ddots & \vdots \\ * & * & & I_{k_m} \end{array} \right) \begin{array}{c} X_1 \\ X_2 \\ \vdots \\ X_m \end{array} \end{array} \qquad (8.3.1)$$

In (8.3.1), $\mathscr{E}(\mu_i)$ designates the columns corresponding to that eigenspace, $X_i$ designates the rows corresponding to that star cell, the symbol $*$ denotes a block matrix of an appropriate size, and $I_k$ denotes the identity matrix of order $k$.

For our present purposes the interpretation of an eigenvector matrix in terms of the *König digraph* is very useful (see Fig. 8.1).

For each $i \in \{1,\ldots,m\}$ the subgraph of $K(E^*)$ induced by vertices corresponding to $\mathscr{E}(\mu_i)$ and $X_i$ (see Fig. 8.1) represents part of a perfect

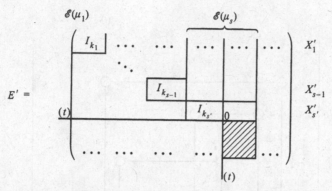

Fig. 8.2. The matrix $E'$.

matching (that is, it consists of $k_i$ independent arcs of weight 1). All other arcs in $K(E^*)$ join vertices corresponding to $\mathscr{E}(\mu_i)$ and $X_j$ $(i \neq j)$.

The rough idea of the first algorithm (which is based on Corollary 7.2.7) is as follows:

start with an eigenvector matrix $E = (E_1 | \cdots | E_m)$ as in (7.2.1), and by means of elementary column operations within each block $E_i$, together with permutations of the rows of $E$, transform it to the form (8.3.1).

We show that there exists a good strategy (with polynomial time bound) for this construction. This (almost greedy) strategy is as follows. For an instance graph, first find an (arbitrary) eigenvector matrix, say $E$, in the form (7.2.1). Suppose next that we have somehow, for some $s$ $(1 \leq s \leq m)$, transformed the blocks $E_1, \ldots, E_{s-1}$ to the forms in (8.3.1), but have failed to do the same for the block $E_s$. In addition, suppose that only one part of $E_s$ (a block of size $k_{s'} \times k_s$; $k_{s'} < k_s$) is transformed as required. For more details see the matrix $E'$, depicted in Fig. 8.2. The reason for failure is the absence of a non-zero element in the shaded area of $E'$, so that no pivot can be brought to the position $(t, t)$, where $t = k_1 + \cdots + k_{s-1} + k_{s'} + 1$. Otherwise, if we can find a pivot in the shaded part of $E'$, we can easily augment $E'$.

In graph-theoretic terms, we have encountered the following situation: all sets $X'_1, \ldots, X'_{s-1}$ so far formed are putative star cells, while $X'_{s'}$ (the $s$-th partial cell) is not, and cannot be extended to a star cell using the

*Canonical star bases*

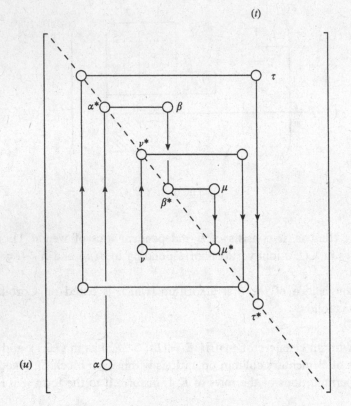

Fig. 8.3. A sequence of non-zero entries of $E'$.

vertices which are not yet classified. Thus at this point we need to make some adjustments, and thereafter extend $X'_{s'}$ by at least one vertex.

To this end, suppose first that we can find in the matrix $E'$ a sequence of non-diagonal non-zero entries say $\alpha, \beta, \ldots, \tau$, whose positions can be most conveniently visualized as in Fig. 8.3. Notice that each entry $v$ of the sequence is reachable from the previous entry $\mu$ in the sequence by moving vertically from $\mu$ to the main diagonal (the entry $\mu^*$), and then moving horizontally to $v$. Also notice that $\tau$ is in the $t$-th column while $\alpha$ is in the $u$-th row for some $u \geq t$. Assume now that the rows of $E'$ which contain the entries from the above sequence are permuted cyclically according to the following rule: for each entry $\mu$, the row containing $\mu$ replaces the row containing $\mu^*$ (to visualize this, see Fig. 8.3). Consequently, the non-zero entries $\alpha, \beta, \ldots, \tau$ are now moved to diagonal

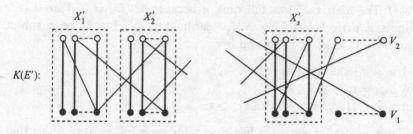

Fig. 8.4. The König digraph $K(E')$

positions (those of entries $\alpha^*, \beta^*, \ldots, \tau^*$, respectively) and accordingly we have a pivot at position $(t, t)$. The problem we might encounter now concerns the blocks along the main diagonal: some of the $s$ unit matrices (including $I_{k_s}$) may be destroyed. We can recover them by means of elementary column operations within the corresponding blocks $E_i$ if and only if these blocks still have full rank. In order to show that this is indeed the case for an appropriate choice of $\alpha, \beta, \ldots, \tau$ we make use of some well-known tools from matching theory (see, for example, [LoPl]).

Let us first interpret the matrix $E'$ in terms of the corresponding König digraph as in Fig. 8.4. Notice that there are two kinds of arcs in this graph. In particular, the boldface arcs correspond to diagonal entries (up to position $t-1$) and form a matching (to be denoted by $M'$). Also notice that the vertices in $V_1$ (resp. $V_2$) corresponding to columns (resp. rows) are coloured black (resp. white). The black vertex $t$ is not adjacent to any white vertex $t'$ ($t' \geq t$), since otherwise we can easily introduce a pivot at position $(t, t)$. Considering again the sequence $\alpha, \beta, \ldots, \tau$ we observe that it corresponds to a matching in $K(E')$. On the other hand, the sequence $\alpha, \alpha^*, \beta, \beta^*, \ldots, \sigma, \sigma^*, \tau$ corresponds to an *augmenting path* with respect to $M'$. Recall that an augmenting path (for which orientation of arcs is ignored) with respect to some matching $M$ is a path whose endvertices are both *free* (i.e. are not incident to edges from $M$) and whose edges are alternately in $\overline{M}$ and $M$. (Here $\overline{M}$ denotes the complement of $M$ with respect to the edge set of the digraph.) The following lemma establishes the existence of a sequence of the type $\alpha, \beta, \ldots, \tau$ described above.

**8.3.1 Lemma** *Given the matching $M'$ in $K(E')$, let $t$ be the vertex in $V_1$ as above (black and free). There exists an augmenting path (with respect to $M'$) with $t$ as an endvertex.*

*Proof* The matrix $E'$ has full rank $n$ because $\det(E') \neq 0$. Thus $K(E')$ admits a perfect matching $N$ (of cardinality $n$). Let $K$ be a subset $\{u_i v_i : i = 1, \ldots, k\}$ of $N$ defined by:

(a) $u_1 = t$ (and hence $v_1 < t$);
(b) $u_{i+1} = v_i$ $(1 \leq i \leq k-1)$;
(c) $v_i < t$ $(1 \leq i \leq k-1)$ and $v_k \geq t$.

Thus $u_i \leq t$ for each $i$, while $v_i < t$ for each $i \neq k$; also notice that $u_i \neq v_i$ for each $i$. Now consider the graph $H = (V_1, V_2, K \cup M')$: the component of $H$ containing the vertex $t \in V_1$ is an augmenting path as required. □

**8.3.2 Remark** A famous theorem of Berge [Berg1] guarantees the existence of an augmenting path (with respect to $M'$), but not necessarily one with a prescribed endvertex. □

**8.3.3 Remark** The existence of a sequence of non-zero non-diagonal entries $\alpha, \beta, \ldots, \tau$ as specified in Fig. 8.3 can also be established as follows, where $(e_{ij}) = E'$. Since $\det(E') \neq 0$, there is a permutation $\sigma$ of $\{1, 2, \ldots, n\}$ such that $\prod_{j=1}^n e_{\sigma(j)j} \neq 0$. Then $\sigma(t) < t, \sigma^{i+1}(t) \neq \sigma^i(t)$ for all $i$ and there exists a positive integer $h$ such that $\sigma^h(t) = t$. Let $k$ be the least positive integer such that $\sigma^k(t) \geq t$. Now we may take $\alpha = e_{\sigma^k(t)\sigma^{k-1}(t)}, \beta = e_{\sigma^{k-1}(t)\sigma^{k-2}(t)}, \ldots, \tau = e_{t\sigma(t)}$. Thus $u = \sigma^k(t)$ and a suitable permutation of columns is obtained when $\sigma^i(t)$ replaces $\sigma^{i-1}(t)$ $(i = 1, \ldots, k)$. □

**8.3.4 Lemma** *Any shortest augmenting path starting at the vertex $t \in V_1$ enables the matrix $E'$ to be augmented.*

*Proof* By Lemma 8.3.1, there is an augmenting path with respect to $M'$, say $P$, which starts at $t \in V_1$. Then the symmetric difference $M' \Delta P$ is a matching in $K(E')$, and $|M' \Delta P| = |M'| + 1$. From our earlier considerations, this yields an augmentation of $E'$ only if the new blocks along the main diagonal (including the $s$-th, i.e. the one being extended) have full rank. We will now show that this is indeed true provided $P$ has shortest length. To prove this claim, let $M_i$ $(1 \leq i \leq s)$ be the part of $M'$ corresponding to the $i$-th diagonal block. (Equivalently the arcs from $M_i$ are incident only with the vertices corresponding to $\mathscr{E}(\mu_i)$ and $X'_i$.) Let $g_1, \ldots, g_r$ be the arcs of $P \cap M_i$ ordered as they are encountered on traversing $P$ from the end vertex $t \in V_1$ to the end vertex $u \in V_2$, and

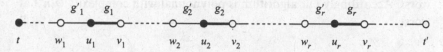

Fig. 8.5. Arcs of the path $P$.

let $g'_1, \ldots, g'_r$ be the arcs of $P$ immediately prior to $g_1, \ldots, g_r$ respectively; see Fig. 8.5. Now notice that $w_1$ is not adjacent to $u_2, \ldots, u_r$; $w_2$ is not adjacent to $u_3, \ldots, u_r$; and so on, since $P$ is a shortest path. After augmentation by $P$, $M_i$ becomes $M'_i$, where $g_i$ is replaced by $g'_i$ for each $i$. Consider now the subgraph of $K(E')$ induced by the vertex set of $M'_i$ ($1 \le i \le s$). This subgraph corresponds to a (lower) triangular matrix with non-zero diagonal entries. $\qquad\square$

**8.3.5 Theorem** *There exists a polynomial-time algorithm for finding a star partition of any graph.*

*Proof* Suppose that our instance graph has $n$ vertices. As is well known from linear algebra, an eigenvector matrix can be obtained in polynomial time with respect to $n$. By [Dem], the complexity is at worst $O(n^3)$. Of course, we suppose here that enough decimal places are taken in representing the real numbers in question, in order to have control over numerical calculations and comparisons. The important point in this respect is the fact that always we have to make only zero versus non-zero decisions.

Now starting from any eigenvector matrix as above, we can easily construct the corresponding König digraph. Thereafter the construction is combinatorial. At each step we have repeatedly ($n$ times) to carry out the following:

(a) find an augmentation path (with prescribed endvertex) in the current König digraph;

(b) update the structure of this digraph to match the diagonal pattern of the eigenvector matrix (after a specified permutation of rows).

In [HoKa] Hopcroft and Karp give a polynomial algorithm for finding a shortest augmenting path beginning at a prescribed vertex $t$ of $V_1$: what is required is step 1 of their Algorithm A applied to the graph obtained from the König digraph by deleting the free vertices in $V_1 \setminus \{t\}$. This has complexity $O(n^2)$, while our task (b) has complexity $O(n^3)$ at worst. Accordingly our algorithm is polynomial with complexity $O(n^4)$ at worst. □

**8.3.6 Remark** The ideas behind this algorithm provide an example of the symbiotic relationship between combinatorics and matrix theory discussed by Brualdi in [Bru]: matrix theory is applied to graphs (cf. [CvDS]) and graph theory to matrices (cf. [Cve11]). More precisely, in the construction of a star partition, we associate with a graph $G$ an adjacency matrix, form the König digraph $K(E)$ of a corresponding eigenvector matrix $E$, and reduce our problem to a matching problem for $K(E)$. □

Next we make some observations on the construction of further star partitions by exchanging elements between cells of a given partition.

Suppose that $X_1 \dot\cup \cdots \dot\cup X_m$ is a star partition and $Y_1 \dot\cup \cdots \dot\cup Y_m$ is a feasible partition for $G$. Let $S_i = Y_i \setminus X_i$, $T_i = X_i \setminus Y_i$ (or equivalently, $X_i = (X_i \setminus S_i) \cup T_i$, $Y_i = (Y_i \setminus T_i) \cup S_i$) and let $E^*(S_i, T_i)$ be the submatrix of an eigenvector matrix (8.3.1) whose rows are indexed by $S_i$ and whose columns are indexed by $T_i$ ($i = 1, \ldots, m$).

**8.3.7 Proposition** *With the above notation, $Y_1 \dot\cup \cdots \dot\cup Y_m$ is a star partition if and only if*

$$\prod_{i=1}^{m} \det E^*(S_i, T_i) \neq 0.$$

*(Here, $\det E^*(S_i, T_i)$ is interpreted as 1 if $S_i = T_i = \emptyset$.)*

*Proof* This follows from Corollary 7.2.7 since $E^*(S_i, T_i)$ is the $i$-th diagonal block in an eigenvector matrix obtained by substituting rows indexed by $S_i$ for those indexed by $T_i$ ($i = 1, \ldots, m$). □

We note also an application of Proposition 8.3.7 in the special case that $|S_i| = |T_i| \leq 1$ for each $i$. If $E^* = (e_{ij})$ then there exists $\sigma \in S_n$ such that $\prod_{j=1}^{n} e_{\sigma(j)j} \neq 0$. Suppose that $\sigma$ has a constituent cycle $\mu = (j_1, j_2, \ldots, j_k)$ such that the elements $j_1, \ldots, j_k$ lie in $k$ different cells of a star partition. Since $e_{\sigma(j)j} \neq 0$ for all $j$ we can generate a second star partition by substituting $\mu(j_h)$ for $j_h$ in the relevant star cell for each $h \in \{1, \ldots, k\}$.

The second algorithm for finding a star partition is based on the matroid intersection problem. As noted in [LoPl] the theory of matroids underlies the construction of matchings such as that related to the eigenvector matrix $E^*$ given by (8.3.1). The matrix $E^*$ is the transition matrix from the standard basis of $\mathbb{R}^n$ to a basis of eigenvectors, and its inverse may be taken to be the transition matrix $(t_{hj})$ of equation (7.1.1). From now on we work with $(t_{hj})$ and appeal to matroid theory to establish the existence of a polynomial algorithm of complexity at most $O(n^5)$ which enables the cells of a star partition to be constructed in succession without backtracking; that is, once the cells $X_1, \ldots, X_s$ $(s < m)$ are constructed they are not subject to subsequent modification.

Recall that $e_j = \sum_{h=1}^{n} t_{hj} x_h$ where $\{x_h : h \in R_i\}$ is a basis for $\mathscr{E}(\mu_i)$ $(i \in \{1, \ldots, m\})$. Here we take $k = k_1$, $R_1 = \{1, \ldots, k\}$ and we write the columns of $(t_{hj})$ as $\begin{pmatrix} a_j \\ b_j \end{pmatrix}$ $(j \in \{1, \ldots, n\})$ where $a_j = (t_{1j}, \ldots, t_{kj})^T$. The Laplacian expansion of $\det(t_{hj})$ determined by $R_1$ guarantees the existence of a $k$-subset $X_1$ of $\{1, \ldots, n\}$ such that both the $k \times k$ matrix $(t_{hj})$ $(h \in R_1, j \in X_1)$ and the $(n-k) \times (n-k)$ matrix $(t_{hj})$ $(h \notin R_1, j \notin X_1)$ are invertible. Thus $X_1$ is an independent set of greatest size in the intersection of two matroids: one is the linear matroid determined by the vectors $a_1, \ldots, a_n$, and the other is the dual of the linear matroid determined by the vectors $b_1, \ldots, b_n$. Now Edmonds [Edm] gives an explicit algorithm for finding a set of maximal cardinality which is independent in each of two matroids on a given finite set; moreover the algorithm is polynomial when for each of the two matroids there is a polynomial algorithm for determining whether a given subset is independent. In our context this last condition is satisfied since there exists a polynomial algorithm for finding the dimension of a subspace spanned by a finite set of vectors. Accordingly $X_1$ can be found in polynomial time. To find $X_2$, we apply the same process to the $(n-k) \times (n-k)$ matrix $(t_{hj})$ $(h \notin R_1, j \notin X_1)$. Repetition for each of $R_3, \ldots, R_{m-1}$ yields a star partition $X_1 \dot\cup \cdots \dot\cup X_m$. Each application of Edmonds' algorithm requires at most $O(n^4)$ steps and so the star partition is obtained in at most $O(n^5)$ steps.

## 8.4 Graphs with distinct eigenvalues

Let $G$ be a graph with $n$ vertices and distinct eigenvalues $\lambda_1, \ldots, \lambda_n$ in decreasing order. Let $X$ be a corresponding eigenvector matrix, say $X = (x_{ij})$ where $(x_{1j}, \ldots, x_{nj})^T$ is an eigenvector $x_j$ corresponding to $\lambda_j$ $(j = 1, \ldots, n)$. From Section 7.2, $\{\pi(1)\} \dot\cup \cdots \dot\cup \{\pi(n)\}$ is a star partition for $G$ if and only if $x_{\pi(j)j} \neq 0$ for all $j \in \{1, \ldots, n\}$. In this situation, CWGVO is trivial because each Gram matrix $W_j$ is just a (positive) scalar. Moreover the partition is orthodox if $(x_{\pi(1)1}^2, \ldots, x_{\pi(n)n}^2)$ is lexicographically maximal, because $P_j e_{\pi(j)}$ is a scalar multiple of $x_j$.

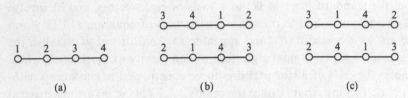

Fig. 8.6. Some labellings of the path $P_4$.

Thus if $X' = (x_{ij}^2)$ then an orthodox star partition may be regarded as a maximal matching in the König digraph $K(X')$, that is, a perfect matching for which the corresponding $n$-tuple of weights is maximal with respect to the lexicographic ordering. There is just one orthodox basis associated with an orthodox star partition $\{\pi(1)\}\cup\cdots\cup\{\pi(n)\}$, namely $\{s_1,\ldots,s_n\}$ where $s_j = P_j e_{\pi(j)}$ $(j = 1,\ldots,n)$. Since $P_j = \|x_j\|^{-2} x_j x_j^T$, we have $s_j = x_{\pi(j)}\|x_j\|^{-2} x_j$ $(j = 1,\ldots,n)$. The corresponding quasi-canonical basis is obtained from the matrix $(s_1|\cdots|s_n)$ by ordering the rows lexicographically. Finally the canonical star basis is found as the quasi-canonical basis which is largest with respect to the lexicographical ordering of columns. We illustrate these ideas in the case of the path $P_4$. An analogous algorithm for graphs with repeated eigenvalues is discussed in [Cve21].

**8.4.1 Example** The path $P_4$ has simple eigenvalues $(\pm\sqrt{5}\pm 1)/2$, here taken in decreasing order. With vertices labelled as in Fig. 8.6(a), a corresponding matrix of unit eigenvectors is

$$X = \begin{pmatrix} b & a & a & b \\ a & b & -b & -a \\ a & -b & -b & a \\ b & -a & a & -b \end{pmatrix} \tag{8.4.1}$$

where $a \approx 0.60$ and $b \approx 0.37$. The orthodox star partitions correspond to the four perfect matchings in $K(X')$ where each edge weight is $a^2$; they are $\{\pi_i(1)\}\cup\{\pi_i(2)\}\cup\{\pi_i(3)\}\cup \{\pi_i(4)\}$ $(i = 1,2,3,4)$, where $\pi_1 = (12)(34)$, $\pi_2 = (1243)$, $\pi_3 = (1342)$ and $\pi_4 = (13)(24)$. The correspondong orthodox bases are given by the following eigenvector matrices:

$$B_1 = \begin{pmatrix} ab & a^2 & a^2 & ab \\ a^2 & ab & -ab & -a^2 \\ a^2 & -ab & -ab & a^2 \\ ab & -a^2 & a^2 & -ab \end{pmatrix}, \quad B_2 = \begin{pmatrix} ab & -a^2 & a^2 & ab \\ a^2 & -ab & -ab & -a^2 \\ a^2 & ab & -ab & a^2 \\ ab & a^2 & a^2 & -ab \end{pmatrix},$$

$$B_3 = \begin{pmatrix} ab & a^2 & a^2 & -ab \\ a^2 & ab & -ab & a^2 \\ a^2 & -ab & -ab & -a^2 \\ ab & -a^2 & a^2 & ab \end{pmatrix}, \quad B_4 = \begin{pmatrix} ab & -a^2 & a^2 & -ab \\ a^2 & -ab & -ab & a^2 \\ a^2 & ab & -ab & -a^2 \\ ab & a^2 & a^2 & ab \end{pmatrix}.$$

The matrices $B_1, B_4$ share the same set of rows, and the same is true of $B_2$ and $B_3$. Thus when rows are ordered lexicographically we obtain just two eigenvector matrices representing quasi-canonical bases, namely

$$C_1 = \begin{pmatrix} a^2 & ab & -ab & -a^2 \\ a^2 & -ab & -ab & a^2 \\ ab & a^2 & a^2 & ab \\ ab & -a^2 & a^2 & -ab \end{pmatrix}, \quad C_2 = \begin{pmatrix} a^2 & ab & -ab & a^2 \\ a^2 & -ab & -ab & -a^2 \\ ab & a^2 & a^2 & -ab \\ ab & -a^2 & a^2 & ab \end{pmatrix}.$$

Note that vertex orderings corresponding to the same quasi-canonical star basis are equivalent in that they differ by an automorphism of the graph: see Fig. 8.6(b) for $C_1$ and Fig. 8.6(c) for $C_2$. The canonical star basis is given by $C_2$, since $C_2 > C_1$ in the lexicographical ordering by columns; accordingly the two total orderings in CVO are those shown in Fig. 8.6(c). $\qquad \square$

The weight of a matching in a weighted graph is defined as the sum of the weights of edges in the matching. In Example 8.4.1 the maximal matchings in $K(X')$ are the perfect matchings of maximal weight. To ensure that this is the case in general we need to scale eigenvectors as follows: we replace $\mathbf{x}_j$ with $\sigma_j \mathbf{x}_j$ $(j = 1, \ldots, n)$, where $\sigma_1, \ldots, \sigma_n$ are chosen so that the difference between any two distinct entries of $\sigma_j \mathbf{x}_j$ exceeds the difference between any two distinct entries of $\sigma_{j+1} \mathbf{x}_{j+1}$ $(j = 1, \ldots, n-1)$. Without loss of generality the vectors $\mathbf{x}_j$ may be taken to be unit vectors, so that $\mathbf{s}_j = x_{\pi(j)j} \mathbf{x}_j$ $(j = 1, \ldots, n)$.

Construction of the canonical star basis may be reduced to finding perfect matchings of maximal weight in a succession of weighted graphs defined as follows. First we delete from $K(X')$ the edges which are contained in no maximal matching to obtain $K'(X')$, which we regard as an unweighted bipartite graph with vertices $r_1, \ldots, r_n$ and $c_1, \ldots, c_n$ corresponding to rows and columns respectively. For each neighbour $s$ of $c_1$ in $K'(X')$, let $K'_s(X')$ be the subgraph of $K'(X')$ obtained by deleting all edges containing $s$ except that joining $s$ to $c_1$. The weighted graph $K''_s(X')$ is obtained from $K'_s(X')$ by assigning the weight $x_{s1}^2$ to the edge $sc_1$ and the weight $x_{sj}x_{ij}$ to the edge $r_i c_j$ $(j = 2, \ldots, n; i = 1, \ldots, n)$. As before, we may rescale columns in the adjacency matrix of $K''_s(X')$ to obtain a weighted graph in which a perfect matching of maximal weight determines a maximal matching in $K_s''(X')$. We select $s$ so that the corresponding sequence of weights $x_{s1}^2, x_{s2}x_{j_22}, \ldots, x_{sn}x_{j_nn}$ is lexicographically maximal. We now repeat the procedure for $c_2$ and each neighbour $s$ of $c_2$ satisfying the above criterion. On repeating the procedure for the vertices $c_3, \ldots, c_n$ in turn, we obtain a perfect matching which determines the canonical star basis.

Let us now consider the complexity of the above procedure The problem of finding a maximal perfect matching in a bipartite graph, also known as the *assignment problem*, can be solved by the *Hungarian algorithm*, of complexity $O(n^3)$ (see, for example, [PaSt]). In order to find the canonical basis, we have to solve several assignment problems, and the number of such tasks is bounded by a polynomial function of $n$. We conclude that the canonical basis for a graph with distinct eigenvalues can be found in polynomial time; hence the isomorphism problem for graphs with distinct eigenvalues can be solved in a polynomial time. This is confirmation, in a special case, of a more general result discussed in the next section.

## 8.5 The maximal clique problem and bounded multiplicities

The procedure for finding the canonical star basis can be designed so that it contains a kind of maximal clique problem. This is especially useful in the case of graphs with eigenvalues of bounded multiplicity. We review the procedure without going into details.

Let $G$ be a graph with distinct eigenvalues $\mu_1, \ldots, \mu_m$, of multiplicities $k_1, \ldots, k_m$ respectively. An orthodox star basis associated with $G$ is characterized by weighted graphs $W_1, \ldots, W_m$ (of orders $k_1, \ldots, k_m$ respectively) in so far as the sequence $W_1, \ldots, W_m$ is lexicographically maximal with respect to the ordering of weighted graphs defined in Section 8.2. Instead of finding several (or all) star bases and selecting the maximal ones among them, we can find maximal sequences $W_1, \ldots, W_m$ and then check whether star bases with such sequences exist.

Let us assume that $G$ is a connected graph with associated spectral decomposition $A = \mu_1 P_1 + \cdots + \mu_m P_m$. Then $\mu_1$ is a simple eigenvalue and $W_1$ is found by inspecting squares of coordinates of an eigenvector corresponding to $\mu_1$. By Theorem 7.4.5 any vertex can form a star cell for $\mu_1$, and so for $W_1$ we merely select a maximal diagonal entry of $P_1$; if the $(j, j)$-entry is chosen we set $X_1' = \{j\}$.

Next we have to select $X_2'$, a putative star cell correspondinging to $\mu_2$.

That means we have to find a principal submatrix $W_2$ of $P_2$ of order $k_2$ so that the weighted graph determined by $W_2$ is maximal. Now we can consider $P_2$ as a weighted graph in which we have to find a clique of order $k_2$ which is maximal with respect to CWGO. The complexity of this decision depends on the order of the clique.

Finding a maximal clique in a weighted graph has been considered in [BaCN]. The problem is essentially similar to the problem of finding a maximal clique in a graph without weights on edges [HaMl] (cf. Remark 8.2.8). Roughly speaking, we have to check all $\binom{n}{k_i}$ principal submatrices of order $k_i$. If $k_i$ is fixed, $\binom{n}{k_i}$ is a polynomial in $n$ of degree $k_i$, i.e. $\binom{n}{k_i} = O(n^{k_i})$. If the size of the clique is not restricted, the problem of finding a maximal clique (the decision version) is known to be NP-complete. Note that for a fixed $c$ $(0 < c < 1)$, $\binom{n}{cn}$ is not polynomially bounded.

Once we have found $X_2'$ such that $W_2$ is maximal we can decide easily whether there exists a star partition $X_1 \dot\cup \cdots \dot\cup X_m$ such that $X_1 = X_1'$ and $X_2 = X_2'$ (cf. Section 7.7). More generally, given any partially constructed partition we can in polynomial time, using algorithms from Section 8.3, extend it to a star partition or establish that this cannot be done.

It should be noted that our reduction of the graph isomorphism problem to certain well-known combinatorial optimization problems does not require the solutions of these problems in the general case; in fact, we have special cases determined by special features of the weight matrices in question (eigenvector and projection matrices). This is important particularly in the case that the general problem is NP-complete, as in the case of the problem of finding a maximal clique. Note that such reductions of the graph isomorphism problem to special cases of NP-hard problems (of unknown complexities) have been identified previously (see [Tim], where a special case of the maximal clique problem occurs).

It has been proved by Babai *et al.* [BaGM] that isomorphism testing for graphs with bounded multiplicities of eigenvalues can be performed in a polynomial time. Using the above ideas we can confirm this result.

If eigenvalue multiplicities are bounded by an absolute constant $a$ then the size of the maximal clique we have to find is limited to $a$ at most. It is known that the problem of finding such a clique has complexity $O(n^a)$. Moreover we can in polynomial time examine and store information on all induced subgraphs whose vertex sets have cardinalities equal to eigenvalue multiplicities. Hence we can find an orthodox star basis in polynomial time. In fact, we can find all orthodox star bases in a polynomial time and go on to find quasi-canonical bases and the canonical basis in a way similar to that described in Section 8.4 for the case of simple eigenvalues.

Ordering vertices in a star cell by means of CWGVO can be carried out in time bounded by a function of the constant $a$, and this has to be carried out at most $n$ times in order to construct an orthodox basis from an orthodox star partiton. We can imagine that for testing isomorphism of graphs in which each eigenvalue has multiplicity at most $a$ we have prepared a (finite) table of automorphism groups and CVO for graphs with at most $a$ vertices. Hence all the necessary maximization procedures can be performed in polynomial time and the result of Babai *et al.* follows.

**8.5.1 Remark** The result can be extended to graphs which together with their induced subgraphs have the property that the multiplicities of all but one eigenvalue are bounded. A stronger result was proved in [BaGM] in that the hereditary nature of the property was not assumed. It is assumed here because the multiplicity of an eigenvalue can increase when passing from graphs to subgraphs, and so the subgraph induced by the star cell corresponding to the eigenvalue of unbounded multiplicity might have more than one eigenvalue whose multiplicity exceeds the prescribed bound. We modify the notion of an orthodox star basis in such a way that the matrix $W_i$, corresponding to the eigenvalue $\mu_i$ whose multiplicity is not bounded, is placed at the end of the sequence of weighted graphs to be considered; in other words, now the sequence $(W_1, \ldots, W_{i-1}, W_{i+1}, \ldots, W_n, W_i)$ should be lexicographically maximal. We readily find in polynomial time star cells corresponding to the matrices $W_1, \ldots, W_{i-1}, W_{i+1}, \ldots, W_n$ while the cell of unbounded size, corresponding to $W_i$, is determined by the vertices which remain. It is also not necessary to order vertices in this star cell; it is sufficient to use the Reconstruction Theorem (Theorem 7.4.1). □

**8.5.2 Remark** Let us note finally that the graph isomorphism problem can also be reduced to the problem of finding a certain kind of matching in an auxiliary bipartite graph where the number of vertices depends on the multiplicities of eigenvalues. The graph in question is the incidence graph between the set of distinct eigenvalues of the original graph $G$

and the set of subsets of $V(G)$ with cardinalities equal to multiplicities of eigenvalues. There is an edge between vertices representing an eigenvalue $\mu$ and a subset $X$ of vertices if and only if $|X|$ equals the multiplicity of $\mu$ and $G - X$ does not have $\mu$ as an eigenvalue. A star partition of $G$ is represented by a matching which satisfies some additional requirements but we shall not go into details. $\quad\square$

## 8.6 Strongly regular graphs

We have seen in Section 8.5 that the problem of finding an orthodox star basis is related to the problem of finding a maximal clique in a weighted graph. The weighted graphs in question are represented by the projection matrices $P_i$, and in the case of strongly regular graphs there are just three such matrices: $P_1$, which corresponds to the index, is a multiple of $J$, while each of $P_2$, $P_3$ has only three distinct entries (see, for example, [CvDGT], p. 28). Indeed, each of $P_2$, $P_3$ has only two distinct off-diagonal entries, and so there are just two modified weight-generated graphs in the sense of Definition 8.2.6. Now, since graphs are ordered by spectral moments and then by canonical star bases (Definition 8.2.5), graphs with a prescribed number of vertices are ordered first by the number of edges (equivalently by the second spectral momnent $s_2$). This implies that in constructing an orthodox star partition of a strongly regular graph, after selecting a singleton star cell corresponding to the index, we seek to partition the remaining vertices into cells $X_2, X_3$ of size $k_2, k_3$, where first $X_2$ induces a subgraph with as many edges as possible, and secondly $X_3$ induces a subgraph with as many edges as possible. In general terms this means that in many cases we seek a bipartition $X_2 \dot\cup X_3$ which is extremal in the sense that the number of edges between $X_2$ and $X_3$ is minimized (cf. [HaHi]). This is a combinatorial problem (the minimal cut problem), known to be in NP even in the case of partitioning into two cells [PaSt].

There is a powerful heuristic by Kernighan and Lin [KeLi] for dividing the graph into two parts with a minimal number of edges between the parts, and the general problem arises in the computer-aided design of electrical circuits.

**8.6.1 Example** Consider again the Petersen graph $P$ (see Section 7.7). The extremal partition of $P - i$ into two parts of cardinalities 5 and 4 yields $C_5$ and $P_4$ (see the star partition $a1$ in Fig. 7.7). The optimal partition for $C_5 - i$ yields $K_2$ and $K_2$. For the path $P_4$ any feasible partition is a

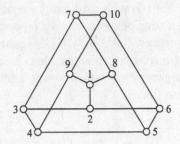

Fig. 8.7. An ordering of components of vectors in an orthodox star basis.

star partition and CVO is given in Example 8.4.1. A possible ordering of components of vectors in an orthodox star basis for the Petersen graph is given in Fig. 8.7. In fact, there are 60 mutually equivalent vertex orderings yielding the same vector ordering in an orthodox star basis because the automorphism group of the Petersen graph has order 120, while the subgroup fixing our star partition setwise has order 2.           □

Further examples are provided by strongly regular graphs with strongly regular decomposition ([HaHi], [Row15]). The Petersen graph represents a special case of this phenomenon. In the next four examples, star cells of a strongly regular graph are again denoted by $X_1, X_2, X_3$, corresponding to the distinct eigenvalues in decreasing order.

**8.6.2 Example** The line graph $L(K_n)$ of the complete graph $K_n$ is strongly regular and its spectrum is given in Example 7.8.15. The graph $L(K_n)$ is transitive and we may take any vertex $x$ to form the star cell $X_1$ belonging to the index $2n-4$. For $X_2$ we can take a clique of order $n-1$ which corresponds to a spanning star in $K_n$. The star cell $X_3$ induces $L(K_{n-1})-x$, and so the star partition is obviously extremal. This decomposition is unique, and so we have constructed the canonical star partition. Note that the unicyclic graph obtained from $X_1 \cup X_2$ (where $X_1, X_2, X_3$ are considered as sets of edges of $K_n$) is extremal among unicyclic graphs with $n$ vertices in the sense that it has largest index (see Section 5.2). The fact that our star partition is extremal in this second sense is perhaps a further reason to believe that our definition of canonical star bases is an appropriate one. □

For the purpose of the next example, we define the *double star* $S_{p,q}$ as the tree with two adjacent vertices of degrees $p$ and $q$, and all other vertices of degree 1.

**8.6.3 Example** The line graph $L(K_{n,n})$ of the complete bipartite graph $K_{n,n}$ is strongly regular with eigenvalues as given in Example 7.8.14. Again we can take any vertex to form $X_1$, while from Theorem 7.8.13 we know that $X_1 \cup X_2$ forms a spanning tree $F$ in $K_{n,n}$. We shall have a maximal number of edges in $L(F)$ if $F$ is a double star $S_{n,n}$ and $L(F)$ is a coalescence of $K_n$ with $K_n$ (at the common vertex $x$, say). The subgraph induced by $X_3$ in $L(K_{n,n})$ is $L(K_{n-1,n-1})$. We take $X_1 = \{y\}$, where $y$ is any vertex from $L(F)$ different from $x$. We conjecture that among the spanning trees of $K_{n,n}$, the double star $K_{n,n}$ has the largest index. □

**8.6.4 Example** The cocktail-party graph $\overline{nK_2}$ is strongly regular with $2n$ vertices and eigenvalues $2n - 2$, $0$ and $-2$ of multiplicities $1$, $n$ and $n - 1$, respectively. It can be partitioned into two copies of $K_n$ and in this case the number of edges between the two parts is minimal. We can take one $K_n$ to form $X_3$ and any vertex from the other copy of $K_n$ to form $X_1$. □

The following general observations can also be made:

(1) Subgraphs induced by the cells of a star partition have smaller vertex degrees than the original graph; for if a vertex $x \in X_i$ is not isolated then $x$ is adjacent to at least one vertex outside the star cell $X_i$ (Theorem 7.3.1).

(2) Lexicographical ordering of regular graphs by the spectral moments $s_3, s_4, s_5$ coincides with lexicographical ordering by the numbers of triangles, quadrangles and pentagons [CvRa2]. This means that in strongly regular graphs the cell $X_2$ should have not only the maximal number of edges but also maximal numbers of triangles, quadrangles and pentagons. These conditions are clearly fulfilled in Examples 8.6.2 and 8.6.4.

It was already noted in [CvRS1] that CVO in connected strongly regular graphs with at least five vertices is reduced to CVO in graphs induced by the two non-trivial canonical star cells. Moreover, the Reconstruction Theorem (Theorem 7.4.1) tells us that for $i \in \{2, 3\}$, a strongly regular graph $G$ is determined by the eigenvalue $\mu_i$ and the graphs $G - E(X_i)$, $G - X_i$. Canonical star bases can therefore help in the classification of strongly regular and, more generally, distance-regular graphs.

**8.6.5 Example** We consider again $G = L(K_n)$ from Example 8.6.2. The graph $G - E(X_3)$, its subgraph $G - X_3$ and the eigenvalue $-2$ determine the whole graph $G$. Note that $E(X_3)$ contains almost all of the edges of $G$. (The graph $L(K_n)$ has $3\binom{n}{3}$ edges while $L(K_{n-1}) - x$ contains

$3\binom{n-1}{3} - (2n-6)$ edges.)                          □

Strongly regular graphs are known to be hard cases for the graph isomorphism problem [Bab2]. This is reflected in our considerations by the fact that eigenvalue multiplicities are not bounded. Let $f$ be the smaller of the multiplicities of the two non-simple eigenvalues. It is known that $f(f+3)/2 \geq n$ and there are cases in which $f = (n-1)/2$, where $n$ is the number of vertices (see, for example, [LiWi], p. 239 and p. 235). (The inequality is known as the *absolute bound*.) We see that $f$ is at least $O(\sqrt{n})$; hence in strongly regular graphs we indeed have a situation in which two eigenvalues have unbounded multiplicities. (Of course, $\binom{n}{c\sqrt{n}}$ is not polynomially bounded and the maximal clique cannot be found by a complete search in polynomial time.)

It is known that the complement of a strongly regular graph is strongly regular, but a canonical star partition of a strongly regular graph is in general not a canonical star partition of its complement. (This follows from the proof of Theorem 7.5.3.) Again, from the proof of Theorem 7.5.4, a canonical star partition for a graph $G$ need not be a canonical star partition for a graph obtained from $G$ by switching.

In the papers dealing with switching, cospectral strongly regular graphs are usually distinguished by the numbers of cliques or cocliques of various orders. For example the graph $L(K_8)$ and the Chang graphs $Ch_1, Ch_2, Ch_3$ defined in Example 1.1.2 can be distinguished by the number of cliques of maximal order: $L(K_8)$ has eight cliques of order 7, $Ch_1$ has eight cliques of order 6, $Ch_2$ has three cliques of order 6 while $Ch_3$ has no cliques of order greater than 5. However, no necessary and sufficient conditions for distinguishing non-isomorphic cospectral strongly regular graphs have been formulated. Now we have tools to handle this problem in the general case.

The canonical star partitions of $L(K_8)$ are described in Examples 8.6.2 and 8.6.5. Consider $Ch_1$, for example. All subgraphs $K_7$ in $L(K_8)$ are destroyed when switching to obtain $Ch_1$, and it is easy to see that $Ch_1$ does not contain an induced subgraph on seven vertices with 21 or 20 edges. However, $Ch_1$ contains induced subgraphs on seven vertices with 19 edges: each is isomorphic to the complement of $K_{2,1} \cup 4K_1$, and the automorphism group of $Ch_1$ acts transitively on these subgraphs. Each such subgraph forms the cell $X_2$ in a canonical star partition. Thus $Ch_1$ and $L(K_8)$ are clearly distinguished by their canonical star partitions. (Computer experiments related to the Chang graphs were performed by means of the package GRAPH and an implementation [Petro] of the Kernighan-Lin heuristic [KeLi].)

# 9

# Miscellaneous results

In this chapter we survey some interesting results which are related directly or indirectly to graph eigenspaces, but which do not fit readily into earlier sections of the book.

## 9.1 Graph structure related to eigenvector components

Here we relate the structure of a graph to some modest information on eigenvectors and associated vectors. In particular, we examine how the sign pattern, or even zero–non-zero pattern, of these vectors influences connectedness. In this respect, we first recall the results of Frobenius (Theorems A.2 and A.3) which imply that a graph is connected if and only if it has a simple largest eigenvalue with a corresponding eigenvector in which all components are of the same (non-zero) sign (Theorem 2.1.3). Fiedler [Fie2] was the first to show that, for a connected graph, further information can be extracted from an eigenvector corresponding to the second largest eigenvalue. Subsequent observations are due to Powers (see [Pow1] and [Pow2]) and some of his results appear below. The sign pattern of eigenvectors is also the context for a heuristic algorithm for a (proper) vertex colouring of graphs given in [AsGi], and this we include at the end of the section.

We start by providing some necessary notation. For any vector $\mathbf{z} = (z_1, \ldots, z_n)^T \in \mathbb{R}^n$, let

$$\mathscr{P}(\mathbf{z}) = \{ i \ : \ z_i > 0 \}, \quad \mathscr{N}(\mathbf{z}) = \{ i \ : \ z_i < 0 \}, \quad \mathscr{O}(\mathbf{z}) = \{ i \ : \ z_i = 0 \}.$$

Assume now that $\mathbf{z}$ is a vector whose $i$-th entry is associated with the vertex $i$ of a graph $G$ whose vertex set is $\{1, \ldots, n\}$. We shall say that the sign of the vertex $i$ is *positive*, *negative*, or *null* (with respect to $\mathbf{z}$) according as $i$ belongs to $\mathscr{P}(\mathbf{z})$, $\mathscr{N}(\mathbf{z})$, or $\mathscr{O}(\mathbf{z})$, respectively. Subsequently we abbreviate $\mathcal{P}(\mathbf{z})$, $\mathcal{N}(\mathbf{z})$, $\mathcal{O}(\mathbf{z})$ to $\mathcal{P}$, $\mathcal{N}$, $\mathcal{O}$ whenever $\mathbf{z}$ is evident from the context. If $U \subseteq \{1, \ldots, n\}$, then $\langle U \rangle$ denotes the subgraph of $G$ induced by the vertices in $U$. For any graph $H$, comp($H$) denotes the number of components of $H$. In contrast to Section 3.2, we write $\mathbf{z}_1 > \mathbf{z}_2$ to mean that each component of $\mathbf{z}_1 - \mathbf{z}_2$ is positive.

The following technical lemma (see [Pow2]), which is a direct consequence of the Interlacing Theorem (see Section 2.4), plays an important role in what follows. As usual, the eigenvalues $\lambda_1, \lambda_2, \ldots, \lambda_n$ of a graph or matrix are assumed to be in non-increasing order.

**9.1.1 Lemma** *Let $B$ be a principal submatrix of the real symmetric matrix $A$, and suppose that $B = \mathrm{diag}(B_1, \ldots, B_k)$, where $B_1, \ldots, B_k$ are irreducible matrices. If $\lambda_1(B_i) \geq \beta$ for each $i$, then $\lambda_k(A) \geq \beta$.*

*Proof* Each eigenvalue of each $B_i$ is an eigenvalue of $B$. In particular, $\lambda_1(B_1), \ldots, \lambda_1(B_k)$ are $k$ eigenvalues of $B$ which are at least as large as $\beta$; hence $\lambda_k(B) \geq \beta$. By Theorem 2.4.1 we have $\lambda_k(A) \geq \lambda_k(B)$, and so $\lambda_k(A) \geq \beta$.      $\square$

Note that it follows from the proof of Lemma 9.1.1 that a strict inequality for any one of the $\lambda_1(B_i)$ yields a strict inequality for $\lambda_k(A)$.

Our first theorem bounds in terms of eigenvalues the number of components in a subgraph induced by non-negative vertices associated with an appropiate vector.

**9.1.2 Theorem** [Pow2] *Let $A$ be the adjacency matrix of a non-trivial connected graph. If $\mathbf{z}$ is a vector such that for some real $\alpha < \lambda_1(A)$,*

$$A\mathbf{z} \geq \alpha \mathbf{z}, \tag{9.1.1}$$

*then*

$$\mathrm{comp}(\langle \mathcal{P} \cup \mathcal{O} \rangle) \;\leq\; \max\{i \,:\, \lambda_i(A) > \alpha\}.$$

*Proof* Suppose that vertices are labelled so that $A$ and $\mathbf{z}$ are of the form

$$A = \begin{pmatrix} B & C \\ C^T & D \end{pmatrix}, \quad \mathbf{z} = \begin{pmatrix} \mathbf{x} \\ -\mathbf{y} \end{pmatrix},$$

where the principal submatrices $B$ and $D$ correspond to non-negative and negative vertices respectively (and similarly for $\mathbf{x}$ and $-\mathbf{y}$). Next, let

$$B = \begin{pmatrix} B_1 & & O \\ & \ddots & \\ O & & B_k \end{pmatrix}, \quad \mathbf{x} = \begin{pmatrix} \mathbf{x_1} \\ \vdots \\ \mathbf{x_k} \end{pmatrix}, \quad C = \begin{pmatrix} C_1 \\ \vdots \\ C_k \end{pmatrix},$$

where $B_1, \ldots, B_k$ are irreducible. The hypothesis $A\mathbf{z} \geq \alpha\mathbf{z}$ implies that $B_i\mathbf{x_i} - C_i\mathbf{y} \geq \alpha\mathbf{x}_i$ for each $i$. Since $A$ is irreducible, no $C_i$ is zero. Thus $C_i\mathbf{y} \geq \mathbf{0}$ with inequality in some element, and hence $B_i\mathbf{x}_i \geq \alpha\mathbf{x_i}$ with strict inequality in some component. Therefore, for each $i \in \{1, 2, \ldots, k\}$,

$$\mathbf{x}_i^T B_i \mathbf{x}_i > \alpha\mathbf{x}_i^T \mathbf{x}_i \qquad (9.1.2)$$

and $\lambda_1(B_i) > \alpha$. From the proof of Lemma 9.1.1 we have $\lambda_k(A) > \alpha$, and so $k \leq \max\{i \ : \ \lambda_i(A) > \alpha\}$. $\qquad\square$

When the scalar $\alpha$ in (9.1.1) is positive, we can deduce a little more from the above proof:

**9.1.3 Corollary** *If $\alpha > 0$ then*

*(i) no component of $\langle \mathscr{P} \cup \mathcal{O} \rangle$ is a singleton,*
*(ii) no component of $\langle \mathscr{P} \cup \mathcal{O} \rangle$ contains only vertices from $\mathcal{O}$.*

*Proof* If the $i$-th component is a singleton, then $B_i = O$ and (9.1.2) is contradicted. If the $i$-th component contains vertices from $\mathcal{O}$ alone, then $\mathbf{x}_i = \mathbf{0}$, and again (9.1.2) is contradicted. $\qquad\square$

**9.1.4 Remark** Theorem 9.1.2 is essentially a graph-theoretical version of a theorem of Fiedler on irreducible symmetric matrices, namely Theorem 2.1 of [Fie2], where the result is proved for $\alpha = \lambda_s(A), s \geq 2$. Note that for $\alpha < \lambda_1(A)$, $\langle \mathscr{P} \cup \mathcal{O} \rangle$ is non-empty, because otherwise $\mathbf{z} < \mathbf{0}$, $A(-\mathbf{z}) \leq \alpha(-\mathbf{z})$ and we have $\lambda_1 \leq \alpha$ as in (3.2.2). $\qquad\square$

The following variant of Theorem 9.1.2 provides an upper bound for the number of components in $\langle \mathscr{P} \rangle$; in the case that $\alpha$ is not an eigenvalue of $A$ the bound is the same as for the number of components in $\langle \mathscr{P} \cup \mathcal{O} \rangle$.

**9.1.5 Theorem** [Pow2] *In the notation of Theorem 9.1.2, with $\alpha$ as in (9.1.1), we have*

$$\mathrm{comp}(\langle \mathscr{P} \rangle) \ \leq \ \max\{i \ : \ \lambda_i(A) \geq \alpha\}.$$

*Proof* We may repeat the arguments for Theorem 9.1.2 with $\mathbf{x} > \mathbf{0}$, $\mathbf{y} \geq \mathbf{0}$ to show that $B_i\mathbf{x_i} - C_i\mathbf{y} \geq \alpha\mathbf{x_i}$. In this situation, $C_i\mathbf{y}$ may be zero and so we have only

$$B_i\mathbf{x_i} \geq \alpha\mathbf{x_i} \qquad (9.1.3)$$

for each $i$. Thus $\lambda_1(B_i) \geq \alpha$, and the conclusion follows from Lemma 9.1.1. $\qquad\square$

The next result provides a spectral bound on the number of components of any induced subgraph of a connected graph.

**9.1.6 Theorem** [Pow2] *Let $G$ be a connected graph, $U$ a proper subset of its vertex set, and $\delta$ the minimum vertex degree in $\langle U \rangle$. Then*

$$\text{comp}(\langle U \rangle) \leq \max\{i \; : \; \lambda_i(G) \geq \delta\}.$$

*Proof* Without loss of generality suppose that the vertices of $G$ are labelled so that $U = \{1, \ldots, m\}$. Define $\mathbf{z}$ by $z_i = 1$ if $i \in U$, and $z_i = 0$ otherwise. Let $A$ be the adjacency matrix of $G$ and let $B$ be the (principal) submatrix of $A$ corresponding to $U$. Then clearly $B\mathbf{j} \geq \delta\mathbf{j}$, where $\mathbf{j}$ is the all-1 vector of appropriate size. If $B = \text{diag}(B_1, \ldots B_k)$, where the $B_i$ are irreducible, then $B_i\mathbf{j} \geq \delta\mathbf{j}$ and thus $\lambda_1(B_i) \geq \delta$ for each $i$. By Lemma 9.1.1, $\lambda_k(A) \geq \delta$, which is equivalent to the assertion of the theorem. $\square$

**9.1.7 Remark** From the above theorem it follows immediately that

$$\text{comp}(H) \leq \max\{i \; : \; \lambda_i(G) \geq 0\}$$

for any induced subgraph $H$ of $G$. Similar results are due to Cvetković (see [CvDS], pp. 88-89). $\square$

In what follows we assume that equality holds in (9.1.1). In this case, both $\mathbf{z}$ and $-\mathbf{z}$ are eigenvectors of $A$, and if now we apply Theorem 9.1.2 to both these vectors when $\alpha = \lambda_2(A)$, we obtain the following result.

**9.1.8. Theorem** [Fie2] *Let $G$ be a connected graph, and let $\mathbf{z}$ be an eigenvector of $G$ corresponding to the second largest eigenvalue. Then both of the subgraphs $\langle \mathscr{P} \cup \mathcal{O} \rangle$, $\langle \mathcal{N} \cup \mathcal{O} \rangle$ are connected.*

The next theorem concerns the graph $\langle \mathscr{P} \cup \mathcal{N} \rangle$.

**9.1.9 Theorem** [Pow2] *Let $A$ be the adjacency matrix of a connected graph, and let $\mathbf{z}$ be an eigenvector of $A$ corresponding to the eigenvalue $\alpha$. Let $s = \min\{i \; : \; \lambda_i(G) = \alpha\}$ and let $m$ be the multiplicity of $\alpha$. Also suppose that $\mathcal{O}$ is non-empty and that it is contained in the set of null vertices for every eigenvector corresponding to $\alpha$ (equivalently, no vertex of $\mathcal{O}$ lies in any star cell corresponding to $\alpha$). Then there are just two possibilities:*

(a) *there are no edges between $\mathscr{P}$ and $\mathcal{N}$, $\alpha \geq 0$ and*

$$m + 1 \leq \text{comp}(\langle \mathscr{P} \cup \mathcal{N} \rangle) \leq s + m - 1;$$

(b) *some edge joins a vertex of $\mathcal{P}$ to a vertex of $\mathcal{N}$ and* $\mathrm{comp}(\langle \mathcal{P} \cup \mathcal{N} \rangle) \leq s + m - 2$.

*Proof* We suppose that the vertices of $G$ are labelled so that

$$A = \begin{pmatrix} A_P & A_{PN} & A_{PO} \\ A_{NP} & A_N & A_{NO} \\ A_{OP} & A_{ON} & A_O \end{pmatrix},$$

where the diagonal blocks are the adjacency matrices of the subgraphs induced by positive, negative and null vertices, respectively. If $\mathbf{z}$ is partitioned accordingly, we have

$$\mathbf{z} = \begin{pmatrix} \mathbf{x} \\ -\mathbf{y} \\ \mathbf{0} \end{pmatrix},$$

where $\mathbf{x}$ and $\mathbf{y}$ are positive vectors. Then from the equation $A\mathbf{z} = \alpha \mathbf{z}$ we deduce that

$$A_P \mathbf{x} - A_{PN}\mathbf{y} = \alpha \mathbf{x}, \quad A_N \mathbf{y} - A_{NP}\mathbf{x} = \alpha \mathbf{y}, \quad A_{OP}\mathbf{x} - A_{ON}\mathbf{y} = \mathbf{0}.$$

Case (a): $A_{NP} = A_{PN}^T = O$. Suppose that

$$B = \begin{pmatrix} A_P & O \\ O & A_N \end{pmatrix} = \begin{pmatrix} B_1 & & O \\ & \ddots & \\ O & & B_k \end{pmatrix}, \quad \begin{pmatrix} \mathbf{x} \\ -\mathbf{y} \end{pmatrix} = \begin{pmatrix} \mathbf{u}_1 \\ \vdots \\ \mathbf{u}_k \end{pmatrix},$$

where each $B_i$ is irreducible. Now for each $i \in \{1, \ldots, k\}$ we have $B_i \mathbf{u}_i = \alpha \mathbf{u}_i$, where $\mathbf{u}_i$ is either positive or negative; thus $\alpha = \lambda_1(B_i)$ and in particular, $\alpha \geq 0$. By Lemma 9.1.1 and Theorem 9.1.2, $\lambda_k(A) \geq \alpha$ and $k \leq s + m - 1$.

Now let $(A_{OP}|A_{ON}) = (C_1|\cdots|C_k)$. Then every eigenvector of $A$ corresponding to $\alpha$ has the form

$$\mathbf{z} = \begin{pmatrix} \mathbf{u}_1 \xi_1 \\ \vdots \\ \mathbf{u}_k \xi_k \\ \mathbf{0} \end{pmatrix},$$

because an eigenvector of $B_i$ corresponding to $\alpha$ is a scalar multiple of $\mathbf{u}_i$. Moreover, $\xi_1, \ldots, \xi_k$ satisfy the equation

$$C_1 \mathbf{u}_1 \xi_1 + \cdots + C_k \mathbf{u}_k \xi_k = \mathbf{0}.$$

No $C_i$ can be zero, and no $\mathbf{u}_i$ can have a zero element. Thus we have a system of equations whose matrix of coefficients has rank at least 1. The nullity of this system is just $m$, the multiplicity of $\alpha$ as an eigenvalue of $A$. Hence $1 + m \le k$.

Case (b): $A_{NP} = A_{PN}^T \ne O$. Suppose that $Q$ is a permutation matrix such that

$$B = \begin{pmatrix} A_P & A_{PN} \\ A_{NP} & A_N \end{pmatrix} = Q \begin{pmatrix} B_1 & & O \\ & \ddots & \\ O & & B_k \end{pmatrix} Q^T, \quad \begin{pmatrix} \mathbf{x} \\ -\mathbf{y} \end{pmatrix} = Q \begin{pmatrix} \mathbf{u}_1 \\ \vdots \\ \mathbf{u}_k \end{pmatrix},$$

where each $B_i$ is irreducible, and $B_i \mathbf{u}_i = \alpha \mathbf{u}_i$. Some of the $\mathbf{u}_i$ must have components of both signs, and without loss of generality we suppose that it is the vectors $\mathbf{u}_1, \dots, \mathbf{u}_h$ which have this property. Then $\alpha$ is not the largest eigenvalue of the corresponding $B_i$, that is, $\lambda_1(B_i) > \alpha$ $(i = 1, \dots, h)$, and so $\lambda_h(B) > \alpha$.

Let $t$ be the multiplicity of $\alpha$ as an eigenvalue of $B$, so that $h \le k \le t$. Since at least $h$ eigenvalues of $B$ exceed $\alpha$, while $B$ has exactly $t$ eigenvalues equal to $\alpha$, we have $\lambda_{h+t}(B) \ge \alpha$; then $\lambda_{h+t}(A) \ge \alpha$ and so $h + t \le s + m - 1$. Now, since $h \ge 1$, we have

$$\text{comp}(\langle \mathcal{P} \cup \mathcal{N} \rangle) = k \le t \le s + m - 1 - h \le s + m - 2.$$

$\square$

Now suppose that $s = 2$ in Theorem 9.1.9. In case (a), the upper and lower bounds coincide. In case (b), it is possible to add to the information provided by Theorem 9.1.8. The situation is described in the following result, which we quote without proof.

**9.1.10 Theorem** [Pow1] *Under the hypotheses of Theorem 9.1.9, one of the following holds when $s = 2$:*

(a) *no edge joins a vertex of $\mathcal{P}$ to one of $\mathcal{N}$, and* $\text{comp}(\langle \mathcal{P} \cup \mathcal{N} \rangle) = m + 1$;
(b) *some edge joins a vertex of $\mathcal{P}$ to one of $\mathcal{N}$, and all of the subgraphs $\langle \mathcal{P} \rangle$, $\langle \mathcal{N} \rangle$, $\langle \mathcal{P} \cup \mathcal{N} \rangle$ are connected.*

In the case that $\mathbf{z}$ is an eigenvector corresponding to an eigenvalue $\alpha < \lambda_1(A)$ one can obtain the following basic upper bound on the size of $\mathcal{O}(\mathbf{z})$. This in turn gives an upper bound for the multiplicity of $\alpha$ (see Corollary 9.1.15).

**9.1.11 Theorem** [Pow2] *Let $A$ be the adjacency matrix of a connected graph* with $n$ vertices, $n > 2$. If $A\mathbf{z} = \alpha\mathbf{z}$, where $\alpha < \lambda_1(A)$, then

$$|\mathcal{O}(\mathbf{z})| \le \begin{cases} n - 2 - 2\alpha & \text{if } \alpha > 0, \\ n - 2 & \text{if } -1 < \alpha \le 0, \\ n - 2|\alpha| & \text{if } \alpha \le -1. \end{cases}$$

*Proof* Let $|\mathcal{O}(\mathbf{z})| = a$, $|\mathcal{P}(\mathbf{z})| = b$, $|\mathcal{N}(\mathbf{z})| = c$. Note that since $\mathbf{z}$ is orthogonal to the principal eigenvector, we have $b \ge 1$ and $c \ge 1$, whence $a \le n - 2$ whatever the value of $\alpha$.

Let $h$ and $k$ be such that $z_h \ge z_i \ge z_k$ for all $i$. On comparing the $h$-th entries of $\alpha\mathbf{z}$ and $A\mathbf{z}$ we obtain $\alpha z_h \le (b - 1)z_h$, and similarly, $\alpha|z_k| \le (c - 1)|z_k|$. It follows that

$$\alpha \le \min\{b, c\} - 1, \tag{9.1.4}$$

and hence that $\alpha \le \frac{1}{2}(b + c) - 1 = \frac{1}{2}(n - a) - 1$. This gives the required bound in the case that $\alpha > 0$. From the equation $A\mathbf{z} = \alpha\mathbf{z}$ we also deduce that $|\alpha|z_h \le c|z_k|$ and $|\alpha||z_k| \le bz_h$. Now we have

$$|\alpha| \le \sqrt{bc} \le \frac{1}{2}(b + c) = \frac{1}{2}(n - a), \tag{9.1.5}$$

and this gives the required bound in the case $\alpha \le -1$. □

**9.1.12 Corollary** [Pow2] *If $G$ is a graph with $n$ vertices then*

$$\lambda_2(G) \le \lfloor \frac{n}{2} \rfloor - 1.$$

*This bound is attained for odd $n$, and is asymptotically sharp for even $n$.*

*Proof* The bound follows from (9.1.4). If $n = 2m + 1$ then the bound is attained in the connected graph constructed from $2K_m \dot\cup K_1$ by adding two edges. If $n = 2m$ and $G_m$ is obtained from $2K_m$ by adding one edge then $\lambda_2(G_m) = \frac{1}{2}(m - 3 + \sqrt{m^2 + 2m - 3})$ and so $\lambda_2(G_m) \sim m - 1$ as $m \longrightarrow \infty$. □

**9.1.13 Remark** In [Pow2], upper bounds for the second largest eigenvalue of a bipartite graph are obtained in similar fashion. □

**9.1.14 Corollary** ([Cons], [Pow2]) *If $G$ is a graph with $n$ vertices then*

$$\lambda_n(G) \ge -\sqrt{\lfloor \frac{n}{2} \rfloor \lceil \frac{n}{2} \rceil}.$$

*This bound is attained for each $n$.*

*Proof* The bound follows form (9.1.5). It is attained in $K_{m,m}$ when $n = 2m$, and in $K_{m,m+1}$ when $n = 2m + 1$. □

**9.1.15 Corollary** [Pow2] *Let $\alpha$ be an eigenvalue of a connected graph with $n$ vertices. If $\alpha$ has multiplicity $m > 1$ then*

$$m \leq \begin{cases} n - 1 - 2\alpha & \text{if } \alpha > 0, \\ n - 1 & \text{if } -1 < \alpha \leq 0, \\ n - 2|\alpha| + 1 & \text{if } \alpha \leq -1. \end{cases}$$

*Proof* This bound follows from Theorem 9.1.11 because there exists an eigenvector $\mathbf{z}$ corresponding to $\alpha$ with $|\mathcal{O}(\mathbf{z})| = m - 1$. □

Finally we indicate (following [AsGi]) how sign patterns in eigenvectors can be used to construct a heuristic algorithm for a proper (vertex) colouring of a graph. Let $\lambda_1 \geq \lambda_2 \geq \cdots \geq \lambda_n$ be the eigenvalues and let $\mathbf{x}_1, \mathbf{x}_2, \ldots, \mathbf{x}_n$ be corresponding orthogonal eigenvectors. Initially vertices are coloured by two colours, corresponding to negative and non-negative components of $\mathbf{x}_n$. This colouring is refined using $\mathbf{x}_{n-1}, \mathbf{x}_{n-2}, \ldots, \mathbf{x}_2$: whenever there are vertices with the same colour but opposite signs in the coordinates of the next eigenvector, a new colour is introduced. If all eigenvectors are considered we obtain a proper colouring with $n$ colours since sign patterns of components of all eigenvectors associated with a given vertex must be different (by the orthogonality condition). Hence, starting from $\mathbf{x}_n$ and including new eigenvectors we must obtain a proper colouring at some point. It is conjectured that eigenvectors of negative eigenvalues will suffice, and it is proved in [AsGi] that in some classes of graphs the algorithm results in proper vertex colourings with a reasonably small number of colours.

## 9.2 Graphs with small second largest eigenvalue

A review of results on the second largest eigenvalue of a graph may be found in the survey paper [CvSi4]. Among the initial observations which can be made are the following (cf. [CvDS], Theorem 6.7). For a connected graph $G$, $\mu_2 \geq -1$, with equality if and only if $G$ is complete (with at least two vertices). There does not exist a connected graph $G$ for which $-1 < \mu_2(G) < 0$. Further, the only connected graphs with $\mu_2 = 0$ are the complete multipartite graphs. Moreover we have seen in

Corollary 9.1.12 that always $\mu_2 \leq \lfloor n/2 \rfloor - 1$, where $n$ is the number of vertices in the graph. Graphs whose second largest eigenvalue is small (in particular, less than or equal to one of the numbers $\frac{1}{3}, \sqrt{2} - 1, \frac{\sqrt{5}-1}{2}, 1, 2$) have recently attracted much attention in the literature (see the third edition of [CvDS] for more details).

Here we indicate briefly the role of graph divisors (see Section 2.4) in finding a collection $\mathcal{F}$ of minimal forbidden subgraphs which characterize those graphs whose second largest eigenvalue does not exceed $\frac{\sqrt{5}-1}{2}$. Thus $\mu_2(G) \leq \frac{\sqrt{5}-1}{2}$ if and only if $G$ has no induced subgraph which is a member of $\mathcal{F}$; and for each graph $H$ in $\mathcal{F}$ we have $\mu_2(H) > \frac{\sqrt{5}-1}{2}$ while $\mu_2(H - j) \leq \frac{\sqrt{5}-1}{2}$ for each $j \in V(H)$. The existence of such a collection $\mathcal{F}$ which is finite was proved in [CvSi3], based on observations from [Sim4]. Except for $2K_2$ and four one-vertex extensions of $P_4$, all other graphs from this collection $\mathcal{F}$ belong to the class $\mathscr{C}$ constructed recursively according to the following rules: (i) $K_1 \in \mathscr{C}$, (ii) if $G \in \mathscr{C}$ then $G \dot{\cup} K_1 \in \mathscr{C}$, (iii) if $G_1, G_2 \in \mathscr{C}$ then $G_1 \nabla G_2 \in \mathscr{C}$.

For example, $((K_3 \nabla \overline{K_2}) \cup K_1) \nabla \overline{K_2} \in \mathscr{C}$, since any complete graph or its complement belongs to $\mathscr{C}$. It is observed in [Sim4] that graphs from $\mathscr{C}$ have a nice representation in the form of weighted rooted trees, where the weights are assigned to the vertices. Given such a tree $T$, with root $u$, the corresponding graph $G$ is obtained as follows:

(i) to each vertex of $T$, say $v$, there corresponds in $G$ an independent set of vertices equal in number to the weight of $v$;
(ii) two vertices from different sets are adjacent in $G$ if and only if the corresponding tree vertices, say $v$ and $w$, are incomparable (in the sense that neither the $u$-$v$ path nor the $u$-$w$ path in $T$ is a subpath of the other).

**9.2.1 Example** For the graph $((K_3 \nabla \overline{K_2}) \cup K_1) \nabla \overline{K_2}$ above the corresponding tree and the graph itself are depicted in Fig. 9.1. Lines between two circles (in Fig. 9.1(b)) indicate that each vertex inside one circle is adjacent to each vertex inside the other.                                                      □

**9.2.2 Remark** The above representation of a graph $G$ from $\mathscr{C}$ is unique provided that (i) all tree vertices (except the root) have non-zero weights, (ii) all internal vertices (vertices other than the root and endvertices) are of degree at least 3, and (iii) the degree of the root is at least 2 if the tree is non-trivial. Such a tree is called the *canonical tree* corresponding to $G$.□

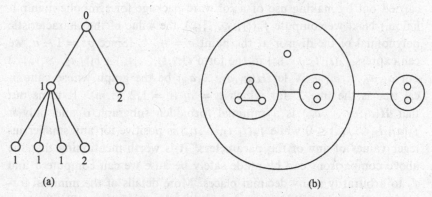

Fig. 9.1. Representations of a graph in $\mathscr{C}$.

It was proved in [Sim4] that for the required collection of minimal forbidden subgraphs, the corresponding collection of canonical trees is finite; the only unresolved problem was how to assign weights to vertices of these trees to obtain the minimal forbidden subgraphs. This problem admits an elegant solution based on the divisor technique. For a graph $G$ in $\mathscr{C}$ we make use of an equitable partition (called in [Sim4] the *natural partition*) whose cells are constructed as follows: if $T$ is the canonical tree of $G$ then

(i) any vertex of $T$, other than the root or an endvertex of weight 1, corresponds to a cell whose vertices are mutually non-adjacent;
(ii) endvertices of weight 1 adjacent to the same internal vertex of $T$ correspond to a cell whose vertices are mutually adjacent.

For example, in Fig. 9.1(b) the vertices within the inner circles constitute the cells of a natural partition.

The next crucial step is the following theorem (where, as in the theory of star partitions, the eigenvalues $-1$ and $0$ play an exceptional role):

**9.2.3 Theorem** [Sim4] *If $G \in \mathscr{C}$, then all eigenvalues of $G$, except possibly $-1$ and $0$, are contained in the spectrum of the divisor corresponding to the natural partition.*

It transpires that we can find the required minimal forbidden subgraphs as follows. We consider a canonical tree $T$ in which the weights are indeterminates, say $t_1, t_2, \ldots, t_m$ (regarded as integer parameters). Then we compute (symbolically) the characteristic polynomial of the divisor determined by the equitable partition defined above. (This step can be carried out by making use of a software package for symbolic manipulation.) Next we compute $f_\sigma(t_1, t_2, \ldots, t_m)$, the value of the characteristic polynomial of the divisor at the point $\sigma = \frac{\sqrt{5}-1}{2}$. (Since $\sigma^2 = 1 - \sigma$, we can express $f_\sigma(t_1, t_2, \ldots, t_m)$ in the form $a(t_1, t_2, \ldots, t_m) + b(t_1, t_2, \ldots, t_m)\sigma$.) For $a_1, a_2, \ldots, a_m \in \mathbb{N}$ let $H(a_1, a_2, \ldots, a_m)$ be the graph whose canonical tree is the tree $T$ for which $t_i = a_i$ ($i = 1, 2, \ldots, m$). It turns out that $H(a_1, a_2, \ldots, a_m)$ is a minimal forbidden subgraph if and only if $f_\sigma(a_1, a_2, \ldots, a_m) \le 0$ while $f_\sigma(t_1, t_2, \ldots, t_m)$ is positive for any smaller integer values of any of the parameters. It is worth mentioning that the above comparisons can be made safely because we can compute $\sigma$ and $f_\sigma$ to arbitrarily many decimal places. More details of the minimal forbidden subgraphs will be provided in a forthcoming paper. To illustrate the advantages of the divisor approach, we mention only that the size of divisor is typically of order 10, whereas the size of a minimal forbidden subgraph might exceed 240 000 000.

## 9.3 Bond order and electron charges

As explained in [Mal] and [CvDS, Chapter 8], for example, graph eigenvalues have a prominent role in Hückel's theory of molecular orbitals, a theory which played an important part in the development of quantum chemistry. In this section we demonstrate a connection with graph angles. A non-saturated hydrocarbon molecule consists of carbon atoms, hydrogen atoms, $\sigma$-electrons and $\pi$-electrons. The $\sigma$-electrons afford bonds between atoms, called *$\sigma$-bonds*, and the corresponding *molecular graph* is the graph whose vertices represent atoms and whose edges represent $\sigma$-bonds. There is just one $\sigma$-electron associated with a hydrogen atom, while there are three $\sigma$-electrons and one $\pi$-electron associated with a carbon atom. Thus in chemical terms a hydrogen atom has valency 1, while a carbon atom has valency 4; but in the molecular graph a hydrogen atom has degree 1 while a carbon atom has degree 3. (Here we follow the description in [Mal].) The subgraph of the molecular graph induced by the carbon atoms is called the *carbon skeleton*. Fig. 9.2 shows the molecular graph and carbon skeleton of a naphthalene molecule.

The $\pi$-electron associated with a carbon atom is treated differently: a solution of Schrödinger's equation determines a probability density function for the location of such a $\pi$-electron, and this function is called an *atomic orbital*. In the Hückel theory, a *molecular orbital* is a linear combination of atomic orbitals for which the total $\pi$-electron energy is a local minimum. It turns out that the eigenvalues $\lambda_1, \lambda_2, \ldots, \lambda_n$ of the

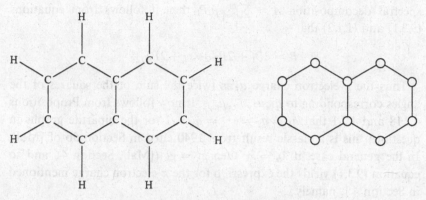

Fig. 9.2. A naphthalene molecule.

carbon skeleton determine the possible $\pi$-electron energy levels, while coordinates of associated orthonormal eigenvectors $(x_{i1}, x_{i2}, \ldots, x_{in})^T$ $(i = 1, 2, \ldots, n)$ determine the corresponding linear combinations of atomic orbitals. The $n$ $\pi$-electrons are ascribed to the $n$ molecular orbitals according to the *Aufbau principle* as described in [Mal]. Suppose that $\lambda_1 \geq \lambda_2 \geq \cdots \geq \lambda_n$ and that $g_k$ is the number of $\pi$-electrons ascribed to the $k$-th molecular orbital. Then *Coulson's bond-order matrix* (or *density matrix*) is the matrix $P = (p_{ij})$, where

$$p_{ij} = \sum_{k=1}^{n} g_k \, x_{ki} \, x_{kj}. \qquad (9.3.1)$$

The number $p_{jj}$ is called the $\pi$-*electron charge* $q_j$ on the $j$-th carbon atom. Note that $\sum_{j=1}^{n} q_j = \sum_{k=1}^{n} g_k = n$. (cf. Section 5b of [Mal]).

For adjacent vertices $i$ and $j$, it transpires that the higher the *bond order* $p_{ij}$ the smaller the *bond length* between the corresponding atoms. (It appears that for distinct non-adjacent vertices, the number $p_{ij}$ has little chemical significance.) It is a consequence of the Aufbau principle that always $g_k \in \{0, 1, 2\}$; and for bipartite molecular graphs without 0 as an eigenvalue (in which case $n$ is even) we have $g_k = 2$ for $k \leq \frac{1}{2}n$, $g_k = 0$ for $k > \frac{1}{2}n$. In the bipartite case,

$$p_{ij} = 2 \sum_{k=1}^{n/2} x_{ki} x_{kj} \qquad (9.3.2)$$

and the number of distinct eigenvalues of the carbon skeleton is even also. If, in the usual notation, the adjacency matrix of the graph has spectral decomposition $A = \sum_{i=1}^{m} \mu_i P_i$ then it follows from equations (9.3.1) and (1.1.2) that

$$P = 2P_1 + 2P_2 + \cdots + 2P_{m/2}.$$

Thus the $\pi$-electron charge $q_j$ is twice the sum of the squares of the angles corresponding to $\mu_1, \mu_2, \ldots, \mu_{m/2}$. It now follows from Propositions 4.2.18 and 4.2.1 that $q_1 = q_2 = \cdots = q_n = 1$ for the bipartite graphs in question: this is a classic result from 1940 cited in Section 5b of [Mal]. In the general case, if $\lambda_h = \lambda_k$ then $g_h = g_k$ ([Mal], Section 4), and so equation (9.3.1) yields the expression for the $\pi$-electron charge mentioned in Section 4.1, namely

$$q_j = \sum_{1=1}^{m} w_i \alpha_{ij}^2,$$

where $w_1, \ldots, w_m$ are the appropriate $g_k$ and the $\alpha_{ij}$ are the angles of the molecular graph.

## 9.4 Certain 3-decompositions of complete graphs

By a 3-decomposition of $K_n$ we mean a set of three isomorphic spanning subgraphs $H_1, H_2, H_3$ such that each edge of $K_n$ lies in exactly one of the $H_i$. For example, it is easy to see that $K_7$ is the edge-disjoint union of three 7-cycles. Is $K_{10}$ (which has 45 edges) the edge-disjoint union of three copies of the Petersen graph (which has 15 edges)? This question was posed by Schwenk in the *American Mathematical Monthly* (Problem 6434(b) of June-July 1983). The following simple argument involving eigenspaces shows that the answer is 'no'. If $K_{10}$ has such a decomposition then

$$A + B + C + I = J \tag{9.4.1}$$

where each of $A, B, C$ is the adjacency matrix of a Petersen graph. Let $\mathcal{W}$ be the orthogonal complement of the all-1 vector $\mathbf{j}$ in $\mathbb{R}^{10}$. Since $\mathscr{E}_A(1)$ and $\mathscr{E}_B(1)$ are five-dimensional subspaces of the nine-dimensional space $\mathcal{W}$, there exists a non-zero vector $\mathbf{x} \in \mathscr{E}_A(1) \cap \mathscr{E}_B(1)$ such that $J\mathbf{x} = \mathbf{0}$. From equation (9.4.1) we have $A\mathbf{x} + B\mathbf{x} + C\mathbf{x} + \mathbf{x} = 0$, whence $C\mathbf{x} = -3\mathbf{x}$. This is a contradiction because $-3$ is not an eigenvalue of $C$.

We can extend the above argument from the Petersen graph to an arbitrary strongly regular graph $G$, and thereby obtain the following result.

**9.4.1 Theorem** [Row2] *Let G be a strongly regular graph with parameters $n, r, e, f$. If $K_n$ is the edge-disjoint union of three spanning subgraphs isomorphic to G then there exists a positive integer k such that, with a consistent choice of sign,*

$$n = (3k \pm 1)^2, \ 3k^2 \pm 2k, \ e = k^2 - 1 \ and \ f = k^2 \pm k.$$

*Proof* The eigenvalues of $G$ are $r, \rho_1, \rho_2$ with multiplicities $1, m_1, m_2$ where

$$\rho_1, \rho_2 = \frac{1}{2}(e - f \pm \sqrt{(e-f)^2 + 4(r-f)})$$

and

$$m_1, m_2 = \frac{1}{2}\{(n-1) \pm \frac{(n-1)(f-e) - 2r}{\sqrt{(e-f)^2 + 4(r-f)}}\}.$$

If $K_n$ has a 3-decomposition as described in the statement of the theorem then a consideration of degrees shows that $n - 1 = 3r$. It follows that $m_1 \neq m_2$ for otherwise $f - e = 2r/(n-1) = \frac{2}{3}$. Equation (9.4.1) holds with each of $A, B, C$ now an adjacency matrix of $G$. On repeating the original argument with $\rho_1$ in place of the eigenvalue 1 we find that $-2\rho_1 - 1$ is an eigenvalue of $C$. This eigenvalue is different from $r$ because a corresponding eigenvector is orthogonal to $\mathbf{j}$; and different from $\rho_1$ because $\rho_1 \neq \frac{1}{3}$. Hence $-2\rho_1 - 1 = \rho_2$ and on expressing $m_1$ in the form

$$m_1 = \frac{1}{2}\{3r - \frac{3r(\rho_1 + \rho_2) + 2r}{\rho_1 - \rho_2}\} \tag{9.4.2}$$

we see that $m_1 = 2r, m_2 = r$. It follows from (9.4.2) that $3(\rho_1 + \rho_2) + 2 = \rho_2 - \rho_1$. Similarly, if $m_2 > m_1$ then $\rho_1$ and $\rho_2$ are interchanged and we have $3(\rho_1 + \rho_2) + 2 = \rho_1 - \rho_2$. Hence always

$$(3e - 3f + 2)^2 = (e - f)^2 + 4(r - f). \tag{9.4.3}$$

Now in any strongly regular graph we have $r(r - e - 1) = (n - r - 1)f$, an equality obtained by counting in two ways the number of walks of length 2. Since here $n - 1 = 3r$ we have $r = 2f + e + 1$ and it follows from (9.4.3) that

$$(f - e + 1)^2 = e + 1.$$

Thus $e$ has the form $k^2 - 1$ and the result follows. □

**9.4.2 Remark** In the terminology of Mesner [Mes] a strongly regular graph which satisfies the conclusions of Theorem 9.4.1 with $n = (3k + 1)^2$

is a graph of negative Latin square type $NL_k(3k+1)$. If $n = (3k-1)^2$ then the graph is called by Bose and Shrikhande [BoSh] a pseudo-net-graph of type $L_k(3k-1)$.                                                        □

**9.4.3 Remark** [Row2] Suppose for definiteness that $m_1 > m_2$ in the proof of Theorem 9.4.1. Our eigenspace argument shows that $\mathscr{E}_A(\rho_1) \cap \mathscr{E}_B(\rho_2) \subseteq \mathscr{E}_C(\rho_2)$. Now $\dim \mathscr{E}_C(\rho_2) = k$ and $\dim(\mathscr{E}_A(\rho_1) \cap \mathscr{E}_B(\rho_1)) = \dim \mathscr{E}_A(\rho_1) + \dim \mathscr{E}_B(\rho_1) - \dim(\mathscr{E}_A(\rho_1) + \mathscr{E}_B(\rho_1)) \geq 2m_1 - (n-1) = k$. Hence $\mathscr{E}_C(\rho_2) = \mathscr{E}_A(\rho_1) \cap \mathscr{E}_B(\rho_1)$. Similarly $\mathscr{E}_B(\rho_2) = \mathscr{E}_C(\rho_1) \cap \mathscr{E}_A(\rho_1)$ and $\mathscr{E}_A(\rho_2) = \mathscr{E}_B(\rho_1) \cap \mathscr{E}_C(\rho_1)$. Since $\mathscr{E}_A(\rho_1) \cap \mathscr{E}_A(\rho_2) = 0$, we have $\mathscr{E}_A(\rho_2) \cap \mathscr{E}_B(\rho_2) = 0$ and, on comparing dimensions, $\mathscr{E}_C(\rho_1) = \mathscr{E}_A(\rho_2) \oplus \mathscr{E}_C(\rho_2)$. Therefore,

$$\mathbb{R}^n = \langle \mathbf{j} \rangle \oplus \mathscr{E}_A(\rho_2) \oplus \mathscr{E}_B(\rho_2) \oplus \mathscr{E}_C(\rho_2). \tag{9.4.4}$$

If $m_2 > m_1$ then $\rho_1$ replaces $\rho_2$ in (9.4.4).

The disposition of the various eigenspaces ensures that the matrices $A, B, C$ are simultaneously diagonalizable and so they commute: in terms of line-colourings of $K_n$ this means that if we use three different colours $c_1, c_2, c_3$ for the three subgraphs isomorphic to $G$ then for any two vertices $u, v$ and any two colours $c_i, c_j$ the number of $u$-$v$ walks of length 2 coloured $c_i, c_j$ is the same as the number of $u$-$v$ walks of length 2 coloured $c_j, c_i$. This is not generally the case for a 3-decomposition of $K_n$ into isomorphic regular subgraphs as may be seen from the following decomposition of $K_7$ into three 7-cycles: if the points of $K_7$ are denoted 1,2,3,4,5,6,7 and the cycles 12345671, 14275361, 13746251 are coloured blue, red, green respectively then the walk 153 is coloured green-red, but there is no walk from 1 to 3 coloured red-green.                    □

The following class of examples illustrating Theorem 9.4.1 is given in [Row2].

**9.4.4 Example** Let $\mathbb{K}$ be a finite field of order $q = p^{2h}$, where $h \in \mathbb{N}$ and $p$ is a prime congruent to 2 mod 3. Let $g$ be a generator for the multiplicative group of $\mathbb{K}$, and let $H = \langle g^3 \rangle$. The subgroup $H$ has index 3 in $\langle g \rangle$ and consists of all the non-zero cubes in $\mathbb{K}$. Since $-1 \in h$ we may define (undirected) graphs $G_i$ ($i = 0, 1, 2$), with vertices the elements of $\mathbb{K}$, as follows: vertices $u$ and $v$ are adjacent in $G_i$ if and only if $u - v \in Hg^i$ ($i = 0, 1, 2$). The map $x \longmapsto xg^i$ ($x \in \mathbb{K}$) is an isomorphism $G_0 \to G_i$, and it follows that $G_0, G_1, G_2$ constitute a 3-decomposition of $K_q$. Moreover $G_0$ is strongly regular because any pair of adjacent vertices may be mapped to $0, g^3$, and any pair of non-adjacent vertices may be mapped to $0, g$, by an automorphism of $\mathbb{K}$. Thus there are infinitely many graphs $G$ which satisfy the hypotheses of Theorem 9.4.1. (The smallest connected example is the *Clebsch graph*, which arises as $G_0$ when $q = 16$.)

We can use the relation between parameters given by Theorem 9.4.1 to find the number of solutions of the Fermat equation $x^3 + y^3 = z^3$ in the field $\mathbb{K}$. Note that $e = k^2 - 1$ where $k = \frac{1}{3}(p^h - 1)$ if $h$ is even and $k = \frac{1}{3}(p^h + 1)$ if $h$ is odd. Now for given $u \in H$, $e$ is the number of solutions $(v, w)$ of the equation $u + v = w$ $(v, w \in H)$. It follows that the number of solutions $(u, v, w)$ of the equation $u + v = w$ $(u, v, w \in H)$ is $e|H|$. Each element of $H$ has three cube roots in $\mathbb{K}$ and so the number of non-trivial solutions $(x, y, z)$ of the equation $x^3 + y^3 = z^3$ $(x, y, z \in \mathbb{K})$ is $f_3(p^{2h})$, where $f_3(p^{2h}) = 27e|H| = (p^{2h} - 1)(p^{2h} - 2(-p)^h - 8)$. $\qquad\square$

**9.4.5 Remark** If $K_n$ is the edge-disjoint union of subgraphs (not necessarily spanning subgraphs) isomorphic to the graph $G$ then (i) $\frac{1}{2}n(n-1)$ is divisible by the number of edges in $G$ and (ii) $n - 1$ is divisible by the greatest common divisor of the degrees of vertices in $G$. An asymptotic converse was proved by R. M. Wilson [Wils]: given a graph $G$, for large enough $n$ satisfying conditions (i) and (ii), $K_n$ is the edge-disjoint union of subgraphs isomorphic to $G$. Note that if $G$ is regular of degree $r$ then conditions (i) and (ii) reduce to the single requirement that $r$ divides $n - 1$. $\qquad\square$

# Appendix A
## Some results from matrix theory

The adjacency matrix of a graph is non-negative in the sense that each of its entries is non-negative. Accordingly the following theorem applies.

**A.1 Theorem** (see, for example, [Gan], vol II, p. 66) *A non-negative matrix always has a non-negative eigenvalue r such that the modulus of any other eigenvalue does not exceed r. To this 'maximal' eigenvalue there corresponds a non-negative eigenvector.*

A matrix $A$ is called *reducible* if there is a permutation matrix $P$ such that the matrix $P^{-1}AP$ is of the form $\begin{pmatrix} X & O \\ Y & Z \end{pmatrix}$, where $X$ and $Z$ are square matrices. Otherwise, $A$ is called *irreducible*. Note that if $A$ is symmetric and reducible then $Y = O$ in the above definition. In particular, the adjacency matrix of a graph $G$ is irreducible if and only if $G$ is connected.

Spectral properties of irreducible non-negative matrices are described by the following theorem of Frobenius.

**A.2 Theorem** (see, for example, [Gan], vol. II, pp. 53-54) *An irreducible non-negative matrix A always has a positive eigenvalue r that is a simple root of the characteristic polynomial. The modulus of any other eigenvalue does not exceed r. To the 'maximal' eigenvalue r there corresponds a positive eigenvector. Moreover, if A has precisely h eigenvalues of modulus r, then these numbers are all distinct and are roots of the equation $\lambda^h - r^h = 0$. More generally: the whole spectrum $\lambda_1, \lambda_2, \ldots, \lambda_n$ of A, regarded as a system of points in the Argand plane, is mapped onto itself by a rotation through the angle $\frac{2\pi}{h}$. If $h > 1$, then by a permutation of rows and the same permutation of columns, A can be put into the following* 'cyclic' form

$$A = \begin{pmatrix} O & A_{12} & O & \dots & O \\ O & O & A_{23} & \dots & O \\ \vdots & & & \ddots & \\ O & O & O & \dots & A_{h-1,h} \\ A_{h1} & O & O & \dots & O \end{pmatrix},$$

where there are square blocks along the main diagonal.

If $h > 1$ in Theorem A.2, the matrix $A$ is called *imprimitive* and $h$ is the *index of imprimitivity*. Otherwise, $A$ is *primitive*. As a converse of Theorem A.2 we have:

**A.3 Theorem** (see, e.g., [Gan], vol II, p. 79) *If the 'maximal' eigenvalue $r$ of a non-negative matrix $A$ is simple and if positive eigenvectors belong to $r$ in both $A$ and $A^T$, then $A$ is irreducible.*

More generally, we have the following result:

**A.4 Theorem** (see, e.g., [Gan], vol. II, p. 78) *To the 'maximal' eigenvalue $r$ of a non-negative matrix $A$ there belongs a positive eigenvector in both $A$ and $A^T$ if and only if $A$ can be represented by a permutation of rows and by the same permutation of columns in block-diagonal form $A = \mathrm{diag}(A_1, \dots, A_s)$, where $A_1, \dots, A_s$ are irreducible matrices each of which has $r$ as its 'maximal' eigenvalue.*

# Appendix B
## A table of graph angles

Table B.1, which is taken from [CvPe3], provides data for all connected graphs on $n$ vertices, $2 \leq n \leq 5$. The graphs are given in the same order as in Table 1 in the Appendix of [CvDS]. The table contains, for each graph, the eigenvalues (first line), the main angles (second line) and the vertex angle sequences, with vertices labelled as in the diagram alongside. Vertices of graphs in Table 1 are ordered in such a way that the corresponding vertex angle sequences are ordered lexicographically. Since similar vertices have the same angle sequence, just one sequence is given for each orbit.

Table B.1. *Eigenvalues, main angles and angles of connected graphs with two to five vertices*

**1.**

| | | |
|---|---|---|
| | 1.0000 | −1.0000 |
| | 1.0000 | 0.0000 |
| 1,2. | 0.7071 | 0.7071 |

**2.**

| | | |
|---|---|---|
| | 2.0000 | −1.0000² |
| | 1.0000 | 0.0000 |
| 1,2,3. | 0.5774 | 0.8165 |

**3.**

| | | | |
|---|---|---|---|
| | 1.4142 | 0.0000 | −1.4142 |
| | 0.9856 | 0.0000 | 0.1691 |
| 1. | 0.7071 | 0.0000 | 0.7071 |
| 2,3. | 0.5000 | 0.7071 | 0.5000 |

**4.**

| | | |
|---|---|---|
| | 3.0000 | −1.0000³ |
| | 1.0000 | 0.0000 |
| 1,2,3,4. | 0.5000 | 0.8660 |

**5.**

| | | | | |
|---|---|---|---|---|
| | 2.5616 | 0.0000 | −1.0000 | −1.5616 |
| | 0.9925 | 0.0000 | 0.0000 | 0.1222 |
| 1,2. | 0.5573 | 0.0000 | 0.7071 | 0.4352 |
| 3,4. | 0.4352 | 0.7071 | 0.0000 | 0.5573 |

**6.**

| | | | | |
|---|---|---|---|---|
| | 2.1701 | 0.3111 | −1.0000 | −1.4812 |
| | 0.9695 | 0.1663 | 0.0000 | 0.1803 |
| 1. | 0.6116 | 0.2536 | 0.0000 | 0.7494 |
| 2,3. | 0.5227 | 0.3682 | 0.7071 | 0.3020 |
| 4. | 0.2818 | 0.8152 | 0.0000 | 0.5059 |

**7.**

| | | | |
|---|---|---|---|
| | 2.0000 | 0.0000² | −2.0000 |
| | 1.0000 | 0.0000 | 0.0000 |
| 1,2,3,4. | 0.5000 | 0.7071 | 0.5000 |

**8.**

| | | | |
|---|---|---|---|
| | 1.7321 | 0.0000² | −1.7321 |
| | 0.9659 | 0.0000 | 0.2588 |
| 1. | 0.7071 | 0.0000 | 0.7071 |
| 2,3,4. | 0.4082 | 0.8165 | 0.4082 |

**9.**

| | | | | |
|---|---|---|---|---|
| | 1.6180 | 0.6180 | −0.6180 | −1.6180 |
| | 0.9732 | 0.0000 | 0.2298 | 0.0000 |
| 1,2. | 0.6015 | 0.3717 | 0.3717 | 0.6015 |
| 3,4. | 0.3717 | 0.6015 | 0.6015 | 0.3717 |

## Table B.1. *Continued*

**10.**

|  |  |  |
|---|---|---|
|  | 4.0000 | $-1.0000^4$ |
|  | 1.0000 | 0.0000 |
| 1,2,3,4,5. | 0.4472 | 0.8944 |

**11.**

|  |  |  |  |  |
|---|---|---|---|---|
|  | 3.6458 | 0.0000 | $-1.0000^2$ | $-1.6458$ |
|  | 0.9957 | 0.0000 | 0.0000 | 0.0930 |
| 1,2,3. | 0.4792 | 0.0000 | 0.8165 | 0.3220 |
| 4,5. | 0.3943 | 0.7071 | 0.0000 | 0.5869 |

**12.**

|  |  |  |  |  |
|---|---|---|---|---|
|  | 3.3234 | 0.3579 | $-1.0000^2$ | $-1.6813$ |
|  | 0.9861 | 0.0837 | 0.0000 | 0.1432 |
| 1,2. | 0.5100 | 0.1378 | 0.7071 | 0.4700 |
| 3,4. | 0.4390 | 0.4294 | 0.7071 | 0.3505 |
| 5. | 0.3069 | 0.7702 | 0.0000 | 0.5590 |

**13.**

|  |  |  |  |  |
|---|---|---|---|---|
|  | 3.2361 | $0.0000^2$ | $-1.2361$ | $-2.0000$ |
|  | 0.9960 | 0.0000 | 0.0898 | 0.0000 |
| 1. | 0.5257 | 0.0000 | 0.8507 | 0.0000 |
| 2,3,4,5. | 0.4253 | 0.7071 | 0.2629 | 0.5000 |

**14.**

|  |  |  |  |  |
|---|---|---|---|---|
|  | 3.0861 | 0.4280 | $-1.0000^2$ | $-1.5141$ |
|  | 0.9567 | 0.2306 | 0.0000 | 0.1774 |
| 1. | 0.5236 | 0.3610 | 0.0000 | 0.7717 |
| 2,3,4. | 0.4820 | 0.2297 | 0.8165 | 0.2196 |
| 5. | 0.1697 | 0.8435 | 0.0000 | 0.5097 |

**15.**

|  |  |  |  |  |
|---|---|---|---|---|
|  | 3.0000 | $0.0000^2$ | $-1.0000$ | $-2.0000$ |
|  | 0.9798 | 0.0000 | 0.0000 | 0.2000 |
| 1,2. | 0.5477 | 0.0000 | 0.7071 | 0.4472 |
| 3,4,5. | 0.3651 | 0.8165 | 0.0000 | 0.4472 |

**16.**

|  |  |  |  |  |  |
|---|---|---|---|---|---|
|  | 2.9354 | 0.6180 | $-0.4626$ | $-1.4728$ | $-1.6180$ |
|  | 0.9839 | 0.0000 | 0.0738 | 0.1629 | 0.0000 |
| 1. | 0.5590 | 0.0000 | 0.3069 | 0.7702 | 0.0000 |
| 2,3. | 0.4700 | 0.3717 | 0.5100 | 0.1378 | 0.6015 |
| 4,5. | 0.3505 | 0.6015 | 0.4390 | 0.4294 | 0.3717 |

## Table B.1. *Continued*

| | | | | | | |
|---|---|---|---|---|---|---|
| **17.** | | 2.8558 | 0.3216 | 0.0000 | −1.0000 | −2.1774 |
| | | 0.9898 | 0.1363 | 0.0000 | 0.0000 | 0.0416 |
| | 1,2. | 0.4912 | 0.3870 | 0.0000 | 0.7071 | 0.3301 |
| | 3,4. | 0.4558 | 0.1312 | 0.7071 | 0.0000 | 0.5244 |
| | 5. | 0.3192 | 0.8161 | 0.0000 | 0.0000 | 0.4817 |
| | | | | | | |
| **18.** | | 2.6855 | 0.3349 | 0.0000 | −1.2713 | −1.7491 |
| | | 0.9602 | 0.1692 | 0.0000 | 0.0486 | 0.2170 |
| | 1. | 0.5825 | 0.2835 | 0.0000 | 0.4008 | 0.6478 |
| | 2. | 0.5237 | 0.3506 | 0.0000 | 0.7611 | 0.1534 |
| | 3,4. | 0.4119 | 0.2004 | 0.7071 | 0.2834 | 0.4581 |
| | 5. | 0.2169 | 0.8464 | 0.0000 | 0.3153 | 0.3704 |
| | | | | | | |
| **19.** | | 2.6412 | 0.7237 | −0.5892 | −1.0000 | −1.7757 |
| | | 0.9550 | 0.1833 | 0.2319 | 0.0000 | 0.0262 |
| | 1,2. | 0.5371 | 0.1655 | 0.1955 | 0.7071 | 0.3820 |
| | 3. | 0.4747 | 0.5030 | 0.3529 | 0.0000 | 0.6301 |
| | 4. | 0.4067 | 0.4573 | 0.6636 | 0.0000 | 0.4303 |
| | 5. | 0.1797 | 0.6950 | 0.5989 | 0.0000 | 0.3549 |
| | | | | | | |
| **20.** | | 2.5616 | 1.0000 | −1.0000[2] | −1.5616 | |
| | | 0.9802 | 0.0000 | 0.0000 | 0.1979 | |
| | 1. | 0.6154 | 0.0000 | 0.0000 | 0.7882 | |
| | 2,3,4,5. | 0.3941 | 0.5000 | 0.7071 | 0.3077 | |
| | | | | | | |
| **21.** | | 2.4812 | 0.6889 | 0.0000 | −1.1701 | −2.0000 |
| | | 0.9850 | 0.1223 | 0.0000 | 0.1220 | 0.0000 |
| | 1,2. | 0.5299 | 0.1793 | 0.5000 | 0.4325 | 0.5000 |
| | 3. | 0.4271 | 0.5207 | 0.0000 | 0.7392 | 0.0000 |
| | 4,5. | 0.3578 | 0.5765 | 0.5000 | 0.1993 | 0.5000 |
| | | | | | | |
| **22.** | | 2.4495 | 0.0000[3] | −2.4495 | | |
| | | 0.9949 | 0.0000 | 0.1005 | | |
| | 1,2. | 0.5000 | 0.7071 | 0.5000 | | |
| | 3,4,5. | 0.4082 | 0.8165 | 0.4082 | | |
| | | | | | | |
| **23.** | | 2.3429 | 0.4707 | 0.0000 | −1.0000 | −1.8136 |
| | | 0.9506 | 0.1587 | 0.0000 | 0.0000 | 0.2667 |
| | 1. | 0.6359 | 0.2414 | 0.0000 | 0.0000 | 0.7331 |
| | 2,3. | 0.4735 | 0.4560 | 0.0000 | 0.7071 | 0.2606 |
| | 4,5. | 0.2714 | 0.5128 | 0.7071 | 0.0000 | 0.4042 |

## Table B.1. *Continued*

**24.**

| | | | | | |
|---|---|---|---|---|---|
| | 2.3028 | 0.6180 | 0.0000 | −1.3028 | −1.6180 |
| | 0.9444 | 0.0000 | 0.2582 | 0.2035 | 0.0000 |
| 1,2. | 0.5651 | 0.3717 | 0.0000 | 0.4250 | 0.6015 |
| 3. | 0.4908 | 0.0000 | 0.5774 | 0.6525 | 0.0000 |
| 4,5. | 0.2454 | 0.6015 | 0.5774 | 0.3263 | 0.3717 |

**25.**

| | | | | | |
|---|---|---|---|---|---|
| | 2.2143 | 1.0000 | −0.5392 | −1.0000 | −1.6751 |
| | 0.9370 | 0.2828 | 0.2021 | 0.0000 | 0.0347 |
| 1. | 0.6037 | 0.0000 | 0.4762 | 0.0000 | 0.6394 |
| 2,3. | 0.4972 | 0.3162 | 0.3094 | 0.7071 | 0.2390 |
| 4. | 0.3425 | 0.6325 | 0.3620 | 0.0000 | 0.5930 |
| 5. | 0.1547 | 0.6325 | 0.6714 | 0.0000 | 0.3540 |

**26.**

| | | | | | |
|---|---|---|---|---|---|
| | 2.1358 | 0.6622 | 0.0000 | −0.6622 | −2.1358 |
| | 0.9762 | 0.0742 | 0.0000 | 0.1835 | 0.0885 |
| 1. | 0.5573 | 0.4352 | 0.0000 | 0.4352 | 0.5573 |
| 2,3. | 0.4647 | 0.1845 | 0.7071 | 0.1845 | 0.4647 |
| 4. | 0.4352 | 0.5573 | 0.0000 | 0.5573 | 0.4352 |
| 5. | 0.2610 | 0.6572 | 0.0000 | 0.6572 | 0.2610 |

**27.**

| | | | |
|---|---|---|---|
| | 2.0000 | $0.6180^2$ | $-1.6180^2$ |
| | 1.0000 | 0.0000 | 0.0000 |
| 1,2,3,4,5. | 0.4472 | 0.6325 | 0.6325 |

**28.**

| | | | |
|---|---|---|---|
| | 2.0000 | $0.0000^3$ | −2.0000 |
| | 0.9487 | 0.0000 | 0.3162 |
| 1. | 0.7071 | 0.0000 | 0.7071 |
| 2,3,4,5. | 0.3536 | 0.8660 | 0.3536 |

**29.**

| | | | | | |
|---|---|---|---|---|---|
| | 1.8478 | 0.7654 | 0.0000 | −0.7654 | −1.8478 |
| | 0.9530 | 0.0785 | 0.0000 | 0.2638 | 0.1267 |
| 1. | 0.6533 | 0.2706 | 0.0000 | 0.2706 | 0.6533 |
| 2. | 0.5000 | 0.5000 | 0.0000 | 0.5000 | 0.5000 |
| 3,4. | 0.3536 | 0.3536 | 0.7071 | 0.3536 | 0.3536 |
| 5. | 0.2706 | 0.6533 | 0.0000 | 0.6533 | 0.2706 |

**30.**

| | | | | | |
|---|---|---|---|---|---|
| | 1.7321 | 1.0000 | 0.0000 | −1.0000 | −1.7321 |
| | 0.9636 | 0.0000 | 0.2582 | 0.0000 | 0.0692 |
| 1. | 0.5774 | 0.0000 | 0.5774 | 0.0000 | 0.5774 |
| 2,3. | 0.5000 | 0.5000 | 0.0000 | 0.5000 | 0.5000 |
| 4,5. | 0.2887 | 0.5000 | 0.5774 | 0.5000 | 0.2887 |

# Bibliography

[Ach]  Acharya B.D., *Spectral criterion for cycle balance in networks*, J. Graph Theory, 4(1980), No. 1, 1-11

[Ahr]  Ahrens J.H., *Paving the chessboard*, J. Combinatorial Theory A, 31(1981), 277-288

[AlMi]  Alon N., Milman V.D., $\lambda_1$, *isoperimetric inequalities for graphs and superconcentrators*, J. Combinatorial Theory B, 38(1985), 73-88

[AsGi]  Aspvall B., Gilbert J.R., *Graph colouring using eigenvalue decomposition*, SIAM J. Alg. Disc. Meth., 5(1984), No.4, 526-538

[Bab1]  Babai L., *Automorphism group and category of cospectral graphs*, Acta Math. Acad. Sci. Hung., 31(1978), 295-306

[Bab2]  Babai L., *On the complexity of canonical labelling of strongly regular graphs*, SIAM J. Computing, 9(1980), 212-216

[Bab3]  Babai L., *Kospektrale Graphen mit vorgegebenen Automorphismengruppen*, Wiss. Z., T.H. Ilmenau, 27(1981), No.5, 31-37

[BaGM]  Babai L., Grigorjev D., Mount D.M., *Isomorphism of graphs with bounded eigenvalue multiplicity*, Proc. 14th Annual ACM Symp. on the Theory of Computing, San Francisco 1982, New York, 1982, 310-324

[Bak1]  Baker G.A., *Drum shapes and isospectral graphs*, BNL 10088, Brookhaven National Laboratory, Long Island, New York, 1966

[Bak2]  Baker G.A., *Drum shapes and isospectral graphs*, J. Math. Phys., 7(1966), 2238-2242

[BaHa]  Balaban A.T., Harary F., *The characteristic polynomial does not uniquely determine the topology of a molecule*, J. Chem. Doc., 11(1971), 258-259

[BaCN]  Balas E., Chvatal V., Nešetril J., *On the maximum weight clique problem*, Mathematics of Operation Research, 12(1987), No.3, 522-535

[BaPa]  Balasubramanian K., Parthasarathy K.R., *In search of a complete invariant for graphs*, Combinatorics and Graph Theory, Proc. 2nd Symp. held at the Indian Statistical Institute, Calcutta, February 25-29, 1980, Lecture Notes in Math. 885, ed. Rao S.B., Springer-Verlag, Berlin, 1981, 42-59

[Bel1]  Bell F.K., *On the maximal index of connected graphs*, Linear Algebra and Appl., 144(1991), 135-151

[Bel2] Bell F.K., *A note on irregularity of graphs*, Linear Algebra and Appl., 161(1992), 45-54

[BeRo1] Bell F.K., Rowlinson P., *The change in index of a graph resulting from the attachment of a pendant edge*, Proc. Royal Soc. Edinburgh, 108A(1988), 67-74

[BeRo2] Bell F.K., Rowlinson P., *On the index of tricyclic Hamiltonian graphs*, Proc. Edinburgh Math. Soc., 33(1990), No. 2, 233-240

[BeRo3] Bell F.K., Rowlinson P., *Certain graphs without zero as an eigenvalue*, Math. Japonica, 48(1993), No. 5, 961-967

[BeSi] Bell F.K., Simić S., *On the index of broken wheels*, Linear and Multilinear Algebra, 39(1995), 137-152.

[Ben] Benson C.T., Jacobs J.B., *On hearing the shape of combinatorial drums*, J. Combinatorial Theory B, 13(1972), 170-178

[Berg1] Berge C., *Two theorems in graph theory*, Proc. Nat. Acad. Sci. U.S.A., 43(1957), 842-844

[Berg2] Berge C., *Théorie des graphes et ses applications*, Dunod, Paris, 1958

[BeFVS] Bermond J.C., Fournier J.-C., Las Vergnas M., Sotteau D. (Eds.), *Problèmes Combinatoire et Théorie des Graphes*, Coll. Int. C.N.R.S., No.260 (Orsay 1976); C.N.R.S. Publ., 1978

[BiMa] Bienenstock E., von der Malsburg C., *A neural network for invariant pattern recognition*, Europhys. Letters, 4(1987), 121-126

[Big] Biggs N.L., *Algebraic Graph Theory* (second edition), Cambridge University Press, Cambridge, 1993

[BoHe] Bondy J.A., Hemminger R.L., *Graph reconstruction – a survey*, J. Graph Theory, 1(1977), 227-268

[BoLa] Bose R.C., Laskar R., *Eigenvalues of the adjacency matrix of tetrahedral graphs*, Aequationes Math., 4(1970), 37-43

[BoMe] Bose R.C., Mesner D.M., *On linear associative algebras corresponding to association schemes of partially balanced designs*, Ann. Math. Statist., 30(1959), 21-36

[BoSh] Bose R.C., Shrikhande S.S., *Graphs in which each pair of vertices is adjacent to the same number d of other vertices*, Studia Sci. Math. Hung., 5(1970), 181-195

[Bot] Bottema O., *Über die Irrfahrt in einem Strassennetz*, Math. Z., 39(1935), 137-145

[BotMe] Botti P., Merris R., *Almost all trees share a complete set of immanantal polynomials*, J. Graph Theory, 17(1993), 467-476

[BrCN] Brouwer A.E., Cohen A.M., Neumaier A., *Distance-Regular Graphs*, Springer-Verlag, Berlin-Heidelberg-New York-London-Paris-Tokyo, 1989

[Bru] Brualdi R.A., *The symbiotic relationship of combinatorics and matrix theory*, Linear Algebra and Appl., 162-164 (1992), 65-105

[BrHo] Brualdi R.A., Hoffman A.J., *On the spectral radius of 0-1 matrices*, Linear Algebra and Appl., 65(1985), 133-146

[BrSo] Brualdi R.A., Solheid E.S., *On the spectral radius of connected graphs*, Publ. Inst. Math. (Beograd) 39(53) (1986), 45-54

[Bruc] Bruck R.H., *Finite nets*, II, Uniqueness and imbedding, Pacific J. Math., 13(1963), 421-457

[BuČCS] Bussemaker F.C., Čobeljić S., Cvetković D., Seidel J.J., *Computer Investigation of Cubic Graphs*,Technological University Eindhoven, T.H. Report 76-WSK-01, 1976

[BuCv] Bussemaker F.C., Cvetković D., *There are exactly 13 connected, cubic, integral graphs*, Univ. Beograd, Publ. Elektrotehn. Fak., Ser. Mat. Fiz., Nos. 544-576(1976), 43-48

[BuCS1] Bussemaker F.C., Cvetković D., Seidel J.J., *Graphs related to exceptional root systems*, Technological University Eindhoven, T.H. Report, 76-WSK-05, 1976

[BuCS2] Bussemaker F.C., Cvetković D., Seidel J.J., *Graphs related to exceptional root systems*, Combinatorics, Vol. I, ed. Haynal A., Sós V., North-Holland, Amsterdam-Oxford-New York, 1978, 185-191

[BuNe] Bussemaker F.C., Neumaier A., *Exceptional graphs with smallest eigenvalue −2 and related problems*, Math. Comput., 59(1992), No. 200, 583-608

[CaGSS] Cameron P.J., Goethals J.M., Seidel J.J., Shult E.E., *Line graphs, root systems, and elliptic geometry*, J. Algebra 43 (1976), 305-327

[Cao] Cao D., *Index function of graphs* (Chinese, English summary), J. East China Norm. Univ. Nat. Sci. Ed., 1987, No.4, 1-8

[Car] Carlitz L., *Enumeration of certain types of sequences*, Math. Nachr., 49(1971), 125-147

[Chan1] Chang L.C., *The uniqueness and non-uniqueness of the triangular association scheme*, Sci. Record, 3(1959), 604-613

[Chan2] Chang L.C., *Association schemes of partially balanced block designs with parameters $v = 28$, $n_1 = 12$, $n_2 = 15$ and $p_{11}^2 = 4$*, Sci. Record, 4(1960), 12-18

[Chao] Chao C.-Y., *A note on the eigenvalues of a graph*, J. Combinatorial Theory B, 10(1971), 301-302

[ChPh] Chaudhari N.S., Phatak D.B., *A new approach for graph isomorphism*, J. Combin. Inform. System Sci., 13(1988), No. 1-2, 1-19

[Che] Chen H.H., Felix Lee, *Numerical representation and identification of graphs*, J. Math. Phys., 22(1981), 2727-2731

[Cher] Chernoff H., *A measure in asymptotic efficiency for tests of a hypothesis based on the sum of observations*, Ann. Math. Statist., 23(1952), 493-507

[Cla] Clapham C.R.J., private communication

[Coa] Coates C.L., *Flow graph solutions of linear algebraic equations*, IRE Trans. Circuit Theory, CT-6 (1959), 170-187

[CoSi] Collatz L., Sinogowitz U., *Spektren endlicher Grafen*, Abh. Math. Sem. Univ. Hamburg, 21(1957), 63-77

[Con] Connor W.S., *The uniqueness of the triangular association scheme*, Ann. Math. Statist., 29(1958), 262-266

[Cons] Constantine G., *Lower bounds on the spectra of symmetric matrices with non-negative entries*, Linear Algebra and Appl. 65(1985), 171-178

[Cou]  Coulson C.A., Streitwieser A., *Dictionary of Electron Calculations*, Pergamon Press, San Francisco, 1965

[Cve1]  Cvetković D., *Spectrum of the graph of n-tuples*, Univ. Beograd, Publ. Elektrotehn. Fak., Ser. Mat. Fiz., Nos. 274-301 (1969), 91-95

[Cve2]  Cvetković D., *Connectedness of the p-sum of graphs*, Univ. Beograd, Publ. Elektrotehn. Fak., Ser. Mat. Fiz., Nos. 274-301 (1969), 96-99

[Cve3]  Cvetković D., *New characterizations of the cubic lattice graph*, Publ. Inst. Math. (Beograd), 10(1970), 195-198

[Cve4]  Cvetković D., *A note on paths in the p-sum of graphs*, Univ. Beograd, Publ. Elektrotehn. Fak., Ser. Mat. Fiz., Nos. 302-319 (1970), 49-51

[Cve5]  Cvetković D., *The generating function for variations with restrictions and paths of the graph and self-complementary graphs*, Univ. Beograd, Publ. Elektrotehn. Fak., Ser. Mat. Fiz., Nos. 320-328(1970), 27-34

[Cve6]  Cvetković D., *Die Zahl der Wege eines Graphen*, Glasnik Mat. Ser. III, 5(1970), 205-210

[Cve7]  Cvetković D., Doctoral Thesis: *Graphs and their spectra*, Univ. Beograd, Publ. Elektrotehn. Fak., Ser. Mat. Fiz., 354(1971), 1-50

[Cve8]  Cvetković D., *On a graph theory problem of M. Koman*, Časopis Pěst. Mat., 98(1973), 233-236

[Cve9]  Cvetković D., *The Boolean operations on graphs – spectrum and connectedness*, V Kongres na Mat. Fiz. i Astr. na Jugoslavija, Ohrid, 14-19 Sept. 1970, Zbornik na Trudovite, Tom I, Dimitrovski D., Sojuz na Društvata na Mat. Fiz. i Astr. na Jugoslavija, Skopje, 1973, 115-119

[Cve10]  Cvetković D., *The main part of the spectrum, divisors and switching of graphs*, Publ. Inst. Math. (Beograd), 23(1978), 31-38

[Cve11]  Cvetković D., *Combinatorial matrix theory, with applications to electrical engineering, chemistry and physics* (Serbo-Croat), Naučna knjiga, Beograd, 1980

[Cve12]  Cvetković D., *Further experiences in computer aided research in graph theory*, Graphs, Hypergraphs and Applications, Proc. Conf. on Graph Theory held in Eyba, October 1984, ed. Sachs H., Teubner, Leipzig, 1985, 27-30

[Cve13]  Cvetković D., *Problems 2, 3*, Graphs, Hypergraphs and Applications, Proc. Conf. on Graph Theory held in Eyba, October 1984, ed. Sachs H., Teubner, Leipzig, 1985, 211

[Cve14]  Cvetković D., *Constructing trees with given eigenvalues and angles*, Linear Algebra and Appl., 105(1988), 1-8

[Cve15]  Cvetković D., *Some graph invariants based on the eigenvectors of the adjacency matrix*, Proc. 8th Yugoslav Sem. on Graph Theory, Novi Sad, 1987, ed. Tošić R., Acketa D., Petrović V., Doroslovački R., University of Novi Sad, Institute of Mathematics, Novi Sad, 1989, 31-42

[Cve16]  Cvetković D., *Some possibilities of constructing graphs with given eigenvalues and angles*, Ars Combinatoria, 29A (1990), 179-187

[Cve17]  Cvetković D., *Some results on graph angles*, Rostock Math. Kolloq., 39(1990), 74-88

[Cve18]  Cvetković D., *Some comments on the eigenspaces of graphs*, Publ. Inst. Math. (Beograd), 50(64) (1991), 24-32

[Cve19] Cvetković D., *Some possible directions in further investigations of graph spectra*, Algebraic Methods in Graph Theory, Vol. I, ed. Lovász L., Sós V.T., North-Holland, Amsterdam, 1981, 47-67

[Cve20] Cvetković D., *Discussing graph theory with a computer II: theorems suggested by the computer*, Publ. Inst. Math. (Beograd), 33(47) (1993), 29-33

[Cve21] Cvetković D., *Star partitions and the graph isomorphism problem*, Linear and Multilinear Algebra, 39(1995), 109-132

[CvDo1] Cvetković D., Doob M., *Root systems, forbidden subgraphs and spectral characterizations of line graphs*, Graph Theory, Proc. 4th Yugoslav Sem. on Graph Theory, Novi Sad, 1983, ed. Cvetković D., Gutman I., Pisanski T., Tošić R., University of Novi Sad, Institute of Mathematics, Novi Sad, 1983, 69-99

[CvDo2] Cvetković D., Doob M., *Some developments in the theory of graph spectra*, Linear and Multilinear Algebra, 18(1985), 153-181

[CvDGT] Cvetković D., Doob M., Gutman I., Torgašev A., *Recent Results in the Theory of Graph Spectra*, North Holland, Amsterdam, 1988

[CvDS] Cvetković D., Doob M., Sachs H., *Spectra of Graphs – Theory and Application*, Deutscher Verlag der Wissenschaften, Berlin, Academic Press, New York, 1980; second edition 1982; Russian edition, Naukova Dumka, Kiev 1984; third edition, Johann Ambrosius Barth Verlag, 1995

[CvDSi] Cvetković D., Doob M., Simić S., *Generalized line graphs*, J. Graph Theory, 5(1981), 385-399

[CvGS] Cvetković D., Gutman I., Simić S., *On self pseudo-inverse graphs*, Univ. Beograd, Publ. Elektrotehn Fak., Ser. Mat. Fiz., Nos. 602-633 (1978), 111-117

[CvGu1] Cvetković D., Gutman I., *The algebraic multiplicity of the number zero in the spectrum of a bipartite graph*, Mat. Vesnik 9(24) (1972), 141-150

[CvGu2] Cvetković D., Gutman I., *On the spectral structure of graphs having the maximal eigenvalue not greater than two*, Publ. Inst. Math. (Beograd), 18(1975), 39-45

[CvKS] Cvetković D., Kraus L., Simić S, *Discussing graph theory with a computer I: implementation of graph-theoretic algorithms*, Univ. Beograd, Publ. Elektrotehn. Fak., Ser. Mat. Fiz., 716(1981), 100-104

[CvLi] Cvetković D., van Lint J.H., *An elementary proof of Lloyd's theorem*, Proc. Kon. Ned. Akad. V. Wet. A, 80(1977), 6-10

[CvLu] Cvetković D., Lučić R., *A new generalization of the p-sum of graphs*, Univ. Beograd, Publ. Elektrotehn. Fak., Ser. Mat. Fiz., Nos. 302-319 (1970), 67-71

[CvPe1] Cvetković D., Petrić M., *Connectedness of the noncomplete extended p-sum of graphs*, Rev. Res. Fac. Sci. Univ. Novi Sad, 13(1983), 345-352

[CvPe2] Cvetković D., Petrić M., *A table of connected graphs on six vertices*, Discrete Math., 50(1984), No.1, 37-49

[CvPe3] Cvetković D., Petrić M., *Tables of graph spectra*, Univ. Beograd, Publ. Elektrotehn. Fak., Ser. Mat., 4(1993), 49-67

[CvRa1] Cvetković D., Radosavljević Z., *A construction of the 68 connected regular graphs non-isomorphic but cospectral to line graphs*, Graph Theory, Proc. 4th Yugoslav Sem. on Graph Theory, ed. Cvetković D., Gutman I.,

Pisanski T., Tošić R., University of Novi Sad, Institute of Mathematics, Novi Sad, 1983, 101-123

[CvRa2] Cvetković D., Radosavljević Z., *A table of regular graphs with at most 10 vertices*, Graph Theory, Proc. 6th Yugoslav Seminar on Graph Theory, Dubrovnik, April 18-19, 1985; ed. Tošić R., Acketa D., Petrović V., University of Novi Sad, Institute of Mathematics, Novi Sad, 1987, 71-105.

[CvRo1] Cvetković D., Rowlinson P., *Spectra of unicyclic graphs*, Graphs and Combinatorics, 3(1987), 7-23

[CvRo2] Cvetković D., Rowlinson P., *Further properties of graph angles*, Scientia (Valparaiso), 1(1988), 41-51

[CvRo3] Cvetković D., Rowlinson P., *On connected graphs with maximal index*, Publ. Inst. Math. (Beograd), 44(58)(1988), 29-34

[CvRo4] Cvetković D., Rowlinson P., *Seeking counterexamples to the reconstruction conjecture for graphs: a research note*, Graph Theory, Proc. 8th Yugoslav Sem. on Graph Theory, Novi Sad, 1987, ed. Tošić R., Acketa D., Petrović V., Doroslovački R., University of Novi Sad, Institute of Mathematics, Novi Sad, 1989, 52-62

[CvRo5] Cvetković D., Rowlinson P., *The largest eigenvalue of a graph – a survey*, Linear and Multilinear Algebra, 28(1990), 3-33

[CvRS1] Cvetković D., Rowlinson P., Simić S., *A study of eigenspaces of graphs*, Linear Algebra and Appl. 182(1993), 45-66

[CvRS2] Cvetković D., Rowlinson P., Simić S., *On some algorithmic investigations of star partitions of graphs*, Discrete Appl. Math., 62(1995), 119-130

[CvSi1] Cvetković D., Simić S., *On enumeration of certain types of sequences*, Univ. Beograd, Publ. Elektrotehn. Fak., Ser. Mat. Fiz., Nos. 421-460 (1973), 159-164

[CvSi2] Cvetković D., Simić S., *Non-complete extended p-sum of graphs, graph angles and star partitions*, Publ. Inst. Math. (Beograd) 53(67) (1993), 4-16

[CvSi3] Cvetković D., Simić S., *On the graphs whose second largest eigenvalue does not exceed* $(\sqrt{5} - 1)/2$, Discrete Math. 138(1995), 213-227

[CvSi4] Cvetković D., Simić S., *The second largest eigenvalue of a graph – a survey*, FILOMAT (Niš), 9(1995), Proc. Conf. on Algebra, Logic and Discrete Math., Niš, 14-16 April 1995, 53-76

[DAGT] D'Amato S.S., Gimarc B.M., Trinajstić N., *Isospectral and subspectral molecules*, Croat. Chem. Acta, 54(1981), No.1, 1-52

[Dem] Demidovich B.P., Maron I.A., *Computational Mathematics*, Mir Publishers, Moscow, 1987

[DeoHS] Deo N., Harary F., Schwenk A.J., *An eigenvector characterization of cospectral graphs having cospectral joins*, Combinatorial Mathematics, Proc. 3rd Internat. Conf., ed. Bloom G.S., Graham R.L., Malkevitch J., Ann. New York Acad. Sci., Vol. 555, New York Acad. Sci., New York, N.Y., 1988, 159-166

[Din] Dinic E.A., Kelmans A.K., Zaitsev M.A., *Non-isomorphic trees with the same T-polynomial*, Inform. Process. Lett., 6(1977), No.3, 3-8

[Djo] Djoković D.Ž., *Isomorphism problem for a special class of graphs*, Acta Math. Acad. Sci. Hung., 21(1970), 267-270

[Doč] Dočev K., *The solution of a difference equation and a combinatorial problem related to it* (Bulgarian), Fiz. Mat. Spis. Bulgar. Akad. Nauk, 6(39)(1963), 284-287

[Doo1] Doob M., *A geometrical interpretation of the least eigenvalue of a line graph.*, Proc. Second Conference on Comb. Math. and Appl., Univ. North Carolina, Chapel Hill, N.C., 1970, 126-135

[Doo2] Doob M., *On characterizing certain graphs with four eigenvalues by their spectra*, Linear Algebra and Appl., 3(1970), 461-482

[Doo3] Doob M., *Graphs with a smaller number of distinct eigenvalues*, Ann. New York Acad. Sci., 175(1970), No. 1, 104-110

[Doo4] Doob M., *On embedding a graph in an isospectral family*, Proc. 2nd Manitoba Conf. on Numerical Mathematics, Utilitas Math., Winnipeg, Man., 1973, 137-142

[Doo5] Doob M., *An interrelation between line graphs, eigenvalues, and matroids*, J. Combinatorial Theory B., 15(1973), 40-50

[Doo6] Doob M., *Seidel switching and cospectral graphs with four distinct eigenvalues*, Ann. New York Acad. Sci., 319(1979), 164-168

[DuMe] Dulmage A.L., Mendelsohn N.S., *Graphs and matrices*, Graph Theory and Theoretical Physics, ed. Harary F., Academic Press, New York, 1967, 167-227

[Edm] Edmonds J., *Matroid Intersection*, Annals of Discrete Mathematics 4, ed. Hammer P.L. *et al.*, Discrete Optimization I, North-Holland, Amsterdam, 1979, 39-49

[Ein] Einbu J.M., *The enumeration of bit-sequences that satisfy local criteria*, Publ. Inst. Math. (Beograd), 27(1980), 51-56

[FaGr] Farrell E.J., Grell J.C., *Some further constructions of cocircuit and cospectral graphs*, Carib. J. Math., 4(1985), 17-28

[Fie1] Fiedler M., *Algebraic connectivity of graphs*, Czechoslovak. Math. J., 23(38) (1973), 298-305

[Fie2] Fiedler M., *A property of eigenvectors of non-negative symmetric matrices and its application to graph theory*, Czechoslovak. Math. J., 25(1975), 619-633

[Fie3] Fiedler M., *Laplacian of graphs and algebraic connectivity*, Combinatorics and Graph Theory, Banach Centre Publ., Vol. 25, 57-70

[Fin] Finck H.-J., *Vollständiges Produkt, chromatische Zahl und charakteristisches Polynom regulärer Graphen II*, Wiss. Z., T.H. Ilmenau, 11(1965), 81-87

[FiSa] Finck H.-J., Sachs H., *Über Beziehungen zwischen Struktur und Spektrum regulärer Graphen*, Wiss. Z., T.H. Ilmenau, 19(1973), 83-99

[Fis] Fisher M., *On hearing the shape of a drum*, J. Combinatorial Theory, 1(1966), 105-125

[Fri1] Friedland S., *The maximal eigenvalue of 0-1 matrices with prescribed number of ones*, Linear Algebra and Appl, 69(1985), 33-69

[Fri2] Friedland S., *The maximal value of the spectral radius of graphs with e edges*, Linear and Multilinear Algebra, 23(1988), 91-93

[Fri3]  Friedland S., *Bounds on the spectral radius of graphs with e edges*, Linear Algebra and Appl., 101(1988), 81-86

[Gan]  Gantmacher F.R., *The Theory of Matrices*, Chelsea Publishing Company, New York, Vol. 1, 1977

[GaJo]  Garey M.R., Johnson D.S., *Computers and Intractability – A Guide to the Theory of NP-Completeness*, W.H. Freeman and Company, San Francisco, 1979

[GoHMK]  Godsil C.D., Holton D.A., McKay B.D., *The spectrum of a graph*, Combinatorial Mathematics V, ed. Little C.H.C., Lecture Notes in Math. 622, Springer-Verlag, Berlin, 1977, 91-117

[GoMK1]  Godsil C.D., McKay B.D., *Some computational results on the spectra of graphs*, Combinatorial Mathematics IV, Proc. 4th Australian Conf. held at the Univ. of Adelaide, Aug. 27-29, 1975, ed. Casse L.R.A., Wallis W.D., Springer-Verlag, Berlin, 1976, 73-92

[GoMK2]  Godsil C.D., McKay B.D., *Feasibility conditions for the existence of walk-regular graphs*, Linear Algebra and Appl., 30(1980), 51-61

[GoMK3]  Godsil C.D., McKay B.D., *Spectral conditions for the reconstructibility of a graph*, J. Combinatorial Theory B, 30(1981), 285-289

[GoMK4]  Godsil C.D., McKay B.D., *Constructing cospectral graphs*, Aequationes Math., 25(1982), 257-268

[Gro]  Grone R., *On the geometry and Laplacian of a graph*, Linear Algebra and Appl., 150(1991), 167-178

[GrMe1]  Grone R., Merris R., *Cutpoints, lobes and the spectra of graphs*, Portugalae Math., 45(1988), 181-188

[GrMe2]  Grone R., Merris R., *Coalescence, majorization, edge valuations and the Laplacian spectra of graphs*, Linear and Multilinear Algebra, 27(1990), 139-146

[GrMe3]  Grone R., Merris R., *The Laplacian spectrum of a graph II*, SIAM J. Discrete Math., 7(1994), 229-237

[GrMS]  Grone R., Merris R., Sunder V.S., *The Laplacian spectrum of a graph*, SIAM J. Matrix Theory, 11(1990), 218-236

[GrMW]  Grone R., Merris R., Watkins W., *Laplacian unimodular equivalence of graphs*, Combinatorial and Graph-Theoretic Problems in Linear Algebra, ed. Brualdi R.A., Friedland S., Klee V., Springer-Verlag, New York, 1993, 175-180

[GrZi]  Grone R., Zimmerman G., *Large eigenvalues of the Laplacian*, Linear and Multilinear Algebra, 28(1990), 45-47

[GuCv]  Gutman I., Cvetković D., *The reconstruction problem for characteristic polynomials of graphs*, Univ. Beograd, Publ. Elektrotehn. Fak., Ser. Mat. Fiz., Nos. 498-541(1975), 45-48

[GüPr]  Günthard Hs.H., Primas H., *Zusammenhang von Graphentheorie und MO-Theorie von Molekeln mit Systemen konjugierter Bindungen*, Helv. Chim. Acta 39 (1956), 1645-1653

[Hae]  Haemers W., *A genralization of the Higman-Sims technique*, Proc. Kon. Ned. Akad. Wet. A, 81(4) (1978), 445-447

[HaHi]  Haemers W.H., Higman D.G., *Strongly regular graphs with strongly regular decomposition*, Linear Algebra and Appl., 114/115(1989), 379-398

[HaMl] Hansen P., Mladenović N., *A comparison of algorithms for the maximum clique problem*, Yugoslav J. Op. Res., 2(1992), 3-11

[Har1] Harary F., *The determinant of the adjacency matrix of a graph*, SIAM Rev., 4(1962), 202-210

[Har2] Harary F., *Graph Theory*, Addison-Wesley, Reading, Mass., 1969

[HaKMR] Harary F., King C., Mowshowitz A., Read R.C., *Cospectral graphs and digraphs*, Bull. London Math. Soc., 3(1971), 321-328

[HaSc] Harary F., Schwenk A.J., *The spectral approach to determining the number of walks in a graph*, Pacific J. Math., 80(1979), 443-449

[Hay] Haynsworth E.V., *Applications of a theorem on partitioned matrices*, J. Res. Nat. Bureau Stand., 62(1959), 73-78

[Hei1] Heilbronner E., *Some comments on cospectral graphs*, Match, 5(1979), 105-133

[Hei2] Heilbronner E., *Das Komposition-Prinzip: Eine anschauliche Methode zur elektronen-theoretischen Behandlung nicht oder niedrig symmetrischer Molekeln in Rahmen der MO-Theorie*, Helv. Chim. Acta, 36(1953), 170-188

[Herm] Hermann E.C., *On the relevance of isospectral nonisomorphic graphs for chemistry*, Match, 19(1986), 43-52

[Hern1] Herndon W.C., *The characteristic polynomial does not uniquely determine molecular topology*, J. Chem. Doc., 14(1974), 150-151

[Hern2] Herndon W.C., *Isospectral molecules*, Tetrahedron Letters, 8(1974), No.8, 671-674

[HeEl1] Herndon W.C., Ellzey M.L. Jr, *Isospectral graphs and molecules*, Tetrahedron, 31(1975), 99-107

[HeEl2] Herndon W.C., Ellzey M.L. Jr, *The construction of isospectral graphs*, Match, 20(1986), 53-79

[HeHi] Hestenes M., Higman D.G., *Rank 3 groups and strongly regular graphs*, Computers in Linear Algebra and Number Theory, SIAM-AMS Proc., Vol. IV, Providence, R.I., 1971, 141-159

[Hof1] Hoffman A.J., *On the exceptional case in a characterization of the arcs of a complete graph*, IBM J. Res. Develop., 4(1960), 487-496

[Hof2] Hoffman A.J., *On the polynomial of a graph*, Amer. Math. Monthly, 70(1963), 30-36

[Hof3] Hoffman A.J., *On the line graph of a projective plane*, Proc. Amer. Math. Soc., 16(1965), 297-302

[Hof4] Hoffman A.J., *Some recent results on spectral properties of graphs*, Beiträge zur Graphentheorie, Leipzig 1968, 75-80

[Hof5] Hoffman A.J., *The change in the least eigenvalue of the adjacency matrix of a graph under imbedding*, SIAM J. Appl. Math., 17(1969), 664-677

[Hof6] Hoffman A.J., *On limit points of spectral radii of non-negative symmetric integral matrices*, Graph Theory and its Applications, Lecture Notes in Math. 303, ed. Alavi Y. et al., Springer-Verlag, New York, 1972, 165-172

[HoRa1] Hoffman A.J., Ray-Chaudhuri D.K., *On the line graph of a finite plane*, Canad. J. Math., 17(1965), 687-694

[HoRa2] Hoffman A.J., Ray-Chaudhuri D.K., *On the line graph of a symmetric balanced incomplete block designs*, Trans. Amer. Math. Soc., 116(1965), 238-252

[HoRa3] Hoffman A.J., Ray-Chaudhuri D.K., *On a spectral characterization of regular line graphs*, unpublished

[HoSm] Hoffman A.J., Smith J.H., *On the spectral radii of topolgically equivalent graphs*, Recent Advances in Graph Theory, ed. Fiedler M., Academia Praha, 1975, 273-281

[HoKa] Hopcroft J.M., Karp R.M., *An $n^{5/2}$ algorithm for maximum matchings in bipartite graphs*, SIAM J. Comput., 2(1973), 225-231

[Hüc] Hückel E., *Quantentheoretische Beitrage zum Benzolproblem*, Z. Phys. 70 (1931), 204-286

[Jia] Jiang Yuan-sheng, *Problem on isospectral molecules*, Sci. Sinica (B), 27(1984), 236-248

[JoLe] Johnson C.R., Leighton F.T., *An efficient linear algebraic algorithm for the determination of isomorphism in pairs of undirected graphs*, J. Res. Nat. Bureau Stand. A., Math. Sci., 80(1976), No. 4, 447-483

[JoNe] Johnson C.R., Newman M., *A note on cospectral graphs*, J. Combinatorial Theory B, 28(1980), No. 1, 96-103

[Kel] Kel'mans A.K., *The properties of the characteristic polynomial of a graph* (Russian), Cybernetics in the Service of Communism (Russian), Vol. 4, Izdat. "Ènergija", Moscow, 1967, 27-41

[KeLi] Kernighan B. W., Lin S., *An efficient procedure for partioning graphs*, Bell Sys. Tech. J., 49(1970), 291-307

[KnMSRT] Knop J.V., Muller W.R., Szymanski K., Randić M., Trinajstić N., *Note on acyclic structures and their self-returning walks*, Croat. Chem. Acta, 56(1983), No. 3, 405-409

[KnMSTKR] Knop J.V., Muller W.R., Szymanski K., Trinajstić N., Kleiner A.F., Randić M., *On irreducible endospectral graphs*, J. Math. Phys., 27(1986), No. 11, 2601-2612

[KoSu] Kolmykov V.A., Subotin V.F., *Spectra of graphs and cospectrality*, Voronezh University, Voronezh, 1983, 23pp. (manuscript no. 6708-83 DEP, VINITI 12. Dec. 1983), 1983

[Kön] König D., *Theorie der Endlichen und Unendlichen Graphen*, Akadem. Verlagsges., Leipzig, 1936

[KrPa1] Krishnamoorthy V., Parthasarathy K.R., *A note on non-isomorphic cospectral digraphs*, J. Combinatorial Theory B, 17(1974), 39-40

[KrPa2] Krishnamoorthy V., Parthasarathy K.R., *Cospectral graphs and digraphs with given automorphism group*, J. Combinatorial Theory B, 19(1975), 204-213

[Kri] Krishnamurthy E.V., *A form invariant multivariable polynomial representation of graphs*, Combinatorics and Graph Theory, Proc. 2nd Symp. held at the Indian Statistical Institute, Calcutta, February 25-29, 1980, Lecture Notes in Math., 885, ed. Rao S.B., Springer-Verlag, Berlin, 1981, 18-32

[Kuh] Kuhn W.W., *Graph isomorphism using vertex adjacency matrix*, Proc. 25th Summer Meeting Canadian Math. Congr., Lakehead University, Thunder

Bay, Ont., June 16-18, 1971, ed. Eames W.R., Stanton R.G., Thomas R.S.D., Lakehead Univ., Thunder Bay, Ont., 1971, 471-476

[LaTi] Lancaster P., Tismenetsky M., *Theory of Matrices* (second edition), Academic Press, New York, 1985.

[Las] Laskar R., *Eigenvalues of the adjacency matrix of cubic lattice graphs*, Pacific J. Math., 29(1969), 623-629

[LeYe] Lee S.-L., Yeh Y.-N., *On eigenvalues and eigenvectors of graphs*, J. Math. Chem. 12(1993), 121-135

[Leh] Lehmer D.H., *Permutations with strongly restricted displacements*, Combinatorial Theory and Its Applications II, ed. Erdös P., Rényi A., Sós V.T., Bolyai Janos Mat. Tarsulat, Budapest, North-Holland Publ. Co., Amsterdam-London, 1970, 755-770

[LiFe] Li Q., Feng K.E., *On the largest eigenvalue of graphs* (Chinese), Acta Math. Appl. Sinica 2(1979), 167-175

[Lih] Lihtenbaum L.M., *Characteristic values of a simple graph* (Russian), Trud 3-go Vses. matem. sezda, tom 1, 1956, 135-136

[LiWZ] Lin Liang-tang, Wang Nan-qin, Zhang Qian-er, *Isospectral molecules*, Acta Univ. Amoiensis Scientiarum Naturalium, 2(1979), 65-75

[Lin] van Lint J.H., *Coding Theory*, Lecture Notes in Math. 201, Springer-Verlag, Berlin-Heidelberg-New York, 1971

[LiWi] van Lint J.H., Wilson R.M., *A Course in Combinatorics*, Cambridge University Press, Cambridge, 1992

[Llo] Lloyd S.P., *Binary block coding*, Bell System Tech. J., 36(1957), 517-535

[LoPe] Lovász L., Pelikan J., *On the eigenvalues of trees*, Periodica Math. Hung., 3 (1973), 175-182

[LoPl] Lovász L., Plummer M.D., *Matching Theory*, Akademiai Kiado, Budapest 1986

[LoSo] Lowe J.P., Soto M.R., *Isospectral graphs, symmetry and perturbation theory*, Match, 20(1986), 21-51

[Maa1] Maas C., *Transportation graphs with optimal spectra*, Methods of Operations Research 53, X. Symp. on Operations Research, Univ. München, August 26-28, 1985, Part I, ed. Beckmann J.M., Gaede K.-W., Ritter K., Schneeweiss H., Verlag Anton Hain, 1986, 289-290

[Maa2] Maas C., *Perturbation results for the adjacency spectrum of a graph*, Z. angew. Math. Mech., 67(1987), 428-430

[Maa3] Maas C., *Transportation in graphs and the admittance spectrum*, Discrete Appl. Math., 16(1987), 31-49

[Mal] Mallion R.B., *Some chemical applications of the eigenvalues and eigenvectors of certain finite, planar graphs*, Applications of Combinatorics, Proc. Conf. on Combinatorics and its Applications, held at the Open University, Buckinghamshire, November 13, 1981, Shiva Mathematics Series 6, ed. Wilson R.J., Shiva Publ. Ltd., Nantwich, Birkhäuser Boston Inc., Cambridge, Mass., 1982, 87-114

[MaMi] Marcus M., Minc H., *A Survey of Matrix Theory and Matrix Inequalities*, Allyn and Bacon, Inc., Boston, Mass., 1964

[Mas] Masuyama M., *A test for graph isomorphism*, Repts. Statist. Appl. Res. Union Japan Sci. Eng., 20(1973), No. 2, 41-64

[McD] McDiarmid C., *On the method of bounded differences*, Surveys in Combinatorics 1989, ed. Siemons J., Cambridge University Press, Cambridge, 148-188

[Mer1] Merris R., *Almost all trees are co-immanantal*, Proc. First Conference of the International Linear Algebra Society (Provo, Utah, 1989), Linear Algebra and Appl., 150(1991), 61-66

[Mer2] Merris R., *Laplacian matrices of graphs: a survey*, Linear Algebra and Appl. 197/198 (1994), 143-176

[Mer3] Merris R., *A survey of graph Laplacians*, Linear and Multilinear Algebra, 39(1995), 19-31

[Mes] Mesner D.M., *A new family of partially balanced incomplete block designs*, Ann. Math. Statist. 38(1967), 571-581

[Mey] Meyer J.F., *Algebraic isomorphism invariants for graphs of automata*, Graph Theory and Computing, including part of the Proc. of a Conf. held at the University of the West Indies, Kingston, Jamaica, January 1969, ed. Read R.C., Academic Press, New York, 1972, 123-152

[Moh1] Mohar B., *Isoperimetric inequalities, growth and the spectrum of graphs*, Linear Algebra and Appl., 103(1988), 119-131

[Moh2] Mohar B., *Isoperimetric numbers of graphs*, J. Combinatorial Theory B, 47(1989), 274-291

[MoOm1] Mohar B., Omladič M., *The spectrum of products of infinite graphs*, Graph Theory, Proc. 6th Yugoslav Sem. on Graph Theory, Dubrovnik, April 18-19, 1985, ed. Tošić R., Acketa D., Petrović V., University of Novi Sad, Institute of Mathematics, Novi Sad, 1987, 135-142

[MoOm2] Mohar B., Omladič M., *Divisors and the spectrum of infinite graphs*, Linear Algebra and Appl., 91(1987), 99-106

[Mow] Mowshowitz A., *The adjacency matrix and the group of a graph*, New Directions in the Theory of Graphs, Proc. 3rd Ann Arbor Conf. on Graph Theory held at the University of Michigan, Oct. 21-23, 1971, ed. Harary F., Academic Press, New York, 1973, 129-148

[Nas] Nash-Williams C.St J.A., *The reconstruction problem*, Selected Topics in Graph Theory, ed. Beineke L.W., Wilson R.J., Academic Press, London-New York-San Francisco, 1978, 205-236

[Neu] Neumaier A., *The second largest eigenvalue of a tree*, Linear Algebra and Appl., 46 (1982), 9-25

[Nor] Nordhaus E. A., *A class of strongly regular graphs*, Proof Techniques in Graph Theory, Proc. 2nd Ann Arbor Graph Theory Conf., ed. Harary F., Academic Press, New York 1969, 119-123

[Nuf] van Nuffelen C., *Rank and domination number*, Graphs and Other Combinatorial Topics, Proc. 3rd Czech. Symp. on Graph Theory, Prague 1982, ed. Fiedler M., Teubner-texte Math., No. 59, Leipzig 1983, 209-211

[Ore] Ore O., *Theory of Graphs*, American Math. Soc., Providence, R.I., 1962

[PaSt] Papadimitrou C.H., Steiglitz K., *Combinatorial Optimization: Algorithms and Complexity*, Prentice-Hall, Englewood Cliffs, N.J., 1982

[Pau] Paulus A.J.L., *Conference matrices and graphs of order 26*, T.H. Eindhoven, T.H. Report 73-WSK-06, 1973

[PeSa1] Petersdorf M., Sachs H., *Über Spektrum, Automorphismengruppe und Teiler eines Graphen*, Wiss. Z., T.H. Ilmenau, 15(1969), 123-128

[PeSa2] Petersdorf M., Sachs H., *Spektrum und Automorphismengruppe eines Graphen*, Combinatorial Theory and Its Applications III, ed. Erdös P., Rényi A., Sós V.T., Bolyai Janos Mat. Tarsulat, Budapest, North-Holland Publ. Co., Amsterdam-London, 1970, 891-907

[Pet1] Petrić M., *Spectral method and the problem of determining the number of walks in a graph*, Master's Thesis, Univ. Belgrade, Fac. Sci., 1980

[Pet2] Petrić M., *A note on the number of walks in a graph*, Univ. Beograd, Publ. Elektrotehn. Fak., Ser. Mat. Fiz., Nos. 716-734 (1981), 83-86

[Pet3] Petrić M., *On generalized direct products of graphs*, Proc. 8th Yugoslav Sem. on Graph Theory, Novi Sad, 1987, ed. Tošić R., Acketa D., Petrović V., Doroslovački R., University of Novi Sad, Institute of Mathematics, Novi Sad, 1989, 99-105

[Pet4] Petrić M., *Connectedness of the generalized direct product of regular digraphs*, Univ. Novi Sad, Zb. Rad. Prirod.-Mat. Fak. Ser. Mat. 21, 2(1991), No. 1, 57-64

[Pet5] Petrić M., *Bipartiteness of certain graph products*, Univ. Novi Sad, Zb. Rad. Prirod.-Mat. Fak. Ser. Mat., submitted

[Petr] Petrović M., *A contribution to the spectral theory of graphs* (Serbo-Croat), Doctoral Thesis, Univ. Belgrade, Fac. Sci., Beograd, 1983

[Petro] Petrović S., *Combinatorial algorithms and heuristics for clustering vertices of a graph and applications to pattern recognition*, (Serbo-Croat), Master's Thesis, Univ. Belgrade,.Fac. Elec. Eng., Beograd, 1988

[Pon] Ponstein J., *Self-avoiding paths and the adjacency matrix of a graph*, SIAM J. Appl. Math., 14(1966), 600-609

[Pow1] Powers D.L., *Structure of a matrix according to its second eigenvector*, Current Trends in Matrix Theory, ed. Uhlig F., Grone R., Elsevier, New York, 1987, 261-266

[Pow2] Powers D.L., *Graph partitioning by eigenvectors*, Linear Algebra and Appl., 101(1988), 121-133

[PoSu] Powers D.L., Sulaiman M.M., *The walk partition and colorations of a graph*, Linear Algebra and Appl., 48(1982), 145-159

[PrDe] Prabhu G.M., Deo N., *On the power of a perturbation for testing nonisomorphism of graphs*, BIT, 24(1984), 302-307

[Ran1] Randić M., *Random walks and their diagnostic value for characterization of atomic environment*, J. Comput. Chem., 1(1980), No. 4, 386-399

[Ran2] Randić M., *On the characteristic equations of the characteristic polynomial*, SIAM J. Alg. Disc. Meth., 6(1985), No. 1, 145-162

[RaKl] Randić M., Kleiner A.F., *On the construction of endospectral graphs*, Combinatorial Mathematics, Proc. 3rd Int. Conf., New York (USA) 1985, Ann. N.Y. Acad. Sci., 555(1989), 320-331

[RaTŽ] Randić M., Trinajstić N., Živković T., *On molecular graphs having identical spectra*, J.C.S. Faraday II, 72(1976), 244-256

[RaWG]  Randić M., Woodworth W.L., Graovac A., *Unusual random walks*, Internat. J. Quantum Chem., 24(1983), 435-452

[Ray]  Ray-Chaudhury D.K., *Characterization of line graphs*, J. Combinatorial Theory 3(1967), 201-214

[ReCo]  Read R.C., Corneil D.G., *The graph isomorphism disease*, J. Graph Theory, 1(1977), 339-363

[RiMW]  Rigby M.J., Mallion R.B., Waller D.A., *On the quest for an isomorphism invariant which characterises finite chemical graphs*, Chem. Phys. Letters, 59(1978), No. 2, 316-320

[Row1]  Rowlinson P., *Simple eigenvalues of intransitive graphs*, Bull. London Math. Soc., 16(1984), 122-126

[Row2]  Rowlinson P., *Certain 3-decompositions of complete graphs with an applications to finite fields*, Proc. Royal Soc. Edinburgh, 99A(1985), 277-281

[Row3]  Rowlinson P., *A deletion-contraction algorithm for the characteristic polynomial of a multigraph*, Proc. Royal Soc. Edinburgh 105A(1987), 153-160

[Row4]  Rowlinson P., *On the maximal index of graphs with a prescribed number of edges*, Linear Algebra and Appl., 110(1988), 43-53

[Row5]  Rowlinson P., *On angles and perturbations of graphs*, Bull. London Math. Soc., 20(1988), 193-197

[Row6]  Rowlinson P., *More on graph perturbations*, Bull. London Math. Soc., 22(1990), 209-216

[Row7]  Rowlinson P., *On the index of certain outerplanar graphs*, Ars Combinatoria 29c(1990), 271-275

[Row8]  Rowlinson P., *Graph perturbations*, Surveys in Combinatorics, ed. A.D. Keedwell, Cambridge University Press, Cambridge, 1991, 187-219

[Row9]  Rowlinson P., *The spectrum of a graph modified by the addition of a vertex*, Univ. Beograd, Publ. Elektrotehn. Fak., Ser. Mat., 3(1992), 67-70

[Row10]  Rowlinson P., *A note on recognition graphs*, unpublished

[Row11]  Rowlinson P., *Eutactic stars and graph spectra*, Combinatorial and Graph-Theoretic Problems in Linear Algebra, ed. Brualdi R.A., Friedland S., Klee V., Springer-Verlag, New York, 1993, 153-164

[Row12]  Rowlinson P., *On Hamiltonian graphs with maximal index*, European J. Combinatorics, 10(1989), 487-497

[Row13]  Rowlinson P., *Dominating sets and eigenvalues of graphs*, Bull. London Math. Soc. 26(1994), 248-254

[Row14]  Rowlinson P., *Graph angles and isospectral molecules*, Univ. Beograd, Publ. Elektrotehn. Fak. Ser. Mat., 2(1991), 61-66

[Row15]  Rowlinson P., *Star partitions and regularity in graphs*, Linear Algebra and Appl., 226-228(1995), 247-265

[Row16]  Rowlinson P., *On the number of simple eigenvalues of a graph*, Proc. Royal Soc. Edinburgh, 94A (1983), 247-250

[RoYu]  Rowlinson P., Yuansheng Y., *Tricyclic hamiltonian graphs with minimal index*, Linear and Multilinear Algebra, 34(1993), No. 3-4, 187-196

[Sac1]  Sachs H., *Beziehungen zwischen den in einem Graphen enthaltenen*

*Kreisen und seinem charakteristischen Polynom*, Publ. Math. (Debrecen) 11(1964), 119-134

[Sac2] Sachs H., *Über Teiler, Faktoren und charakteristische Polynome von Graphen*. Teil I, Wiss. Z., T.H. Ilmenau, 12(1966), 7-12

[Sac3] Sachs H., *Über Teiler, Faktoren und charakteristische Polynome von Graphen*. Teil II, Wiss. Z., T.H. Ilmenau, 13(1967), 405- 412

[Sac4] Sachs H., *On a theorem connecting the factors of a regular graph with the eigenvectors of its line graph*, Combinatorics I, II, Proc. V Hungarian Coll. on Combinatorics, Keszthely 1976, ed. Hajnal A., Sós V.T., North-Holland, Amsterdam, 1978, 947-957

[Sch1] Schwenk A.J., *Almost all trees are cospectral*, New Directions in the Theory of Graphs, Proc. 3rd Ann Arbor Conf. on Graph Theory held at the University of Michigan, Oct. 21-23, 1971, ed. Harary F., Academic Press, New York, 1973, 275-307

[Sch2] Schwenk A.J., *Computing the characteristic polynomial of a graph*, Graphs and Combinatorics, ed. Bari R., Harary F., Springer-Verlag, Berlin-Heidelberg-New York, 1974, 153-172

[Sch3] Schwenk A.J., *Removal-cospectral sets of vertices in a graph*, Proc. 10th South Eastern Conference on Combinatorics, Graph Theory and Computing, 1979, Congr. Num., XXIII-XXIV, Utilitas Math., Winnipeg, Man., 1979, 849-860

[Sch4] Schwenk A.J., *Spectral reconstruction problems*, Ann. N. Y. Acad. Sci., 328(1978), 183-189

[Sed] Sedláček J., *Über Inzidenzmatrizen gerichteter Graphen* (Czech, Russian and German summaries), Časopis Pěst. Mat., 84(1959), 303-316

[Sei1] Seidel J.J., *Strongly regular graphs with* $(-1, 1, 0)$-*adjacency matrix having eigenvalue 3*, Linear Algebra and Appl., 1(1968), 281-298

[Sei2] Seidel J.J., *Graphs and two-graphs*, Proc. 5th South Eastern Conf. on Combinatorics, Graph Theory and Computing, Boca Raton (Fla), 1974, Congr. Num. X, Utilitas Math., Winnipeg, Man., 1974, 125-143

[Sei3] Seidel J.J., *Eutactic stars*, Combinatorics, ed. Hajnal A., Sós V.T., North-Holland, Amsterdam-Oxford-New York, 1978, 983-999

[SeFK] Seno A., Fukunaga K., Kasai T., *An algorithm for graph isomorphism*, Bull. Univ. Osaka Prefecture Ser. A, 26(1977), No. 2, 61-74

[She] Shee S.C., *A note on the C-product of graphs*, Nanta Math., 7(1974), 105-108

[Shr] Shrikhande S.S., *The uniqueness of the $L_2$ association scheme*, Ann. Math. Statist. 30(1959), 781-798

[Sie] Siemons J., *Automorphism groups of graphs*, Arch. Math. 41 (1982), 379-384

[Sim1] Simić S., *Some results on the largest eigenvalue of a graph*, Ars Combinatoria, 24A (1987), 211-219

[Sim2] Simić S., *On the largest eigenvalue of unicyclic graphs*, Publ. Inst. Math. (Beograd), 42(56) (1988), 13-19

[Sim3] Simić S., *On the largest eigenvalue of bicyclic graphs*, Publ. Inst. Math. (Beograd), 46(60) (1989), 1-6

[Sim4] Simić S., *Some notes on graphs whose second largest eigenvalue is less than* $(\sqrt{5} - 1)/2$, Linear and Multilinear Algebra, 39(1995), 59-71

[SiKo] Simić S., Kocić V., *On the largest eigenvalue of some homeomorphic graphs*, Publ. Inst. Math. (Beograd), 40(54) (1986), 3-9

[SiMe] Simmons H.E., Merrifield R.E., *Paraspectral molecular pairs*, Chem. Phys. Letters, 62(1979), 235-237

[Sla] Slater P.J., *Dominating and reference sets in a graph*, J. Math. Phys. Sci. 22(1988), 445-455

[Šok] Šokarovski R., *A generalized direct product of graphs*, Publ. Inst. Math. (Beograd), 22(36)(1977), 267-269

[Sta] Stanley R.P., *A bound on the spectral radius of graphs with e edges*, Linear Algebra and Appl., 87(1987), 267-269

[StMa] Stewartson K., Maechter R.T., *On hearing the shape of a drum – further results*, Proc. Cambridge Phil. Soc., 69(1971), 353-363

[Sto1] Stockmeyer P.K., *The falsity of the reconstruction conjecture for tournaments*, J. Graph Theory, 1(1977), 19-25

[Sto2] Stockmeyer P.K., *Which reconstruction results are significant?*, The Theory and Application of Graphs, Proc. 4th Internat. Conf. on Theory and Applications of Graphs, ed. Chartrand G. *et al.*, John Wiley, New York 1981, 543-555

[Tim] Timkovski V.G., *Some recognition problems that are connected with the isomorphism of graphs* (Russian), Kibernetika (Kiev), 1988, no. 2, 13-18, 133; English translation in Cybernetics, 24(1988), no. 2, 153-160

[Turn1] Turner J., *Point-symmetric graphs with a prime number of points*, J. Combinatorial Theory, 1(1966), 136-145

[Turn2] Turner J., *Generalized matrix functions and the graph isomorphism problem*, SIAM J. Appl. Math., 16(1968), 520-526

[Tut] Tutte W.T., *All the king's horses, A guide to reconstruction*, Graph Theory and Related Topics, Proc. Conf. held at the University of Waterloo, Waterloo, Ont. 1977, ed. Bondy J.A., Murty U.S.R., Academic Press, New York, 1979, 15-33

[Vah] Vahovski E.V., *O harakterističeskih čislah matricah sosedstva dla neosobennih grafov*, Sibir. Matem. Žurn., 6(1965), No. 1, 44-49

[Ver] de Verdière Y.C., *Sur un nouvel invariant des graphes et un critère de planarité*, J. Combinatorial Theory B, 50(1990), 11-21

[Wan] Wang N.S., *On the specific properties of the characteristic polynomial of a tree* (Chinese, English summary), J. Lanzhou Railway Coll. 5(1986), 89-94

[Wei] Wei T.H., *The algebraic foundations of ranking theory*, Thesis, Cambridge Univ., 1952

[WeSt] Weinstein A., Stenger W., *Methods of Intermediate Problems for Eigenvalues*, Academic Press, New York, 1972

[WiKe] Wild U., Keller J., Günthard Hs.H., *Symmetry properties of the Hückel matrix*, Theoret. Chim. Acta, 14(1969), 383-395

[Wil] Wilf H.S., *Spectral bounds for the clique and independence numbers of graphs*, J. Combinatorial Theory B, 40(1986), 113-117

[Wils]  Wilson R.M., *Decomposition of complete graphs into subgraphs isomorphic to a given graph*, Proc. 5th British Combinatorial Conf., 1975, ed. Nash-Williams C.St J.A., Sheehan J., Congr. Num. XV, Utilitas Math., Winnipeg, Man., 1976, 647-659

[Yap]  Yap H.P., *The characteristic polynomial of the adjacency matrix of a multi-digraph*, Nanta Math., 8(1975), No.1, 41-46

[ZhZZ]  Zhang F.J., Zhang Z.N., Zhang Y.H., *Some theorems about the largest eigenvalue of graphs*, (Chinese, English summary), J. Xinjiang Univ. Nat. Sci. (1984), No. 3, 84-90

[Živ]  Živković T., *On graph isomorphism and graph automorphism*, J. Math. Chem., 8(1991), Nos. 1-3, 19-37

[ZiTR]  Živković T., Trinajstić N., Randić M., *On conjugated molecules with identical topological spectra*, Mol. Phys., 30(1975), 517-533

# Index